# Water Quality Assessment of River Basins

# Water Quality Assessment of River Basins

Guest Editor

**Qianqian Zhang**

MDPI

Basel • Beijing • Wuhan • Barcelona • Belgrade • Novi Sad • Cluj • Manchester

*Guest Editor*
Qianqian Zhang
Institute of Hydrogeology
and Environmental Geology
Chinese Academy of
Geological Sciences
Shijiazhuang City
China

*Editorial Office*
MDPI AG
Grosspeteranlage 5
4052 Basel, Switzerland

This is a reprint of the Special Issue, published open access by the journal *Water* (ISSN 2073-4441), freely accessible at: https://www.mdpi.com/journal/water/special_issues/2Q35A0SU2F.

For citation purposes, cite each article independently as indicated on the article page online and as indicated below:

| Lastname, A.A.; Lastname, B.B. Article Title. *Journal Name* **Year**, *Volume Number*, Page Range. |
|---|

**ISBN 978-3-7258-4088-5 (Hbk)**
**ISBN 978-3-7258-4087-8 (PDF)**
**https://doi.org/10.3390/books978-3-7258-4087-8**

# Contents

# Preface

In the context of significant global environmental changes and increasing human activities, the water quality of rivers is directly related to the health of ecosystems, the sustainable development of human society, and the rational use of water resources. Based on this, we have carefully planned and published this Book, titled "Water Quality Assessment of River Basins", which aims to compile cutting-edge technologies and practical strategies for assessing the quality of various water bodies, providing a comprehensive and in-depth reference for researchers and practitioners in relevant fields.

The scope of this reprint is extensive, covering the entire process ranging from basic water quality monitoring practices to complex data analysis methods. We have elaborated on the application of these tools in the context of water environment management, including river water quality assessment, tracking of pollutant sources, research on the migration and transformation processes of pollutants, development and application of water quality models, and evaluation of the ecological impacts of water quality changes. These contents not only help us to better understand the dynamic changes in water quality within river basins but also provide a solid scientific basis for formulating effective water quality protection measures.

The target audience of this reprint is broad, including, but not limited to, scholars engaged in water environment research, policymakers who formulate water policies, and other professionals interested in the water quality of river basins. We believe that this reprint will provide them with valuable reference information, which will help them make wiser decisions in their respective fields.

This reprint compiles 14 high-quality articles from around the world, authored by distinguished experts in their respective fields. Their research not only showcases the latest advancements in the field of water quality assessment but also offers valuable insights for future research directions. We sincerely thank all the authors for their active participation and contributions; without their efforts, the publication of this Book would not have been possible.

Finally, we extend our gratitude to all individuals and institutions that have provided assistance and support for the publication of this Book. Special thanks go to the editorial team of MDPI, whose professionalism and hard work have ensured the smooth publication of this reprint. We are also grateful to the reviewers, whose valuable comments and suggestions have significantly enhanced the quality of this rerpint.

**Qianqian Zhang**
*Guest Editor*

*water*

MDPI

*Editorial*

# Water Quality Assessment of River Basins: New Insights and Practical Solutions

**Qianqian Zhang**

Hebei and China Geological Survey Key Laboratory of Groundwater Remediation, Institute of Hydrogeology and Environmental Geology, Chinese Academy of Geological Sciences, Shijiazhuang 050061, China; zhangqianqian@mail.cgs.gov.cn

## 1. Introduction

Human activities and environmental changes have severely compromised water quality in river basins, posing challenges to ecosystems, public health, and sustainable development [1,2]. Issues such as industrial pollution, agricultural runoff, urbanization, and climate change necessitate a comprehensive and effective water quality assessment [3,4]. Over the past few decades, significant advancements in analytical techniques, monitoring technologies, and modeling approaches have been made [5,6]. However, many challenges remain due to the complexity of river basin systems and the dynamic nature of water quality [7,8]. This Special Issue on "Water Quality Assessment of River Basins" presents the latest research findings and practical applications, exploring various aspects of water quality assessment, including novel monitoring methods, advanced analytical techniques, integrated hydrological and water quality models, and innovative management strategies. The contributions reflect the diverse and evolving nature of water quality challenges in river basins, highlighting the importance of interdisciplinary approaches and the need for practical and sustainable solutions.

## 2. Main Contribution of This Special Issue

After a rigorous peer-review process, thirteen papers have been selected for publication in this Special Issue. The contributions and implications of these papers are discussed below.

Wang et al. (Contribution 1) investigates the features and driving factors of shallow groundwater quality in arid areas, specifically focusing on the Beichuan River Basin in Northwest China. The authors utilized hydrochemical analysis and multivariate statistics, including principal component analysis (PCA) and correlation analysis, to identify key factors influencing groundwater quality. The study found that the dissolution of rock salt primarily contributes to the presence of $Na^+$ and $Cl^-$ ions, while the weathering of silicate and carbonate rocks is the main origin of $Ca^{2+}$, $Mg^{2+}$, and $HCO_3^-$ ions. Additionally, the dissolution of evaporite rocks is identified as the principal source of $SO_4^{2-}$. Human activities, particularly sewage discharge and fertilization, significantly contribute to nitrate contamination. The findings suggest that rock weathering and industrial activities are the main controlling factors during the high-flow season, while the hydrochemistry of groundwater during the low-flow season is mainly influenced by the weathering of silicate rocks, evaporite rocks, and rock salt.

Rogéliz-Prada and Nogales (Contribution 2) introduce AFAR-WQS, an open-source MATLAB™ toolbox for rapid water quality simulation in large basins. It integrates assimilation factors with graph theory and a DFS algorithm to simulate 13 water quality

 https://doi.org/10.3390/w17081207

determinants across complex networks. AFAR-WQS can process up to 30,000 segments in 163 s, enabling real-time evaluations. Its scalable, object-oriented design supports customization and efficiency. Case studies show its utility in prioritizing sanitation investments and fostering stakeholder collaboration. AFAR-WQS bridges simplified and complex models, supporting adaptive management under hydrological uncertainty and climate change. Freely available on GitHub, it offers a transformative approach to integrated water resource management.

Song et al. (Contribution 3) applied eDNA metabarcoding and a multi-species biotic integrity index (Mt-IBI) to assess aquatic ecosystem health in the Irtysh River Basin. They analyzed eDNA from 52 sites, finding that dissolved oxygen, total nitrogen, and elevation significantly influence community structure, while alien fish species negatively impact health. The Mt-IBI showed high sensitivity to ecological changes, providing a robust tool for early degradation detection and guiding conservation efforts. The study highlights the importance of maintaining environmental parameters and controlling invasive species.

Gao et al. (Contribution 4) examined the distribution, sources, and risks of dissolved heavy metals (DHMs) in the Chaobai River, an urban river in the Beijing–Tianjin–Hebei region. Using ICP-MS and PCA, they found significant spatial variation in metal concentrations, with V, As, and Mo being most abundant. Sources include industrial pollution, natural processes, agricultural activities, and urban influences. The study also used a water quality index (WQI) and Monte Carlo simulation to assess health risks, particularly for As in vulnerable populations. The findings underscore the need for targeted water management policies to mitigate heavy metal contamination in urban rivers.

Garcia et al. (Contribution 5) examined the spatial–temporal variability of water quality in a rural property enrolled in a Payment for Environmental Services (PES) program in Campinas, Brazil. The study monitored physicochemical indicators of water quality at seven sampling points over five months, revealing significant variability in dissolved oxygen, biochemical oxygen demand, pH, total phosphorus, and total nitrogen. The results highlight the influence of both spatial location and seasonal changes in water quality parameters. The study underscores the importance of long-term monitoring to differentiate between natural seasonal variations and improvements resulting from PES program interventions. The findings provide valuable insights into the effective implementation and evaluation of PES programs in similar regions.

Paudel et al. (Contribution 6) investigated the hydrochemical dynamics and water quality of the Ghodaghodi Lake Complex in Nepal, analyzing 49 samples across pre- and post-monsoon periods. They found significant seasonal differences in key indicators, with higher TDS levels pre-monsoon due to evaporation and lower levels post-monsoon due to dilution. The dominant water type was $Ca^{2+}$-$HCO_3^-$, indicating carbonate weathering. The study confirmed the lakes' excellent irrigation quality and emphasized the importance of sustainable wetland management for biodiversity and alignment with the UN Sustainable Development Goals.

Ougrad et al. (Contribution 7) assessed health risks from trace elements in Dayat Roumi Lake, Morocco, through seasonal sampling at six stations. Results showed that Pb and Se concentrations exceeded WHO limits, with contamination sources identified as natural geological origins and anthropogenic inputs. The Water Quality Index (WQI) indicates poor water quality, posing significant risks to human and animal health, especially for children. The study emphasizes the need for effective management strategies to protect this vital ecosystem.

Bhattarai et al. (Contribution 8) examined the impact of freshwater inflow on coastal water quality in the Western Mississippi Sound (WMSS) using an integrated SWAT-vEFDC modeling approach. The study focused on major tributaries like the Jourdan and Wolf

Rivers, revealing significant impacts on total nitrogen (TN) and total phosphorus (TP) concentrations at inflow points, diminishing with distance. A sensitivity analysis during a tropical depression showed that extreme weather had a greater impact on water quality than a 25% variation in nutrient load. The study highlights the importance of integrated modeling for effective coastal water quality management.

Zhang et al. (Contribution 9) examined the spatiotemporal variations and sources of nitrogen in Baiyangdian Lake, China. While monitoring during three hydrological periods (normal, flood, and dry), the study analyzed 165 water quality data points. Results showed significant seasonal and spatial variations in nitrogen concentrations, with the highest total nitrogen (TN) levels in the dry season, exceeding national standards at 31.3% of sites. Northern areas and sites near inflowing rivers were most impacted, likely due to point-source pollution and agricultural runoff. Domestic sewage and manure were identified as the main sources of nitrogen pollution, with agricultural activities also contributing. Despite these challenges, Baiyangdian Lake's water quality is relatively good compared to similar lakes, indicating effective management measures. The findings provide a scientific basis for ongoing management and pollution prevention efforts.

Rezouki et al. (Contribution 10) investigated the ecological hazards in the Inaouen Wadi and its tributaries in Morocco by analyzing the presence of potentially toxic elements (PTEs) in sediments. The research collected sediment samples from 12 locations in 2019 and measured concentrations of Cd, Pb, Cr, Ag, Al, Cu, Fe, and Zn using ICP–AES. The study utilized multiple indices, including the enrichment factor (EF), geo-accumulation index (Igeo), potential ecological hazard index (RI), and modified ecological risk index (MRI), to evaluate sediment contamination levels. Results indicated significant contamination by Pb, Cd, Fe, and Zn, particularly at sites near urban discharges in Taza and Oued-Amlil. The EF and Igeo values revealed anthropogenic sources of Fe and Pb, while the RI and MRI highlighted potential ecological risks at several stations. The study underscores the urgent need for improved pollution management practices to mitigate these environmental risks.

Ren and Liu (Contribution 11) investigated the hydrochemical characteristics and controlling factors of the lower Yellow River, revealing significant spatiotemporal variations influenced by both natural and anthropogenic factors. The study identified rock weathering and evaporation as primary drivers, with notable impacts from agricultural fertilization, industrial emissions, and wastewater discharge. Nitrate levels increased during the flood season due to rainfall flushing, while other components showed higher concentrations in the low-flow season due to dilution. Spatial variations were more pronounced in the low-flow season, influenced by human activities, whereas the high-flow season showed less fluctuation due to the Xiaolangdi Reservoir's dilution effect. Principal component analysis highlighted carbonate weathering, evaporite dissolution, and industrial activities as key factors during the low-flow season, while agricultural activities and domestic sewage played significant roles during the flood season. These findings offer valuable insights into water resource management and protection in the region.

Gutiérrez-Rial et al. (Contribution 12) investigated microplastic (MP) pollution in river ecosystems, focusing on the effects of land use and biotic indices on MP distribution. The research was conducted in four river catchments in the NW Iberian Peninsula, analyzing MP concentrations in both water and sediments. The results indicate that urbanization is the primary driver of MP pollution in water, while population density significantly correlates with sediment pollution levels. The study also explores the potential impact of MPs on benthic macroinvertebrate communities, revealing that higher MP concentrations in sediments are associated with lower values of certain biotic indices (e.g., IASPT and EPT), suggesting a negative impact on sensitive taxa. This work highlights the importance of land use and population factors in MP pollution dynamics and suggests that existing biotic indices may serve as indicators of MP pollution in riverine environments.

Fashagba et al. (Contribution 13) evaluated the water quality of the Keddara Dam in Algeria using two water quality index (WQI) methods: the Canadian Council of Ministers of the Environment (CCME) WQI and the Weighted Arithmetic Method (WAM) WQI. The research analyzed 11 water quality parameters (temperature, pH, conductivity, turbidity, total suspended solids, full alkalimetric title, hydrometric title, nitrite, nitrate, ammonium, and phosphate ions) based on data collected from December 2018 to June 2021. The results indicate that the water quality of the Keddara Dam is generally suitable for agricultural and municipal use, with CCME WQI values ranging from 81.92 (acceptable) to 95.08 (excellent) and WAM WQI values ranging from 9.52 to 17.77 (excellent). However, conductivity and turbidity values exceeded WHO permissible limits. The study highlights the importance of using WQI methods to assess and manage water quality in arid and semi-arid regions.

## 3. Conclusions

The papers published in this Special Issue highlight the multifaceted challenges and innovative solutions in river basin water quality assessment. The first key finding is the critical role of understanding and managing the complex interactions between natural processes and human activities. For instance, rock weathering and industrial activities are identified as significant drivers of groundwater quality in arid regions (Contribution 1), while domestic sewage and agricultural runoff are major sources of nitrogen pollution in Baiyangdian Lake (Contribution 9). Additionally, urbanization and population density are primary drivers of microplastic pollution in river ecosystems (Contribution 12). The hydrochemical characteristics of the lower Yellow River reveal significant spatiotemporal variations influenced by both natural and anthropogenic factors (Contribution 11). The presence of potentially toxic elements in sediments of the Inaouen Wadi and its tributaries in Morocco underscores the need for improved pollution management practices (Contribution 10). Finally, the evaluation of the water quality of the Keddara Dam in Algeria using water quality indices demonstrates its suitability for agricultural and municipal use while highlighting the need for further treatment to meet WHO standards (Contribution 13).

The second main result is the importance of advanced tools and techniques in water quality assessment. The introduction of AFAR-WQS, a MATLAB™ toolbox for rapid water quality simulation, demonstrates the potential of numerical modeling to support decision-making in large basins (Contribution 2). Similarly, the application of eDNA metabarcoding combined with a multi-species biotic integrity index (Mt-IBI) provides a sensitive tool for assessing aquatic ecosystem health in the Irtysh River Basin (Contribution 3). These tools enhance our ability to monitor and manage water quality effectively.

A third finding is the significance of long-term monitoring and adaptive management strategies. Long-term monitoring is essential to differentiate natural seasonal variations from improvements due to interventions such as Payment for Environmental Services (PES) programs in Brazil (Contribution 5). Integrated modeling approaches, such as those used to assess the impact of freshwater inflow on coastal water quality in the Western Mississippi Sound (Contribution 8), further emphasize the need for adaptive management to address dynamic water quality issues.

Other important findings include the significant spatial heterogeneity of heavy metals in the Chaobai River, highlighting the need for targeted water management policies (Contribution 4); the importance of sustainable wetland management for biodiversity conservation in the Ramsar-listed Ghodaghodi Lake Complex in Nepal (Contribution 6); and the health risks associated with trace elements in surface water of Dayat Roumi Lake, Morocco, emphasizing the need for effective management strategies (Contribution 7).

In conclusion, the collection of papers in this Special Issue underscores the importance of integrated and adaptive management strategies for river basin water quality assessment. The studies provide valuable insights into the sources and impacts of pollution, the effectiveness of management interventions, and the potential of new technologies to support sustainable water resource management. We hope that these contributions will stimulate further interdisciplinary research and practical actions to improve water quality in river basins worldwide.

**Funding:** This research received no external funding.

**Conflicts of Interest:** The author declares no conflicts of interest.

**List of Contributions:**

1. Wang, L.; Yang, N.; Zhao, Y.; Zhang, Q. Research on the Features and Driving Factors of Shallow Groundwater Quality in Arid Areas, Northwest China. *Water* **2025**, *17*, 934. https://doi.org/10.3390/w17070934.
2. Rogéliz-Prada, C.A.; Nogales, J. AFAR-WQS: A Quick and Simple Toolbox for Water Quality Simulation. *Water* **2025**, *17*, 672. https://doi.org/10.3390/w17050672.
3. Song, T.; Zi, F.; Huang, Y.; Fang, L.; Zhang, Y.; Liu, Y.; Chang, J.; Li, J. Assessment of Aquatic Ecosystem Health in the Irtysh River Basin Using eDNA Metabarcoding. *Water* **2025**, *17*, 246. https://doi.org/10.3390/w17020246.
4. Gao, X.; Han, G.; Zhang, S.; Zeng, J. Sources, Water Quality, and Potential Risk Assessment of Heavy Metal Contamination in Typical Megacity River: Insights from Monte Carlo Simulation. *Water* **2025**, *17*, 224. https://doi.org/10.3390/w17020224.
5. Garcia, J.M.; Longo, R.M.; Nunes, A.N.; Gomes, R.C. Spatial-Temporal Monitoring of Water Quality in Rural Property Enrolled in a Program for Payment for Environmental Water Services (PES-Water)—A Case Study in Brazil. *Water* **2024**, *16*, 3673. https://doi.org/10.3390/w16243673.
6. Paudel, G.; Pant, R.R.; Joshi, T.R.; Saqr, A.M.; Đurin, B.; Cetl, V.; Kamble, P.N.; Bishwakarma, K. Hydrochemical Dynamics and Water Quality Assessment of the Ramsar-Listed Ghodaghodi Lake Complex: Unveiling the Water-Environment Nexus. *Water* **2024**, *16*, 3373. https://doi.org/10.3390/w16233373.
7. Ougrad, I.; Elassassi, Z.; Mrabet, A.; Mssillou, I.; Lim, A.; Shahat, A.A.; Rezouki, S.; Moubchir, T. Health Risk Assessment of Trace Elements in Surface Water from Dayat Roumi Lake, Morocco. *Water* **2024**, *16*, 3231. https://doi.org/10.3390/w16223231.
8. Bhattarai, S.; Parajuli, P.; Linhoss, A. Integrated Modeling Approach to Assess Freshwater Inflow Impact on Coastal Water Quality. *Water* **2024**, *16*, 3012. https://doi.org/10.3390/w16213012.
9. Zhang, Q.; Xu, S.; Yang, L. Spatiotemporal Variation Characteristics and Source Identification of Nitrogen in the Baiyangdian Lake Water, China. *Water* **2024**, *16*, 2969. https://doi.org/10.3390/w16202969.
10. Rezouki, S.; Moubchir, T.; El Hanafi, L.; Flouchi, R.; Zahir, I.; Alzain, M.N.; El Guerrouj, B.; Noman, O.; Shahat, A.A.; Allali, A. Assessment of Ecological Hazards in the Inaouen Wadi and Its Tributaries Using the Presence of Potentially Toxic Elements in Its Sediments. *Water* **2024**, *16*, 2936. https://doi.org/10.3390/w16202936.
11. Ren, C.; Liu, L. Under the Strong Influence of Human Activities: The Patterns and Controlling Factors of River Water Chemistry Changes—A Case Study of the Lower Yellow River. *Water* **2024**, *16*, 1886. https://doi.org/10.3390/w16131886.
12. Gutiérrez-Rial, D.; Villar, I.; Álvarez-Troncoso, R.; Soto, B.; Mato, S.; Garrido, J. Assessment of Microplastic Pollution in River Ecosystems: Effect of Land Use and Biotic Indices. *Water* **2024**, *16*, 1369. https://doi.org/10.3390/w16101369.
13. Fashagba, T.S.; Bessedik, M.; ElSayed, N.B.; Abdelbaki, C.; Kumar, N. Evaluating the Water Quality of the Keddara Dam (Algeria) Using Water Quality Indices. *Water* **2024**, *16*, 1291. https://doi.org/10.3390/w16091291.

## References

1. Zhu, L.; Liu, J.T.; Yang, M.N.; Zhang, Y.X.; Wen, D.P. Evolutionary trend of water cycle in Beichuan River Basin of China under the influence of vegetation restoration. *J. Groundw. Sci. Eng.* **2021**, *9*, 202–211. [CrossRef]
2. Palmer, M.A.; Reidy Liermann, C.A.; Nilsson, C.; Flörke, M.; Alcamo, J.; Lake, P.S.; Bond, N. Climate change and the world's river basins: Anticipating management options. *Front. Ecol. Environ.* **2008**, *6*, 81–89. [CrossRef]
3. Zhai, T.L.; Zhang, Q.Q.; Wang, L.; Wang, H.W. Temporal and spatial variations hydrochemical components and driving factors in baiyangdian lake in the northern plain of China. *J. Groundw. Sci. Eng.* **2024**, *12*, 293–308. [CrossRef]
4. Jin, T.; Zhang, X.; Wang, T.; Liang, J.; Ma, W.; Xie, J. Spatiotemporal impacts of climate change and human activities on blue and green water resources in northwest river basins of China. *Ecol. Indic.* **2024**, *160*, 111823. [CrossRef]
5. Gandhimathi, G.; Chellaswamy, C.; Thiruvalar Selvan, P. Comprehensive river water quality monitoring using convolutional neural networks and gated recurrent units: A case study along the Vaigai River. *J. Environ. Manag.* **2024**, *365*, 121567. [CrossRef]
6. Behmel, S.; Damour, M.; Ludwig, R.; Rodriguez, M.J. Water quality monitoring strategies—A review and future perspectives. *Sci. Total Environ.* **2016**, *571*, 1312–1329. [CrossRef] [PubMed]
7. Makumbura, R.K.; Mampitiya, L.; Rathnayake, N.; Meddage, D.P.P.; Henna, S.; Dang, T.L.; Rathnayake, U. Advancing water quality assessment and prediction using machine learning models, coupled with explainable artificial intelligence (XAI) techniques like shapley additive explanations (SHAP) for interpreting the black-box nature. *Results Eng.* **2024**, *23*, 102831. [CrossRef]
8. Paudel, G.; Pant, R.R.; Joshi, T.R.; Saqr, A.M.; Đurin, B.; Cetl, V.; Kamble, P.N.; Bishwakarma, K. Hydrochemical Dynamics and Water Quality Assessment of the Ramsar-Listed Ghodaghodi Lake Complex: Unveiling the Water-Environment Nexus. *Water* **2024**, *16*, 3373. [CrossRef]

water

MDPI

Article

# Research on the Features and Driving Factors of Shallow Groundwater Quality in Arid Areas, Northwest China

Long Wang [1], Nan Yang [2], Yang Zhao [2] and Qianqian Zhang [1,*]

1 Hebei and China Geological Survey Key Laboratory of Groundwater Remediation, Institute of Hydrogeology and Environmental Geology, Chinese Academy of Geological Sciences, Shijiazhuang 050061, China; wanglikang@mail.cgs.gov.cn
2 Shandong Coalfield Geological Planning Research Institute, Jinan 250000, China; tayn85@126.com (N.Y.); 19105385699@163.com (Y.Z.)
* Correspondence: zhangqianqian@mail.cgs.gov.cn

**Abstract:** Given the increasing threat of groundwater pollution, comprehending the trends and influencing factors of groundwater quality variation is essential for effective mitigation strategies. This study addresses groundwater quality variations in the Beichuan River, a critical area in China's arid region. Using hydrochemical analysis and multivariate statistics, we identified key factors influencing groundwater quality. Groundwater is mildly alkaline, with $HCO_3^-$-Ca as the dominant hydrochemical type. The concentrations of major ions increase during the high-flow period due to rainfall effects. The dissolution of rock salt primarily contributes to the presence of $Na^+$ and $Cl^-$ ions. Meanwhile, the weathering of silicate and carbonate rocks is the main origin of $Ca^{2+}$, $Mg^{2+}$, and $HCO_3^-$ ions. Additionally, the dissolution of evaporite rocks is identified as the principal source of $SO_4^{2-}$. Human activities, particularly sewage discharge and fertilization, significantly contribute to nitrate contamination. Principal component analysis revealed that the weathering of rocks and industrial activities are the main controlling factors during the high-flow season, while the hydrochemistry of groundwater during the low-flow season is mainly influenced by the weathering of silicate rocks, evaporite rocks, and rock salt. Our findings provide a scientific basis for preventing groundwater quality deterioration and ecological environmental protection in arid regions.

**Keywords:** arid areas; groundwater; sources; controlling factors; anthropogenic activities

## 1. Introduction

As a vital resource for human survival, groundwater also holds a crucial role in sustaining sustainable social development and ecological stability [1]. In recent years, as industrialization and urbanization have made rapid progress, the impact of human activities on groundwater quality has increased annually, especially in arid areas with restricted groundwater recharge sources, where human activities may have a more severe effect on groundwater [2–4]. Considering the escalating impact on groundwater quality, it is highly vital to determine the characteristics and controlling factors of groundwater quality variation in arid areas [5–7].

The quality of groundwater is primarily impacted by both natural factors, including topography, lithology, and the conditions of groundwater recharge, and human factors, such as land use, agricultural practices, and industrial activities [8,9]. Presently, researchers both domestically and internationally have explored the characteristics of groundwater



https://doi.org/10.3390/w17070934

quality variation to some extent. For instance, Bozdağ [10] applied saturation index methods and multivariate statistical analysis to investigate the hydrochemical controlling factors of groundwater in Turkey's Konya Plain. The findings indicated that water–rock interactions are the predominant factors influencing the hydrochemical composition in this region. Zhang et al. [11] examined the hydrochemical evolution patterns of groundwater in the Hutuo River region and suggested that the chemical composition of groundwater is impacted not only by water–rock interactions and cation exchange but also by domestic sewage. In addition, researchers found that the determinants of groundwater quality include mineral dissolution, interaction with mine water, contamination from domestic wastewater and septic tanks, and pollution from agricultural fertilizers [12]. Nevertheless, investigations into the characteristics and driving factors of groundwater quality fluctuations in arid regions are rather sparse, and the underlying mechanisms causing groundwater quality degradation in these areas remain to be fully elucidated.

The Beichuan River Basin is situated in China's arid northwest region, where groundwater serves as the primary water supply. Nowadays, the rapid progress of industrialization and urbanization has led to an increasing impact of human activities on groundwater quality, resulting in a deteriorating trend of groundwater quality [13]. Previous research in this area has predominantly concentrated on the features of surface water quality variation, contamination sources, and prevention strategies [14,15]. However, studies on the features and influential factors of groundwater quality variation remain inadequate. Groundwater in arid areas has limited recharge sources and significant evaporation, making the groundwater system more vulnerable and more sensitive to human activities. The patterns and driving factors of groundwater quality variation in these regions may differ from those in other areas. Consequently, this study comprehensively gathered groundwater samples from the Beichuan River Basin and employed hydrochemical diagrams along with multivariate statistical methods to thoroughly examine the patterns of hydrochemical change and the controlling factors of groundwater across different periods. It is anticipated that the findings of this research will offer a significant scientific foundation for the sustainable utilization of groundwater resources and ecological environment conservation in regions with similar geological backgrounds

## 2. Materials and Methods

### 2.1. Overview of the Study Area

The Beichuan River Basin spans two major geomorphological units: the Qilian Mountain Plateau and the Huang–Liangzhong Plateau [16]. The topography of the area is characterized by higher elevations in the northwest and lower elevations in the southeast, forming a narrow and elongated shape extending from north to south. The region experiences a semi-arid continental climate, characterized by a mean annual rain volume of 367.5 mm, which is primarily concentrated between May and September.

### 2.2. Geological and Hydrogeological Conditions

The geological strata within this area are composed of formations from the Neogene (Lower Tertiary), Upper Tertiary, and Quaternary systems. The Quaternary system is primarily composed of alluvial, fluvioglacial, and glacial deposits of sand and gravel. The types of groundwater include porous aquifer water in loose rocks, fractured porous water in clastic rocks, and karstic fracture water in carbonate rocks, each reflecting distinct hydrogeological characteristics. Topography plays a crucial role in determining the interaction between groundwater and surface water. In narrow valley areas, groundwater flow is obstructed, causing the water table to rise and groundwater to discharge to the surface. In wider valley areas, surface water infiltrates extensively, strongly recharging the groundwater. The primary sources of groundwater recharge include rainfall and snowmelt, as well as the percolation of surface water into the subsurface. The main exit routes for groundwater are pumping, spring outflow, and evaporation, which collectively influence the groundwater balance.

### 2.3. Sample Collection and Testing

A total of 21 groundwater samples were collected during the low-flow season (July 2024) and the high-flow season (October 2024). Sampling sites were predominantly situated along both banks of the Beichuan River, with sampling depths varying between 5 and 25 m below the ground surface. The spatial distributions of these sampling sites are illustrated in Figure 1.

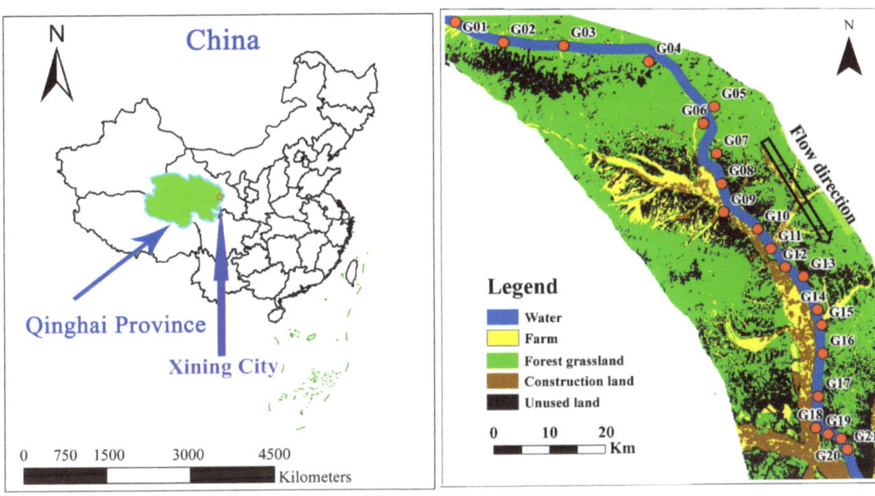


**Figure 1.** Geographical location map of the Beichuan River Basin.

In this study, a 500 mL polyethylene plastic bottle was used for sampling. Before sampling, the bottle was thoroughly cleaned with distilled water. During sampling, the collection bottle was thoroughly rinsed three times using the groundwater collected.

Prior to collecting the water samples, the wells were purged to remove stagnant water. The collected samples were subsequently split into two parts. One part was used for the analysis of cations ($Na^+$, $K^+$, $Ca^{2+}$, and $Mg^{2+}$), with high-purity nitric acid added in situ to acidify the sample to pH < 2. The other part was reserved for the analysis of anions ($HCO_3^-$, $SO_4^{2-}$, $Cl^-$, and $NO_3^-$). All samples were filled to capacity in the containers without any air bubbles to prevent contamination and stored in a dark environment at low temperatures to maintain the sample's integrity. Testing was conducted within a week after collection to ensure accurate results.

The pH of the water was measured on site using a portable water quality multiparameter analyzer (Hach-HQ40D, produced by Hach Company Water Quality Analyzer (Shanghai) Co., Ltd., Shanghai, China). The total dissolved solids (TDSs) were determined via the drying method. $Na^+$, $K^+$, $Ca^{2+}$, and $Mg^{2+}$ were measured using an atomic absorption spectrophotometer (AA240FS/GTA120, produced by Agilent Technologies Co., Ltd., Shanghai, China), and the detection limit values for the respective analytes were 0.01 mg/L, 0.05 mg/L, 4.0 mg/L, and 3.0 mg/L. The $SO_4^{2-}$, $Cl^-$, and $NO_3^-$ were analyzed using ion chromatography (CIC-D180, manufactured by Qingdao Shenghan Chromatography Technology Co., Ltd., Qingdao, China), with detection limits of 0.75 mg/L, 1.0 mg/L, and

0.664 mg/L, respectively. $HCO_3^-$ was determined via titration with hydrochloric acid, with a detection limit of 5.0 mg/L.

### 2.4. Data Analysis Methods

Water Quality Index (WQI)

In our research, the WQI served as a tool for evaluating the groundwater quality within the Beichuan River Basin. The calculation of the WQI involved allocating a specific weight ($W_i$) to each parameter based on its critical role relative to the overall potability of groundwater. The criteria for water quality were primarily based on China's Grade III standards for groundwater [17]. In cases where these standards did not include a particular parameter, the guidelines from the World Health Organization (WHO) published in 2011 were utilized. The weights ($W_i$) and their corresponding relative weights ($RW_i$) for each parameter are detailed in Supplementary Table S1. The WQI values obtained were then categorized into five distinct quality levels: excellent (WQI < 50), good ($50 \leq$ WQI $\leq 100$), poor ($100 <$ WQI $\leq 200$), very poor ($200 <$ WQI $\leq 300$), and unsuitable for drinking (WQI > 300).

The WQI was calculated as follows:

$$RW_i = \frac{W_i}{\sum_{i=1}^{n} W_i} \tag{1}$$

$$Q_i = \frac{C_i}{S_i} \times 100 \tag{2}$$

$$SI_i = W_i \times Q_i \tag{3}$$

$$WQI = \sum SI_i \tag{4}$$

In the formula, $Q_i$ denotes the quality rating; $C_i$ and $Si$ denote the concentration (mg/L) and the water quality standard of each indicator, respectively; $SI_i$ is the sub-index of the $i$-th indicator.

ArcGIS10.5 was employed in this study to generate the map of sampling point distribution for the study area, providing a visual representation of the sampling locations. Ion ratio diagrams were drawn using Origin 2022. SPSS25.0 was employed in this study to conduct correlation assessment and principal component evaluation on the data of water quality.

## 3. Results and Discussion

### 3.1. Pattern of Groundwater Quality in the Beichuan River Basin

To analyze the water quality characteristics during the two hydrological periods, statistical analyses were conducted on the range of values, means, and exceedance rates of various water quality indicators (Table 1). The results indicate that in the high-flow season, groundwater pH levels fluctuated between 6.60 and 8.31, averaging 7.53, which suggests a mildly alkaline to neutral nature. In contrast, during the low-flow season, pH values oscillated from 6.80 to 8.20, with a mean of 7.64, consistently reflecting a weakly alkaline condition. In the high-flow season, the levels of cations were ordered as follows: $Na^+ > Ca^{2+} > Mg^{2+} > K^+$. In the low-flow season, the ranking was as follows: $Ca^{2+} > Na^+ > Mg^{2+} > K^+$. For anions, the content ranking was consistent in both seasons: $SO_4^{2-} > HCO_3^- > Cl^- > NO_3^-$. During the high-flow season, the average levels of $Na^+$, $Mg^{2+}$, $Cl^-$, $HCO_3^-$, $NO_3^-$, and TDS were elevated compared to the low-flow season. In contrast, the average levels of pH, $K^+$, $Ca^{2+}$, and $SO_4^{2-}$ were greater during the low-flow season than in the high-flow season. TDS, $SO_4^{2-}$, $Na^+$, and $Cl^-$ were recognized as the key determinants of groundwater quality, with excess rates during the high-flow season being

35.3%, 33.3%, 14.3%, and 4.76%, respectively, and during the low-flow season, they were 35.3%, 28.6%, 19.1%, and 4.76%.

**Table 1.** Statistical analysis of the water sample index of the study area.

| Parameters | Range | | Average | | Exceedance Rate/(%) | | Class III Groundwater Quality Standard * |
|---|---|---|---|---|---|---|---|
| | HFS | LFS | HFS | LFS | HFS | LFS | |
| pH | 6.60–8.31 | 6.80–8.20 | 7.53 | 7.64 | 0 | 0 | 6.50–8.50 |
| $K^+$ | 1.44–24.2 | 1.51–27.8 | 6.13 | 6.86 | – | – | – |
| $Na^+$ | 3.88–1514 | 4.38–452 | 136 | 90.0 | 14.3% | 19.1% | 200 |
| $Ca^{2+}$ | 12.9–354 | 38.3–375 | 96.8 | 114 | – | – | – |
| $Mg^{2+}$ | 4.40–211 | 0.97–126 | 41.0 | 33.6 | – | – | – |
| $SO_4^{2-}$ | 9.30–3210 | 20.1–1692 | 373 | 299 | 33.3% | 28.6% | 250 |
| $Cl^-$ | 2.81–307 | 5.60–332 | 50.90 | 57.5 | 4.76% | 4.76% | 250 |
| $HCO_3^-$ | 47.7–576 | 91.6–514 | 278 | 276 | – | – | – |
| $NO_3^-$ | 1.23–82.4 | 1.77–66.8 | 25.0 | 23.8 | 0 | 0 | 88.6 |
| TDS | 65.0–5189 | 126–2803 | 869 | 764 | 35.3% | 35.3% | 1000 |

Note: * Refers to the Class III water quality standard in the "Groundwater Quality Standard" [17]. The units of all water quality parameters are $mg \cdot L^{-1}$, except for pH, which is dimensionless. HFS: High-flow season; LFS: low-flow season.

### 3.2. Water Quality Assessment

Table 2 presents the water quality index (WQI) classifications for groundwater quality during the high- and low-flow seasons along the Beichuan River. During the high-flow season, the quality distribution of the samples was as follows: 61.9% were rated as Class I, 19.0% as Class II, 9.52% as Class III, 4.76% as Class IV, and 4.76% were deemed Class V. In contrast, during the low-flow season, the classification shifted to 51.7% Class I, 28.6% Class II, and 6.3% Class III. Overall, the proportion of poor-quality groundwater was higher in the high-flow season compared to the low-flow season, likely due to the infiltration of groundwater by surface runoff carrying pollutants.

**Table 2.** Groundwater quality index classification for the two seasons in Beichuan River.

| WQI Type | Grade | High-Flow Season | | Low-Flow Season | |
|---|---|---|---|---|---|
| | | Number | Rate (%) | Number | Rate (%) |
| Excellent water | I | 13 | 61.9 | 12 | 57.1 |
| Good water | II | 4 | 19.0 | 6 | 28.6 |
| Poor water | III | 2 | 9.52 | 3 | 14.3 |
| Very poor water | IV | 1 | 4.76 | 0 | 0 |
| Water unsuitable for drinking purposes | V | 1 | 4.76 | 0 | 0 |

### 3.3. Analysis of Groundwater Chemical Types

The Piper diagram is widely utilized for depicting the hydrochemical type of groundwater [18]. As depicted in Figure 2, during the high-flow season, the primary hydrochemical categories of groundwater included the $HCO_3$–Ca type (38.1%), $HCO_3$–Ca·Mg type (19.0%), and $HCO_3 \cdot SO_4$–Ca·Na·Mg type (14.3%). During the low-flow season, the dominant hydrochemical categories were $HCO_3$–Ca type (38.1%), $HCO_3 \cdot SO_4$–Ca type, $SO_4$–Na·Ca type, and $HCO_3 \cdot SO_4$–Ca·Na·Mg type (each accounting for 14.3%). Overall, the proportion of sulfate-type water is higher during the high-flow season than during the dry season. This increase is primarily attributed to the enhanced leaching of evaporite rocks during the high-flow season. This facilitated the accumulation of $SO_4^{2-}$ in groundwater [19].

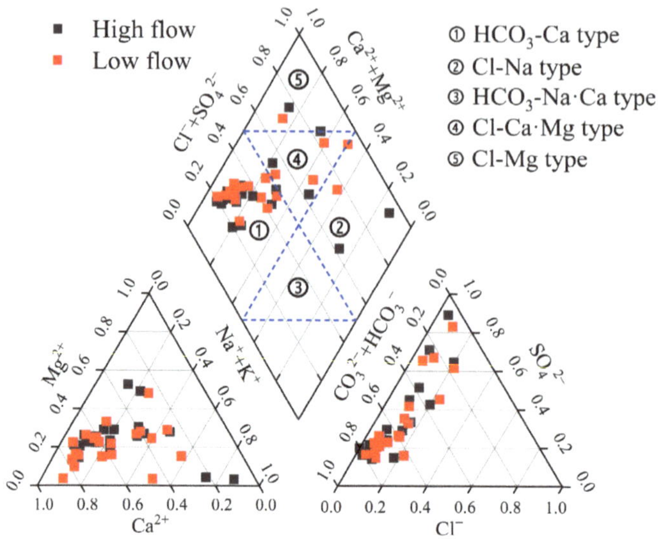


**Figure 2.** Piper triplex of the Beichuan River Basin.

### 3.4. *Controlling Factors and Origins of Groundwater Ions*

#### 3.4.1. Identify the Controlling Factors

The chemical composition of natural water is primarily impacted by rock weathering, atmospheric precipitation, and evaporation [20]. The atmospheric precipitation control zone is typified by low TDS concentrations (<10 mg/L) and elevated $[Na^+]/[Na^+ + Ca^{2+}]$ and $[Cl^-]/[Cl^- + HCO_3^-]$ ratios (typically between 0.5 and 1), and it is positioned at the Gibbs diagram's lower right corner. The rock weathering control zone, with TDS values generally between 70 and 300 mg/L and $[Na^+]/[Na^+ + Ca^{2+}]$ and $[Cl^-]/[Cl^- + HCO_3^-]$ ratios both below 0.5, is located in the middle left section. The evaporation concentration control zone, characterized by high TDS (>300 mg/L) and $[Na^+]/[Na^+ + Ca^{2+}]$ and $[Cl^-]/[Cl^- + HCO_3^-]$ ratios ranging from 0.5 to 1, is situated at the diagram's upper right corner [21].

Figure 3a,b show that the sampling sites are primarily located in areas controlled by rock weathering and evaporative concentration processes. This trend suggests that they are the dominant factors influencing the hydrochemical characteristics of groundwater in this area. However, the Gibbs diagram is not sufficient for distinguishing the impact of human activities. Subsequent studies will concentrate on examining how human activities impact groundwater chemistry.

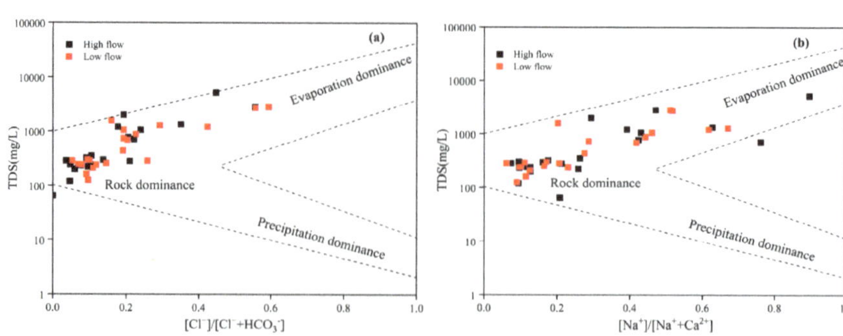


**Figure 3.** (**a**,**b**) Gibbs chart of the Beichuan River basin.

### 3.4.2. Water–Rock Interaction

Water–rock interactions significantly influence the chemical composition of groundwater. Studies have shown that the breakdown and dissolution of carbonate rocks mainly produce $Ca^{2+}$, $Mg^{2+}$, and $HCO_3^-$. Silicate rocks, through analogous processes, yield $Ca^{2+}$, $Mg^{2+}$, $Na^+$, $K^+$, and $HCO_3^-$. In addition, the weathering and dissolution of evaporite rocks lead to the formation of $Ca^{2+}$, $Mg^{2+}$, $Na^+$, $K^+$, $SO_4^{2-}$, and $Cl^-$. With these findings in mind, ion ratio diagrams can be effectively employed to identify the sources of major ions in the groundwater of the study area [22].

The relationship between $[Ca^{2+}/Na^+]$ and $[HCO_3^-/Na^+]$ can be used to identify the effects of the weathering and dissolution of carbonate, silicate, and evaporite rocks on the chemical composition of groundwater [23]. Researchers found that the endmember ratios for silicate rocks are $[Ca^{2+}/Na^+] = 0.35 \pm 0.15$ and $[HCO_3^-/Na^+] = 2 \pm 1$; for carbonate rocks, they are $[Ca^{2+}/Na^+] = 50$ and $[HCO_3^-/Na^+] = 120$; for evaporite rocks, they are $[Ca^{2+}/Na^+] = 0.25 \pm 0.15$ and $[HCO_3^-/Na^+] = 0.01 \pm 0.01$ [24]. Figure 4a reveals that, in both high-flow and low-flow seasons, the majority of groundwater sampling points cluster in the central region between the silicate and carbonate rock endmembers, with a few points nearing the evaporite dissolution zone. This suggests that $Ca^{2+}$, $Mg^{2+}$, $Na^+$, $K^+$ and $HCO_3^-$ in groundwater predominantly stem from the weathering and dissolution of silicate and carbonate rocks, while minor quantities of $Na^+$, $K^+$, and $Ca^{2+}$ could be sourced from evaporite rocks like rock salt and gypsum [25]. Considering the hydrogeological context, groundwater resides in sand and gravel layers, where the dissolution of minerals such as quartz and feldspar serves as the primary origin of $Ca^{2+}$, $Mg^{2+}$, $Na^+$, $K^+$, and $HCO_3^-$.

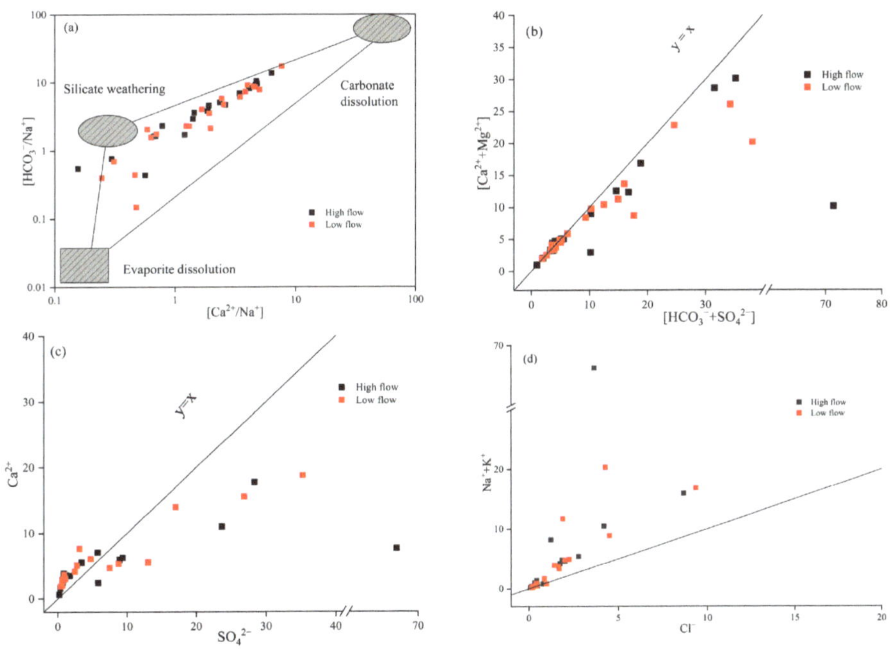


**Figure 4.** Ion ratio diagram: (**a**) $[Ca^{2+}/Na^+]$ vs. $[HCO_3^-/Na^+]$; (**b**) $[HCO_3^- + SO_4^{2-}]$ vs. $[Ca^{2+} + Mg^{2+}]$; (**c**) $[SO_4^{2-}]$ vs. $[Ca^{2+}]$; (**d**) $[Cl^-]$ vs. $[Na^+ + K^+]$.

Figure 4b indicates that during both seasons, the groundwater sampling points are predominantly situated above the y = x line, indicating that the dissolution of silicate and evaporite rocks is dominant, providing $K^+$, $Na^+$, $Ca^{2+}$, $Mg^{2+}$, and $SO_4^{2-}$, while the dissolution of carbonate rocks mainly supplies $HCO_3^-$ and a portion of $Ca^{2+}$ and $Mg^{2+}$.

Figure 4c, which is the $[Ca^{2+}]/[SO_4^{2-}]$ ratio diagram, shows that in both seasons, 71.4% of samples are positioned above the y = x line, where a strong correlation between $Ca^{2+}$ and $SO_4^{2-}$ is evident. This implies that the dissolution of gypsum is the primary source of $SO_4^{2-}$ [26], whereas excess $Ca^{2+}$ primarily originates from the weathering and dissolution of silicate or carbonate rocks. Conversely, 28.6% of samples fall below the y = x line, indicating additional sources of $SO_4^{2-}$ beyond gypsum dissolution, such as industrial activities and domestic sewage [27].

The $[Na^+ + K^+]/[Cl^-]$ milliequivalent ratio serves as an indicator of the primary sources of $Na^+$ and $Cl^-$ in the water's chemical processes. As depicted in Figure 4d, during both seasons, all water samples are positioned above the y = x line and display a general correlation. This suggests that $Na^+$ and $Cl^-$ share a common origin, specifically the dissolution of rock salt. The milliequivalent ratio of $[Na^+ + K^+]$ typically exceeds that of $Cl^-$, which is primarily attributed to the dissolution of silicate rocks. Additionally, domestic sewage discharge may contribute to elevated $Na^+$ concentrations [28].

### 3.5. Human Activity Impact

Human activities significantly influence the variation in groundwater quality by altering the chemical composition through the discharge of wastewater, waste, and exhaust gases [29]. In regions with high levels of human activity, the levels of $NO_3^-$, $Cl^-$, $SO_4^{2-}$, and $Na^+$ in groundwater are prone to substantial increases. Specifically, $SO_4^{2-}$ is predominantly influenced by industrial operations, whereas $NO_3^-$, $Cl^-$, and $Na^+$ are more affected by agricultural practices, such as the use of fertilizers and manure, and domestic wastewater [30]. Figure 5a shows that during both seasons, 90.5% and 85.7% of the sampling points, respectively, exhibited higher $[SO_4^{2-}]/[Ca^{2+}]$ ratios than $[NO_3^-]/[Ca^{2+}]$. Specifically, during the high-flow season, the $[SO_4^{2-}]/[Ca^{2+}]$ ratio ranged from 0.201 to 8.74, while the $[NO_3^-]/[Ca^{2+}]$ ratio ranged from 0.001 to 0.241. During the low-flow season, the $[SO_4^{2-}]/[Ca^{2+}]$ ratio ranged from 0.220 to 2.34, and the $[NO_3^-]/[Ca^{2+}]$ ratio ranged from 0.001 to 0.210. Combined with the distribution of sampling points on the diagram, this suggests that industrial activities are the primary source of human impact on groundwater, followed by agricultural activities. According to the survey results, large aluminum and thermal power plants are located in the southern part of the study area, and the discharge of wastewater from these plants may lead to increased $SO_4^{2-}$ concentrations in groundwater [31].

The $[NO_3^-]/[Cl^-]$ ratio in relation to $[Cl^-]$ concentrations is typically employed to identify the sources of $NO_3^-$. Usually, elevated $[Cl^-]$ levels coupled with low $[NO_3^-]/[Cl^-]$ ratios point to domestic sewage and manure as the main contributors of $NO_3^-$. Conversely, low $[Cl^-]$ levels with high $[NO_3^-]/[Cl^-]$ ratios imply that $NO_3^-$ is predominantly from agricultural activities. When both $[Cl^-]$ and $[NO_3^-]/[Cl^-]$ ratios are low, soil nitrogen is the key source of $NO_3^-$ [32]. As shown in Figure 5b, during the high- and low-flow periods, 52.4% and 61.9% of the sampling points, respectively, exhibited higher $[Cl^-]$ than $[NO_3^-]/[Cl^-]$. Specifically, in the high-flow period, the content of $[Cl^-]$ varied between 0.081 and 8.66, while the $[NO_3^-]/[Cl^-]$ ratio fluctuated from 0.021 to 0.652. In contrast, during the low-flow season, $[Cl^-]$ concentrations spanned from 0.161 to 9.35, with the $[NO_3^-]/[Cl^-]$ ratio ranging similarly from 0.021 to 0.652. Collectively, these data suggest that $NO_3^-$ in groundwater is predominantly sourced from manure and domestic sewage, with their influence on groundwater quality being more pronounced in the low-flow season than in the high-flow season.

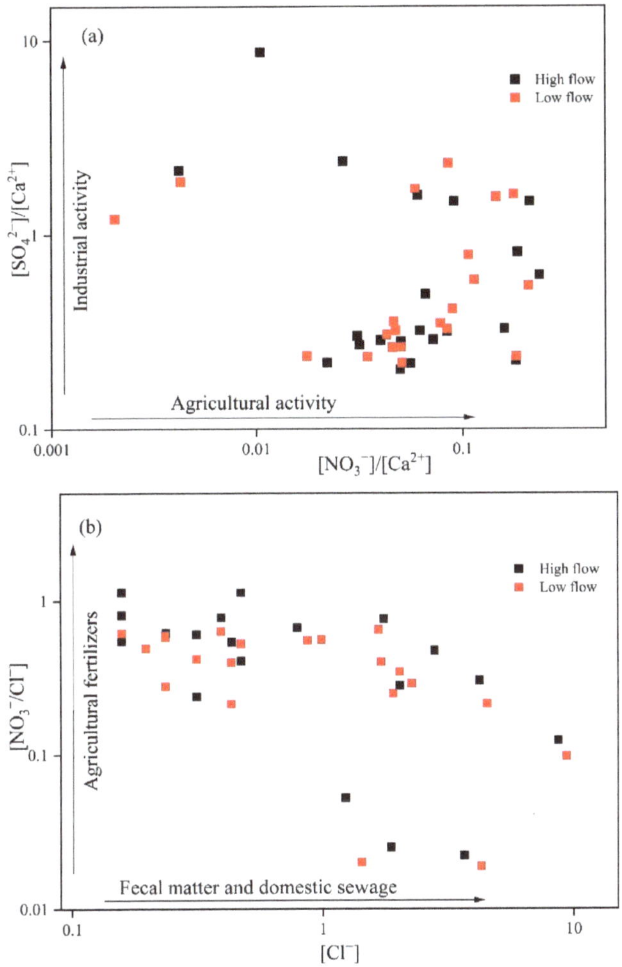


**Figure 5.** Ion ratio diagram: (**a**) $[NO_3^-]/[Ca^{2+}]$ vs. $[SO_4^{2-}]/[Ca^{2+}]$; (**b**) $[Cl^-]$ vs. $[NO_3^-]/[Cl^-]$.

Considering the geographical overview of the study area and the layout of the sampling points, it is evident that most groundwater sampling sites are situated in the southern region, an area characterized by intensive human activities and a high population density. Under such conditions, manure and domestic sewage emerge as the key source of $NO_3^-$ in groundwater [33]. The low-flow season (July) coincides with the concentrated period of atmospheric precipitation in the region, and surface runoff collects manure and domestic sewage, resulting in higher $Cl^-$ concentrations in groundwater during the low-flow season (57.5 mg/L) compared to the high-flow season (50.90 mg/L).

### *3.6. Identification of Major Controlling Factors of Groundwater Hydrochemistry*

#### 3.6.1. Correlation Analysis

Pearson correlation analysis was utilized to examine the relationships between groundwater indicators, offering valuable insights for a more detailed investigation into the origins of each hydrochemical component. As shown in Figure 6, throughout both seasons, pH exhibited negative correlations with $Na^+$, $Ca^{2+}$, $Mg^{2+}$, $Cl^-$, $SO_4^{2-}$, $HCO_3^-$, $NO_3^-$, and TDS (correlation coefficients ranged from −0.510 to −0.12). The increase in ion concentrations is generally caused by enhanced water–rock interactions, while acidic conditions (low pH) favor the occurrence of water–rock interactions [34], leading to negative correlations between pH and most ion concentrations.

High flow

| | $K^+$ | $Na^+$ | $Ca^{2+}$ | $Mg^{2+}$ | $Cl^-$ | $SO_4^{2-}$ | $HCO_3^-$ | $NO_3^-$ | TDS | pH |
|---|---|---|---|---|---|---|---|---|---|---|
| $K^+$ | 1.0 | 0.41 | 0.31 | 0.52 | 0.39 | 0.46 | 0.64 | 0.16 | 0.51 | -0.16 |
| $Na^+$ | | 1.0 | 0.39 | 0.13 | 0.50 | 0.95 | 0.18 | 0.016 | 0.93 | -0.12 |
| $Ca^{2+}$ | | | 1.0 | 0.82 | 0.90 | 0.60 | 0.67 | 0.47 | 0.68 | -0.46 |
| $Mg^{2+}$ | | | | 1.0 | 0.63 | 0.41 | 0.77 | 0.34 | 0.48 | -0.37 |
| $Cl^-$ | | | | | 1.0 | 0.61 | 0.61 | 0.61 | 0.71 | -0.43 |
| $SO_4^{2-}$ | | | | | | 1.0 | 0.31 | 0.020 | 0.99 | -0.18 |
| $HCO_3^-$ | | | | | | | 1.0 | 0.66 | 0.43 | -0.50 |
| $NO_3^-$ | | | | | | | | 1.0 | 0.16 | -0.51 |
| TDS | | | | | | | | | 1.0 | -0.26 |
| pH | | | | | | | | | | 1.0 |

Low flow

| | $K^+$ | $Na^+$ | $Ca^{2+}$ | $Mg^{2+}$ | $Cl^-$ | $SO_4^{2-}$ | $HCO_3^-$ | $NO_3^-$ | TDS | pH |
|---|---|---|---|---|---|---|---|---|---|---|
| $K^+$ | 1.0 | 0.57 | 0.61 | 0.48 | 0.31 | 0.65 | 0.44 | 0.10 | 0.64 | 0.094 |
| $Na^+$ | | 1.0 | 0.83 | 0.49 | 0.85 | 0.94 | 0.33 | 0.35 | 0.95 | -0.12 |
| $Ca^{2+}$ | | | 1.0 | 0.57 | 0.73 | 0.95 | 0.43 | 0.14 | 0.94 | -0.18 |
| $Mg^{2+}$ | | | | 1.0 | 0.71 | 0.55 | 0.84 | 0.54 | 0.67 | -0.27 |
| $Cl^-$ | | | | | 1.0 | 0.78 | 0.51 | 0.58 | 0.86 | -0.29 |
| $SO_4^{2-}$ | | | | | | 1.0 | 0.32 | 0.14 | 0.98 | -0.070 |
| $HCO_3^-$ | | | | | | | 1.0 | 0.66 | 0.48 | -0.40 |
| $NO_3^-$ | | | | | | | | 1.0 | 0.32 | -0.32 |
| TDS | | | | | | | | | 1.0 | -0.16 |
| pH | | | | | | | | | | 1.0 |

**Figure 6.** Correlation among hydrochemical components.

During the high-flow season, $Na^+$ exhibited significant positive correlations with $SO_4^{2-}$, TDS, $Ca^{2+}$, $Cl^-$, $Mg^{2+}$, $HCO_3^-$, and $NO_3^-$. Based on the previous analysis, the typical origins of $Na^+$, $SO_4^{2-}$, and TDS include the dissolution of rock salt and evaporite rocks, in addition to industrial activities. The common sources of $Ca^{2+}$, $Cl^-$, and $Mg^{2+}$ are the dissolution of silicate rocks and domestic sewage. The common sources of $NO_3^-$, $HCO_3^-$, and $Cl^-$ are agricultural fertilization and the dissolution of carbonate rocks. During the low-flow season, $Na^+$ exhibited significant positive correlations with TDS, $SO_4^{2-}$, $Cl^-$, and $Ca^{2+}$, indicating that their common sources are primarily the dissolution of rock salt and evaporite rocks, as well as domestic and industrial wastewater discharges. During the low-flow season, a notable positive correlation was observed between $Mg^{2+}$ and both $HCO_3^-$ and $NO_3^-$; this indicates that these ions primarily originate from the weathering of silicate and carbonate and agricultural activities.

3.6.2. Principal Component Analysis

To elucidate the dominant factors influencing groundwater hydrochemistry across various periods, a total of 10 key water quality parameters were chosen for detailed analysis ($K^+$, $Na^+$, $Ca^{2+}$, $Mg^{2+}$, $Cl^-$, $SO_4^{2-}$, $HCO_3^-$, $NO_3^-$, TDS, and pH) and conducted principal component analysis (PCA) on groundwater samples from both seasons. Before entering the data into the model, we performed the Kaiser–Meyer–Olkin (KMO) and Bartlett's sphericity tests. The test results indicated that the KMO values for groundwater were 0.776 during the high-flow season and 0.545 during the low-flow season. Correspondingly, the Bartlett's sphericity test values were 277 and 313 ($p < 0.001$) for the high- and low-flow seasons, respectively. These results demonstrate that the hydrochemical data are suitable for PCA analysis. In both seasons, two principal components were extracted based on eigenvalues exceeding 1, which accounted for 79.8% and 78.9% of the total variance, respectively.

In Table 3, it is indicated that during the high-flow season, PC1 accounted for 51.5% of the total variance. PC1 displayed strong positive correlations with $SO_4^{2-}$, TDS, $K^+$, $Na^+$, $Cl^-$, $Mg^{2+}$, $HCO_3^-$, and $Ca^{2+}$. Based on prior analysis, $Ca^{2+}$, $Mg^{2+}$, and $SO_4^{2-}$ in the region's groundwater primarily stem from the dissolution of silicate and evaporite rocks. $HCO_3^-$ is mainly derived from carbonate rock dissolution, and $Na^+$ and $Cl^-$ are predominantly from rock salt dissolution. Moreover, industrial activities in the area have a notable effect on $SO_4^{2-}$ [35]. Thus, PC1 reflects the joint influence of rock weathering and industrial activities on the hydrochemical properties of groundwater. PC2 explained 28.3% of the total variance and showed strong positive correlations with $NO_3^-$. Additionally, moderate positive correlations were observed with $Cl^-$, $HCO_3^-$, and $Ca^{2+}$. As previously discussed, nitrate primarily originates from sewage and agricultural fertilizers. Studies have shown that domestic sewage contains elevated levels of $Cl^-$ [36]. Hence, PC2 highlights the impact of sewage and fertilizers on groundwater hydrochemistry.

**Table 3.** The results of PCA of groundwater for different seasons in the study area.

| Parameters | High Flow | | Low Flow | |
|---|---|---|---|---|
| | PC1 | PC2 | PC1 | PC2 |
| $SO_4^{2-}$ | 0.921 | 0.316 | 0.903 | 0.376 |
| TDS | 0.885 | 0.396 | 0.951 | 0.233 |
| $K^+$ | 0.859 | −0.106 | 0.786 | 0.015 |
| $Na^+$ | 0.846 | 0.380 | 0.902 | 0.164 |
| $Cl^-$ | 0.756 | 0.596 | 0.725 | 0.637 |
| $Mg^{2+}$ | 0.714 | 0.344 | 0.556 | 0.657 |
| $HCO_3^-$ | 0.702 | 0.549 | 0.371 | 0.812 |
| $Ca^{2+}$ | 0.652 | 0.535 | 0.916 | 0.127 |
| $NO_3^-$ | 0.161 | 0.854 | 0.136 | 0.837 |
| pH | −0.191 | −0.790 | 0.059 | −0.684 |
| Eigenvalue | 6.83 | 1.14 | 6.23 | 1.66 |
| Contribution Rate (%) | 51.5 | 28.3 | 49.9 | 29.0 |
| Cumulative Contribution Rate (%) | 51.5 | 79.8 | 49.9 | 78.9 |

During the low-flow period, PC1 accounted for 49.9% of the total variance, and PC1 exhibited strong positive correlations with $SO_4^{2-}$, TDS, $K^+$, $Na^+$, $Cl^-$, and $Ca^{2+}$. Therefore, PC1 represents the influence of the weathering of silicate rocks, evaporite rocks, and rock salt on groundwater hydrochemistry. PC2 explained 29.0% of the total variance, and PC2 exhibited strong positive correlations with $NO_3^-$ and $HCO_3^-$, while $Ca^{2+}$ has a moderate correlation with PC2. Therefore, PC2 represents the influence of domestic sewage, agricultural fertilizers, and the dissolution of carbonate rocks on groundwater hydrochemistry.

## 4. Conclusions

This research in-depth analyzed the patterns and determinants of groundwater quality variations in the Beichuan River Basin. The findings suggested that the chemical composition of groundwater is predominantly influenced by precipitation, showing marked differences between the high- and low-flow seasons. Water–rock interactions, sewage, and chemical fertilizer are the main factors affecting groundwater quality. The average contents of $Na^+$, $Mg^{2+}$, $Cl^-$, $HCO_3^-$, $NO_3^-$, and TDS are higher during the high-flow season compared to the low-flow season, primarily due to the scouring effect of rainfall. Human activities, particularly domestic sewage discharge and agricultural fertilization, are the primary contributors to nitrate contamination in groundwater. Principal component analysis revealed that rock weathering and industrial activities are the main controlling factors for variations in groundwater quality in the high-flow season, while the hydrochemistry of groundwater is mainly influenced by the weathering of silicate rocks, evaporite rocks, and rock salt in the low-flow season.

To address the rational exploitation of groundwater and ecological conservation in this area, we propose several solutions: Firstly, stricter management of sewage and agricultural runoff is needed to reduce the discharge of pollutants into water bodies. Secondly, the promotion of water-saving technologies and irrigation systems can improve water use efficiency. Additionally, comprehensive aquifer assessments should be conducted to better understand groundwater dynamics and inform protective measures. The findings of this research offer a scientific foundation for addressing groundwater quality deterioration in arid regions and provide valuable guidance for the sustainable management of water resources and ecological conservation efforts in such areas.

**Supplementary Materials:** The following supporting information can be downloaded at: https://www.mdpi.com/article/10.3390/w17070934/s1, Table S1: Relative weight of physicochemical parameters and water quality standard (all units of the parameters are in mg/L, except pH).

**Author Contributions:** L.W.: Investigation, methodology, software, data curation, and writing—original draft preparation. N.Y.: Investigation, data curation, and software. Y.Z.: Investigation and software. Q.Z.: Supervision, methodology, and writing—reviewing and editing. All authors have read and agreed to the published version of the manuscript.

**Funding:** This work was supported by the Natural Science Foundation of China (Grant No. 42377232), the Open Foundation of the Key Laboratory of Coupling Process and Effect of Natural Resources Elements (No. 2024KFKT015), the Natural Science Foundation of Hebei Province of China (Grant No. D2022504015), and the Fundamental Research Funds for the Chinese Academy of Geological Sciences (No. YK202310).

**Data Availability Statement:** Data is contained within the article or Supplementary Materials.

**Acknowledgments:** The authors gratefully acknowledge the editor and anonymous reviewers for their valuable comments on this manuscript. The authors also appreciate the financial support from the different organizations.

**Conflicts of Interest:** The authors declare no conflicts of interest.

## References

1. Wei, S.; Zhang, Y.; Cai, Z.; Bi, D.; Wei, H.; Zheng, X.; Man, X. Evaluation of groundwater quality and health risk assessment in Dawen River Basin, North China. *Environ. Res.* **2025**, *264*, 120292. [CrossRef] [PubMed]
2. Qin, R.; Wu, Y.; Xu, Z.; Xie, D.; Zhang, C. Assessing the impact of natural and anthropogenic activities on groundwater quality in coastal alluvial aquifers of the lower Liaohe River Plain, NE China. *Appl. Geochem.* **2013**, *31*, 142. [CrossRef]
3. Yang, M.; Zhu, L.; Liu, J.; Zhang, Y.; Zhou, B. Influence of water conservancy project on runoff in the source region of the Yellow River and wetland changes in the Lakeside Zone, China. *J. Groundw. Sci. Eng.* **2023**, *11*, 333–346. [CrossRef]

4. Torres-Martínez, J.A.; Mahlknecht, J.; Kumar, M.; Loge, F.J.; Kaown, D. Advancing groundwater quality predictions: Machine learning challenges and solutions. *Sci. Total Environ.* **2024**, *949*, 174973. [CrossRef]
5. El-Rawy, M.; Wahba, M.; Fathi, H.; Alshehri, F.; Abdalla, F.; El Attar, R.M. Assessment of groundwater quality in arid regions utilizing principal component analysis, GIS, and machine learning techniques. *Mar. Pollut. Bull.* **2024**, *205*, 116645. [CrossRef]
6. Chandra, G.V.; Ghosh, P.K. Groundwater quality in high-sulfur coal mining region of India: Spatial distribution, source control, and health risk assessment. *J. Environ. Manag.* **2024**, *368*, 122281. [CrossRef]
7. Deng, Y.; Lu, Y.; Du, X.; Ye, X.; Feng, J. Identifying spatial patterns and driving factors of anthropogenic impacts on the groundwater environment based on groundwater chemical kinetics. *J. Clean. Prod.* **2025**, *486*, 144436. [CrossRef]
8. Huang, G.; Sun, J.; Zhang, Y.; Chen, Z.; Liu, F. Impact of anthropogenic and natural processes on the evolution of groundwater chemistry in a rapidly urbanized coastal area, South China. *Sci. Total Environ.* **2013**, *463*, 209. [CrossRef]
9. Lin, C.Y.; Abdullah, M.H.; Praveena, S.M.; Yahaya, A.H.B.; Musta, B. Delineation of temporal variability and governing factors influencing the spatial variability of shallow groundwater chemistry in a tropical sedimentary island. *J. Hydrol.* **2012**, *432*, 26. [CrossRef]
10. Bozdağ, A. Combining AHP with GIS for assessment of irrigation water quality in Çumra irrigation district (Konya), Central Anatolia, Turkey. *Environ. Earth Sci.* **2015**, *73*, 8217–8236. [CrossRef]
11. Zhang, Q.Q.; Wang, H.W.; Wang, Y.C.; Yang, M.N.; Zhu, L. Groundwater quality assessment and pollution source apportionment in an intensely exploited region of northern China. *Environ. Sci. Pollut. Res.* **2017**, *24*, 16639–16650. [CrossRef] [PubMed]
12. Gu, H.; Chi, B.; Li, H.; Jiang, J.; Qin, W.; He, W. Assessment of groundwater quality and identification of contaminant sources of Liujiang basin in Qinhuangdao, North China. *Environ. Earth Sci.* **2015**, *73*, 6477. [CrossRef]
13. Zhou, B.; Wang, H.; Zhang, Q. Assessment of the evolution of groundwater chemistry and its controlling factors in the Huangshui River Basin of northwestern China, using hydrochemistry and multivariate statistical techniques. *Int. J. Environ. Res. Public Health* **2021**, *18*, 7551. [CrossRef] [PubMed]
14. Qiu, Y.; Lu, C.; Xu, Z.; Wang, Y.Q. Spatio-temporal variation characteristics and water pollution sources in the Huangshui River Basin. *Acta Sci. Circumstantiae* **2017**, *37*, 2829–2837. (In Chinese) [CrossRef]
15. Wang, X.; He, K.; Dong, Z. Effects of climate change and human activities on runoff in the Beichuan River Basin in the northeastern Tibetan Plateau, China. *CATENA* **2019**, *176*, 81–93. [CrossRef]
16. Zhu, L.; Liu, J.T.; Yang, M.N.; Zhang, Y.X.; Wen, D.P. Evolutionary trend of water cycle in Beichuan River Basin of China under the influence of vegetation restoration. *J. Groundw. Sci. Eng.* **2021**, *9*, 202–211. [CrossRef]
17. Ministry of Natural Resources of the People's Republic of China (MNRPRC). *Standard for Groundwater Quality, (GB/T14848-2017)*; General Administration of Quality Supervision, Inspection and Quarantine of the People' s Republic of China: Beijing, China, 2017.
18. Sunkari, E.D.; Seidu, J.; Ewusi, A. Hydrogeochemical evolution and assessment of groundwater quality in the Togo and Dahomeyan aquifers, Greater Accra Region, Ghana. *Environ. Res.* **2022**, *208*, 112679. [CrossRef]
19. Mao, H.; Wang, G.; Rao, Z.; Liao, F.; Shi, Z.; Huang, X.; Chen, X.; Yang, Y. Deciphering spatial pattern of groundwater chemistry and nitrogen pollution in Poyang Lake Basin (eastern China) using self-organizing map and multivariate statistics. *J. Clean. Prod.* **2021**, *329*, 129697. [CrossRef]
20. Li, P.; Wu, J.; Qian, H. Assessment of groundwater quality for irrigation purposes and identification of hydrogeochemical evolution mechanisms in Pengyang County, China. *Environ. Earth Sci.* **2013**, *69*, 2211. [CrossRef]
21. Tang, J.P.; Zhang, Q.; Hu, Y. Groundwater chemical characteristics and analysis of their controlling factors in an alluvial fan of Jianjiang River. *Environ. Sci.* **2019**, *40*, 3089. (In Chinese) [CrossRef]
22. Li, Z.J.; Yang, Q.C.; Yang, Y.S. Isotopic and geochemical interpretation of groundwater under the influences of anthropogenic activities. *J. Hydrol.* **2019**, *576*, 685–697. [CrossRef]
23. Wang, Q.; Dong, S.; Wang, H.; Yang, J.; Huang, H.; Dong, X.; Yu, B. Hydrogeochemical processes and groundwater quality assessment for different aquifers in the Caojiatan coal mine of Ordos Basin, northwestern China. *Environ. Earth Sci.* **2020**, *79*, 199. [CrossRef]
24. Hartmann, J.; Jansen, N.; Dürr, H.H.; Kempe, S.; Köhler, P. Global $CO_2$-consumption by chemical weathering: What is the contribution of highly active weathering regions? *Glob. Planet. Change* **2009**, *69*, 185–194. [CrossRef]
25. Jin, Z.; Xue, Q.; Chen, L.; Jin, M.; Li, F. Using dual isotopes to evaluate sources and transformations of nitrate in the West Lake watershed, eastern China. *J. Contam. Hydrol.* **2015**, *177*, 64–75. [CrossRef]
26. Wang, L.; Zhang, Q.; Wang, H. Rapid urbanization has changed the driving factors of groundwater chemical evolution in the large groundwater depression funnel area of northern China. *Water* **2023**, *15*, 2917. [CrossRef]
27. Ye, H.; Han, Z.; Wu, P.; Zha, X.; Li, X.; Hou, E.; Peñuelas, J. Disentangling sources and transformation mechanisms of nitrogen, sulfate, and carbon in water of a Karst Critical Zone. *Sci. Total Environ.* **2024**, *922*, 171310. [CrossRef]

28. Nayak, A.; Matta, G.; Prasad Uniyal, D.; Kumar, A.; Kumar, P.; Pant, G. Assessment of potentially toxic elements in groundwater through interpolation, pollution indices, and chemometric techniques in Dehradun in Uttarakhand State. *Environ. Sci. Pollut. Res.* **2024**, *31*, 36241–36263. [CrossRef]
29. Yang, J.; Ye, M.; Tang, Z.; Jiao, T.; Song, X.; Pei, Y.; Liu, H. Using cluster analysis for understanding spatial and temporal patterns and controlling factors of groundwater geochemistry in a regional aquifer. *J. Hydrol.* **2020**, *583*, 124594. [CrossRef]
30. Liu, W.; Jiang, H.; Guo, X.; Li, Y.; Xu, Z. Time-series monitoring of river hydrochemistry and multiple isotope signals in the Yarlung Tsangpo River reveals a hydrological domination of fluvial nitrate fluxes in the Tibetan Plateau. *Water Res.* **2022**, *225*, 119098. [CrossRef]
31. Liu, C.Q.; Lang, Y.C.; Satake, H.; Wu, J.; Li, S.L. Identification of anthropogenic and natural inputs of sulfate and chloride into the karstic ground water of Guiyang, SW China: Combined $\delta^{37}$Cl and $\delta^{34}$S approach. *Environ. Sci. Technol.* **2008**, *42*, 5421–5427. [CrossRef]
32. Li, S.L.; Liu, C.Q.; Li, J.; Liu, X.; Chetelat, B.; Wang, B. Assessment of the sources of nitrate in the Changjiang River, China using a nitrogen and oxygen isotopic approach. *Environ. Sci. Technol.* **2010**, *44*, 1573–1578. [CrossRef] [PubMed]
33. Wu, Y.; Wang, J.; Liu, Z.; Li, C.; Niu, Y.; Jiang, X. Seasonal nitrate input drives the spatiotemporal variability of regional surface water-groundwater interactions, nitrate sources and transformations. *J. Hydrol.* **2025**, *655*, 132973. [CrossRef]
34. Kim, K.H.; Yun, S.T.; Choi, B.Y.; Chae, G.T.; Joo, Y.; Kim, K.; Kim, H.S. Hydrochemical and multivariate statistical interpretations of spatial controls of nitrate concentrations in a shallow alluvial aquifer around oxbow lakes (Osong area, central Korea). *J. Contam. Hydrol.* **2009**, *107*, 114–127. [CrossRef]
35. Zhang, D.; Li, X.D.; Zhao, Z.Q.; Liu, C.Q. Using dual isotopic data to track the sources and behaviors of dissolved sulfate in the western North China Plain. *Appl. Geochem.* **2015**, *52*, 43–56. [CrossRef]
36. Li, Y.; Zhang, Z.; Fei, Y.; Chen, H.; Qian, Y.; Yu, D. Investigation of quality and pollution characteristics of groundwater in the Hutuo River alluvial plain, North China Plain. *Environ. Earth Sci.* **2016**, *75*, 581. [CrossRef]

water

MDPI

*Article*

# Sources, Water Quality, and Potential Risk Assessment of Heavy Metal Contamination in Typical Megacity River: Insights from Monte Carlo Simulation

**Xi Gao [1,2,3], Guilin Han [1,2,3,*], Shitong Zhang [1,2,3] and Jie Zeng [1,2,3]**

1 State Key Laboratory of Geomicrobiology and Environmental Changes, China University of Geosciences (Beijing), Beijing 100083, China; gaoxi0716@email.cugb.edu.cn (X.G.); stongzhang0103@email.cugb.edu.cn (S.Z.); zengjie@cugb.edu.cn (J.Z.)
2 Institute of Earth Sciences, China University of Geosciences (Beijing), Beijing 100083, China
3 Frontiers Science Center for Deep-Time Digital Earth, China University of Geosciences (Beijing), Beijing 100083, China
* Correspondence: hanguilin@cugb.edu.cn; Tel.: +86-10-8232-3536

**Abstract:** Due to the intense human activities and rapid development of economy, dissolved heavy metals (DHMs) pose a significant threat to urban river ecosystems. Therefore, the distribution, sources, and potential risks of DHMs in the Chaobai River (typical urban river) were investigated via ICP-MS in detail. Results revealed considerable spatial heterogeneity of heavy metals with various concentrations from the upper to lower reach. Principal component analysis (PCA) revealed that V, Ni, As, Mo, and Pb mainly originated from a mixing process of industrial input and natural process, Cr and Cu were mainly derived from urban activities, and Zn was mainly influenced by agriculture activities. Furthermore, land use types within the buffer zone near sampling points were innovatively analyzed, revealing strong correlations between DHMs and regional land use patterns. Monte Carlo simulations were employed to assess the differentiated non-carcinogenic and carcinogenic risks associated with DHMs across four age groups. This study provided scientific references for the sustainable management of urban rivers and aquatic systems in such a megacity region.

**Keywords:** heavy metal pollution; Monte Carlo simulation; urban river; Beijing-Tianjin-Hebei region; land use; risk assessment

## 1. Introduction

Urban river water plays a vital role in socioeconomic development, providing essential ecological functions such as the transportation of terrestrial loads, aquaculture development, and ecological conservation [1–3]. However, urban rivers are highly vulnerable to external pollution due to its high sensitivity and quick response to various human activities [4–6]. Therefore, comprehensive studies on urban river water have become a pressing global need and have garnered significant attention in recent years [7–10]. Megacity represents the most urbanized city area with a high concentration of human activities. The river that drains through megacity not only provides diversified facilities for the life of urban residents, but also carries essential information on megacity pollution [11,12], since urban wastewater poses a unique and serious threat to the urban water environment [13,14]. Therefore, systematic investigations and environmental evaluation of urban water contamination levels are crucial for effective management and sustainable development.

Dissolved heavy metals (DHMs) are significant and harmful pollutants in rivers worldwide [15,16], and have received a rich body of worldwide attention according to their significant characteristics of non-degradation, persistence, bio-accumulation, and biological toxicity [17–19]. DHMs profoundly affect human health and urban water quality in the long run, even though their content is relatively low [20], and this situation may worsen in a megacity river system [21–23]. The Beijing-Tianjin-Hebei (BTH) region, one of China's most densely populated and industrialized areas, is particularly affected by DHM contamination in its river systems [24,25]. Rapid urbanization, industrial emissions, agricultural runoff, and untreated wastewater discharge have led to the accumulation of DHMs such as Cd, Pb, and As in local water bodies [23]. These pollutants pose significant environmental and health risks, affecting both aquatic ecosystems and human populations. In response, various mitigation measures, including stricter environmental regulations, improvements in wastewater treatment, and pollution control technologies, have been implemented in the BTH region [26–28]. Despite these efforts, challenges remain in effectively monitoring and reducing DHMs levels across the BTH region. Inadequate enforcement of environmental laws, combined with the complexity of pollution sources and the region's industrial activities, hampers progress. Therefore, there is a need for more comprehensive DHM pollution investigation and the better integration of industrial, agricultural, and urban management practices to ensure sustainable urban water quality [29–35]. For the first time, an age grouping method based on the Monte Carlo simulation was proposed to accurately evaluate the differences of non-carcinogenic and carcinogenic risks among four age groups. This approach provides new insights for precise health risk management concerning heavy metal exposure.

Previous works have provided extensive pollution information on world's largest river systems (e.g., the Yangtze River, the Yellow River) with various lithology and natural landscapes [36,37]; however, the systematic research on DHMs pollution in urban river draining through megacities remains relatively scarce [38,39]. The BTH region is the most significant agglomeration of technology and economy in northern China with two megacities (Beijing city and Tianjin city), which has special representativeness in the sustainable development of society, economy, and environment [40]. The Chaobai River is recognized as the main water supply for urban survival in the BTH region, such as for drinking purposes and the shipping industry. In recent years, with highly concentrated urbanization, a large number of industrial and domestic sewage directly or indirectly enter the Chaobai River, resulting in poor water quality [41]. However, the limited studies of riverine pollutants in the Chaobai River are mostly centered on organic components, for example, polycyclic aromatic hydrocarbons and polychlorinated biphenyls, while the risk evaluation of DHMs in this typical megacity river is still unclear. Therefore, this study systematically collected the river water in this typical megacity river to (1) determine the content and distribution pattern of DHMs in the megacity river, (2) identify the potential DHM sources by examining their relationships with land use patterns, and (3) evaluate the health risks of DHMs in urban water via the probabilistic model. This age-group-based risk assessment method could benefit local government by proposing effective management strategies to target different risk groups more accurately and enhance public health protection.

## 2. Materials and Methods

### 2.1. Study Area

The Chaobai River covers an area of 19,354 $km^2$ and provides water supply for the BTH metropolitan region, the most famous urban region in northern China. The Chaobai River originates from the northern Yanshan mountain, draining through Hebei Province, Beijing, and Tianjin, and finally flows into the Bohai Bay [42]. The study region is affected by temperate monsoon together with the mean temperature of 12~13 °C and an evaporation amount of 1100~1800 mm [43]. The rainfall amount increases from northwest to southeast with an average of 488.9 mm annually, and the precipitation amount from June to August generally occupies 65~75% of total precipitation [44].

More than 40 hydropower stations are constructed upon the Chaobai River basin, and the capacity of the famous Miyun reservoir is $4.38 \times 10^9$ $m^3$. Many diversion channels and dams have also been constructed. The total storage capacity of the water conservancy projects within the basin exceeds $4.8 \times 10^9$ $m^3$, and the capacity of water diversion exceeds $3.0 \times 10^8$ $m^3$ with the maximum irrigation area of $2.5 \times 10^5$ $hm^2$. The land use types include farmland, grassland, forests, water bodies, and construction land, in which the two most significant land use types are farmland and construction land. By 2016, the total population in the Chaobai River basin increased rapidly to 3.28 million. The social economy developed rapidly with the dramatically changed land use structure. The Chaobai River basin has made great efforts in developing urban construction, industrial parks, and real estate, resulting in a rapid expansion rate of construction land. Meanwhile, a large amount of cultivated land has been expropriated for the construction of real estate, development zones, and industrial parks, leading to a serious loss of cultivated land [44].

### 2.2. Sampling and Chemical Analysis

Samples were collected from downstream to upstream (marked as CB1~CB11) in the Chaobai River (Figure 1). All instruments were pre-cleaned in the laboratory and moistened via river water three times in the field. The basic physiochemical parameters were analyzed by YSI monitoring meter for water quality (Xylem Inc., Yellow Spring, OH, USA). After the filtering process through membrane, all samples were acidified until $pH < 2$ and sealed in polyethylene bottles, and then kept in the fridge for further analysis. DHMs (V, Cr, Ni, Cu, Zn, As, Mo, and Pb) were measured via ICP-MS (PerkinElmer, USA) at the IGSNRR, Chinese Academy of Sciences. To maintain the quality of accuracy, duplicates, standard references, and procedure, blanks were applied reasonably during the whole analyzing process [23]. The results of repeated samples indicated acceptable repetitiveness during the whole analysis. The MDA levels of V, Cr, Ni, Cu, Zn, As, Mo, and Pb were 0.2 ng/L, 0.6 ng/L, 2 ng/L, 0.3 ng/L, 1 ng/L, 0.6 ng/L, 0.4 ng/L, and 0.1 ng/L, respectively. The standard reference material (GSB04-1767-2004) was applied to ensure the quality of DHM analysis. A mixed standard sample of DHMs were prepared, and the concentration of V, Cr, Ni, Cu, Zn, As, Mo, and Pb in the mixed standard sample was 120:5:40:40:60:80:160:1. In this study, the recovery rate of each heavy metal was higher than 98%.

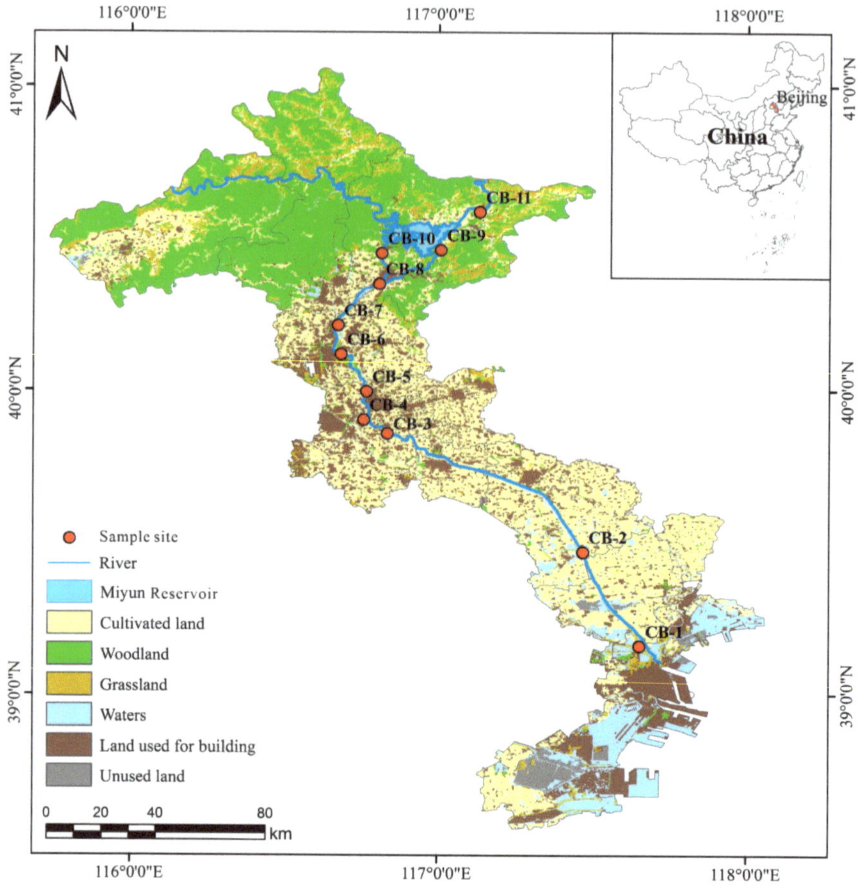


**Figure 1.** Sampling sites and land use types in the Chaobai River basin.

### 2.3. Assessment Methods

#### 2.3.1. Water Quality Index

Water quality index (WQI) is an essential method to evaluate water quality through multiple variables [29] WQI can be calculated via the equation below.

$$WQI = \Sigma\,[W_i \times (C_i/S_i) \times 100] \quad (1)$$

Here, $W_i$ represents the relative weight ($W_i = w_i/\Sigma w_i$, $w_i$ = 1~5) of water parameters on both human health and water quality; $C_i$ is the corresponding content of the parameters contained in the calculation; $S_i$ represents the drinking water standard by WHO (2011). The water quality in WQI analysis is categorized into 5 categories: "Excellent" means WQI < 50, "Good" represents the WQI range of 50~100, "Poor" means the WQI range of 100~200, "Very poor" represents the WQI range of 200~300, and "Undrinkable" means WQI > 300 [39].

#### 2.3.2. Health Risk Assessment

To determine the exposure degree and solidification tendency of toxic elements in the human body, the health risk evaluation was applied via the Monte Carlo simulation to evaluate DHMs in this typical Chaobai River [45]. In this study, the main consideration is that the health risks of DHMs in aquatic environments typically come from direct intake and fish intake. Since the government prohibits swimming near the river, we ignore the health risks of skin absorption of river water to urban residents. Detailed calculations can be found in Supplementary Materials.

2.3.3. Statistical Methods

Correlation analysis and principal component analysis (PCA) were applied in this study to identify and quantify possible DHM sources. Monte Carlo simulation was conducted by Oracle crystal ball 11.1.34160 (10,000 repetitions) via the applied parameters in Tables S1–S3; 30 m resolution land use datasets were extracted to 8 types by ArcGIS10.6. Furthermore, the influence of upper reach on lower reach should be firstly excluded in urban river research. To do so, the individual DHM content of each site was divided by detected metal concentrations to obtain the relative proportions. Through successive subtraction of the proportions from the nearest upper point, the potential influence of the upper reach could be eliminated. The land use proportions could thus be identified via this method in Table S4.

## 3. Results and Discussion

### 3.1. *Metal Content in Urban River*

The physicochemical parameters of the Chaobai River are summarized in Table 1. Water temperature varied between 23.7 and 31.0 °C. The river water samples were relatively alkaline with the pH of 8.36 ± 0.47. Alkaline river water may easily speed up the physicochemical process of absorption and oxidation in DHMs [46]. The pH value in the Chaobai River water did not show significantly lower values and had no significant correlation (Table 2) with DHMs in the correlation matrix. Therefore, riverine pH may not be a major factor affecting DHM concentrations in the Chaobai River. TDS values decreased from upstream to downstream with a range of 154~591 mg/L. EC values widely varied from 277 to 1050 μS/cm and decreased from downstream to upstream. The average content (in μg/L) of the selected DHMs in the Chaobai River declined as follows: Mo (3.12 μg/L) > V (2.31 μg/L) > As (1.65 μg/L) > Zn (1.19 μg/L) > Ni (0.83 μg/L) > Cu (0.82 μg/L) > Cr (0.12 μg/L) > Pb (0.02 μg/L). Mo, V, As, and Zn were the four most abundant DHMs in the Chaobai River. All DHM content was within the allowable range for drinking water recommended by the drinking water guidelines in the World Health Organization (WHO) and China, except for V, because of a lacking limited value. Cu, Zn, Pb, and As also met the class I of China's surface water standard. As the concentration (1.65) is the closest to the threshold of China's drinking water guidelines, this represents potential pollution at the corresponding sampling point and high human emissions (such as industrial and agricultural emissions) in the region. Moreover, V, Zn, As, and Mo content was higher than that the world average, implying the influence of increasing population and intensive domestic inputs. However, when comparing with the average value in other typical urban rivers of the world (Table 1), the average concentration of dissolved Cr (0.12 μg/L), Ni (0.83 μg/L), Cu (0.82 μg/L), and Zn (1.19 μg/L) in the Chaobai River was generally below that in the Seine River, Thames River, and Pearl River, except As (1.65 μg/L) and Mo, which exhibited relatively higher average values than those in the Pearl River (1.65 μg/L for As and 2.41 μg/L for Mo) [47,48]. These results indicated less man-made disturbance on the DHMs of the Chaobai River water via urbanization than other typical urban rivers in the world, even though the Chaobai River flows through two megacities. The lower concentrations of heavy metals in the Chaobai River may also reflect the positive effects of the ecological water replenishment policies implemented by the regional government over the past two years. Meanwhile, when comparing these results with historical data published in the same area in 2018 and 2020, Cu (0.82 μg/L), Zn (1.19 μg/L), and Ni (0.83 μg/L) content in 2021 (this study) was relatively lower than that in 2018 and 2020 (Cu = 1.07~1.79 μg/L, Zn = 2.91~114.5 μg/L, Ni = 1.64~2.40 μg/L) [49].

**Table 1.** Physiochemical parameters of the Chaobai River and other world rivers with drinking water guidelines.

| | Unit | Chaobai River | | | | | Seine River [1] | Yellow River [1] | Thames River [2] | Pearl River [3] | World Mean [1] | Drinking Water Guidelines | | Water Standard [6] |
|---|---|---|---|---|---|---|---|---|---|---|---|---|---|---|
| | | Min | Max | Mean | SD | Median | | | | | | China [4] | WHO [5] | |
| **V** | μg/L | 0.82 | 4.17 | 2.31 | 0.98 | 2.26 | | | | 1.51 | 0.71 | | | |
| **Cr** | μg/L | 0.05 | 0.44 | 0.12 | 0.11 | 0.09 | 11.46 | | 0.42 | 15.94 | 0.7 | 50 | 50 | |
| **Ni** | μg/L | 0.24 | 1.95 | 0.83 | 0.54 | 0.55 | 5.06 | 0.30–0.59 | 2.94 | 28.57 | 0.80 | 20 | 70 | |
| **Cu** | μg/L | 0.59 | 1.46 | 0.82 | 0.28 | 0.67 | 3.53 | 0.96–1.6 | 4.38 | 13.96 | 1.48 | 1000 | 2000 | 10 |
| **Zn** | μg/L | 0.46 | 2.91 | 1.19 | 0.73 | 0.92 | 4.98 | 0.07–0.32 | 8.09 | 61.84 | 0.6 | 1000 | | 50 |
| **As** | μg/L | 0.69 | 3.52 | 1.65 | 0.86 | 1.43 | 2.71 | 2 | 2.03 | 0.15 | 0.62 | 10 | 10 | 50 |
| **Mo** | μg/L | 1.36 | 4.67 | 3.12 | 1.12 | 3.67 | | | 35.12 | 2.41 | 0.42 | 70 | 70 | |
| **Pb** | μg/L | 0.01 | 0.08 | 0.02 | 0.02 | 0.01 | 0.22 | 0.01–4.1 | 0.40 | 0.58 | 0.079 | 10 | 10 | 10 |
| **T** | °C | 23.7 | 31.0 | 27.8 | 2.6 | 28.0 | | | | | | | | |
| **pH** [1] | | 7.70 | 9.30 | 8.40 | 0.50 | 8.30 | | 8.3 | | | | 6.5–8.5 | | |
| **EC** [1] | μS/cm | 277.0 | 1050 | 577.6 | 245.2 | 560.0 | | | | | | | | |
| **TDS** [1] | mg/L | 154.0 | 591.0 | 323.7 | 138.3 | 315.0 | | 460 | | | | 100 | | |

Notes: [1] DHM concentrations in the Seine River and world average [50]; [2] DHM concentrations in Thames River [47]; [3] DHM concentrations in Pearl River [48]; [4] China drinking water standards (GB 5749-2006) [45]; [5] WHO drinking water guidelines [25]; [6] limited values of Grade I in Chinese surface water standard [51].

**Table 2.** The correlation coefficients of eight heavy metals in the Chaobai River.

| | T | EC | pH | TDS | V | Cr | Ni | Cu | Zn | As | Mo | Pb |
|---|---|---|---|---|---|---|---|---|---|---|---|---|
| T | 1 | | | | | | | | | | | |
| EC | −0.12 | 1 | | | | | | | | | | |
| pH | 0.70 * | −0.62 | 1 | | | | | | | | | |
| TDS | −0.15 | 1.00 ** | −0.60 | 1 | | | | | | | | |
| V | 0.40 | 0.57 | −0.02 | 0.57 | 1 | | | | | | | |
| Cr | −0.39 | −0.01 | −0.17 | −0.01 | −0.12 | 1 | | | | | | |
| Ni | 0.39 | 0.62 * | −0.12 | 0.63 * | 0.81 ** | −0.10 | 1 | | | | | |
| Cu | −0.46 | 0.04 | −0.35 | 0.04 | 0.21 | 0.57 | −0.12 | 1 | | | | |
| Zn | 0.24 | 0.38 | 0.01 | 0.39 | 0.36 | −0.01 | 0.74 ** | −0.19 | 1 | | | |
| As | 0.42 | 0.67 * | −0.09 | 0.67 * | 0.91 ** | −0.34 | 0.84 ** | −0.13 | 0.32 | 1 | | |
| Mo | 0.87 ** | 0.20 | 0.48 | 0.20 | 0.72 * | −0.47 | 0.69 * | −0.36 | 0.42 | 0.76 ** | 1 | |
| Pb | 0.10 | 0.49 | −0.32 | 0.48 | 0.70 * | −0.17 | 0.44 | 0.21 | −0.22 | 0.76 ** | 0.38 | 1 |

Notes: * Correlation is significant at $p < 0.05$ level. ** Correlation is significant at $p < 0.01$ level.

The concentrations of eight DHMs in the Chaobai River are presented in Figure 2. Overall, the distribution for each DHM exhibited wide variations. Dissolved V, As, and Mo concentrations exhibited an upward trend from the upper to lower reach. It is remarkable that Cr concentrations increased significantly at site CB-8, which may be attributed to the discharge of wastewater from the manufacturing industry in the upper reach and the inflow of the Chaohe River and Baihe River. The distribution trend of Ni, Mo, and As concentrations in the Chaobai River were roughly similar, suggesting the same origin. V, Ni, As, and Mo had higher concentrations in Tianjin city, whereas Cu had higher concentrations in Beijing city. Overall, the spatial heterogeneity observed in the distribution of DHMs can be attributed to varying anthropogenic processes, such as agriculture, industrial activities, and economic development, across different regions.

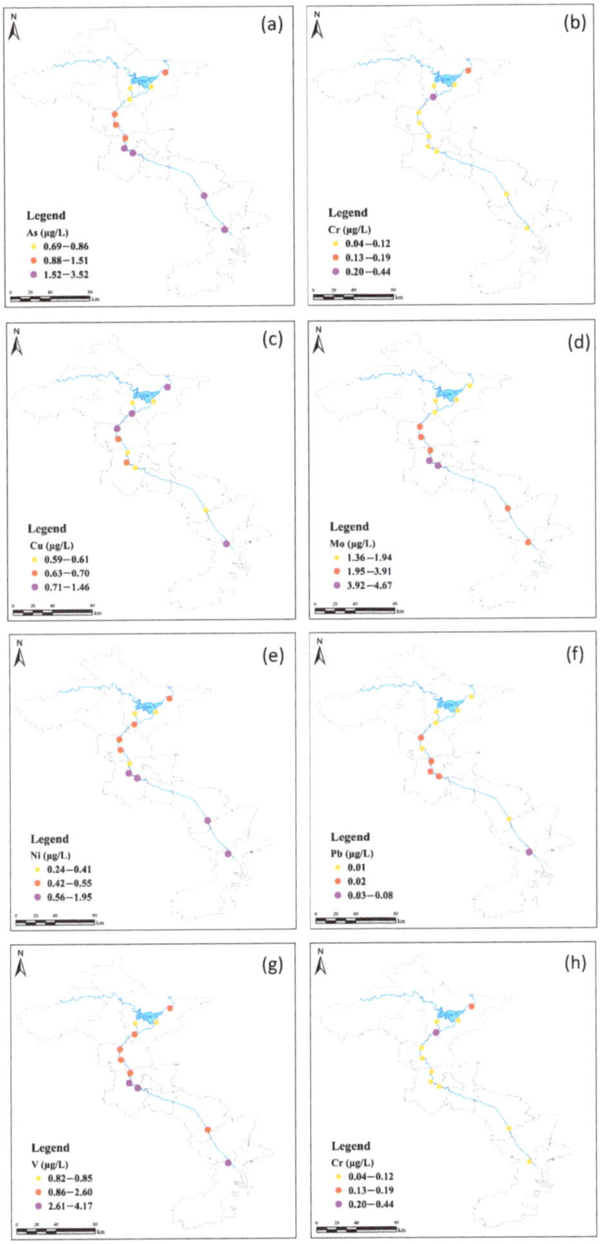


**Figure 2.** Spatial variation of (**a**) As (**b**) Cr (**c**) Cu (**d**) Mn (**e**) Ni (**f**) Pb (**g**) V (**h**) Zn concentrations in the Chaobai River.

### 3.2. Source Identification

Pearson correlation analysis was applied to analyze associations among heavy metals and other parameters of the river water (Table 2). The significant positive relationship of Mo and T (0.870, $p < 0.01$) in the study area indicated possible correlations between the solubility of Mo and the temperature. Moderate positive correlations were extracted between EC and Ni (0.623, $p < 0.05$), As (0.669, $p < 0.05$), and TDS and Ni (0.625, $p < 0.05$),

As (0.669, $p < 0.05$), indicating that Ni and As were closely associated with physicochemical parameters such as TDS and EC.

Principal component analysis (PCA) was performed to further identify the sources of dissolved heavy metals in the Chaobai River, with the results presented in Figure 3 [52,53]. Three principal components (PCs) existed in the Chaobai River in Table 3. PC1 had the primary loadings of V, Ni, As, Mo, and Pb, and contributed 51.7% of total variance, demonstrating similar sources of the five DHMs. The Beijing-Tianjin-Hebei (BTH) region is a major coal-consuming area in China, and during coal combustion, trace elements such as V, As, and Pb are often released into the environment [51,54]. Insufficient pollution control measures could lead to the atmospheric deposition or direct discharge of V, As, and Pb into water bodies, which was much more severe near large combustion facilities like thermal power plants [23,27,55,56]. The geological structure of the Beijing-Tianjin-Hebei region is complex, and nickel-containing minerals such as pyroxene and nickel iron ore are widely distributed in rocks within the basin, which are especially released during the weathering process of ultrabasic rocks and diabase [57–59]. Therefore, Ni may originate from natural sources. In addition to the Yanjiao national high-tech Industrial Development Zone, the Dachang Chaobai River Economic Zone, and the demonstration zone of emerging industry and the industrial estate of environmental protection, there are also several small furniture manufacturing and metal production enterprises not included in scientific statistics that may provide additional contributions to V, Ni, As, Mo, and Pb concentrations [60]. Therefore, PC1 represents a mixed origin influenced by industrial (thermal) pollution and natural processes.

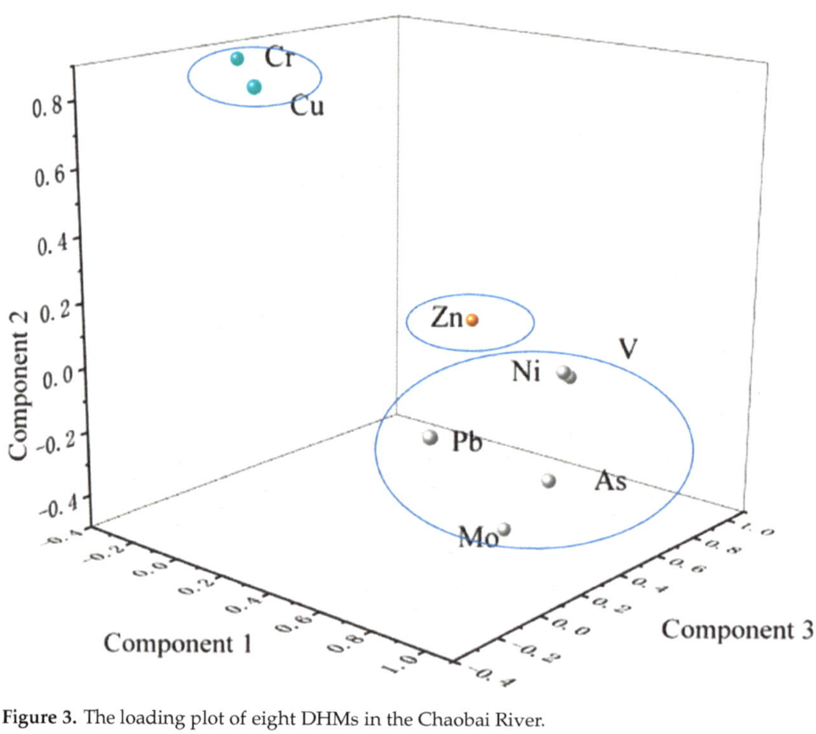


**Figure 3.** The loading plot of eight DHMs in the Chaobai River.

**Table 3.** Varimax-rotated component matrix for eight DHMs in the Chaobai River.

| Variable | PC1 | PC2 | PC3 |
|---|---|---|---|
| Eigenvalues | 4.14 | 1.75 | 1.37 |
| Variance (%) | 51.7 | 21.9 | 17.2 |
| V | 0.92 | 0.33 | 0.09 |
| Cr | −0.23 | 0.13 | 0.88 |
| Ni | 0.67 | 0.71 | −0.06 |
| Cu | 0.17 | −0.20 | 0.88 |
| Zn | 0.05 | 0.98 | −0.05 |
| As | 0.92 | 0.27 | −0.22 |
| Mo | 0.64 | 0.42 | −0.48 |
| Pb | 0.92 | −0.29 | 0.05 |

The PC2 contributed 21.9% of the total variance with the significant loading of Zn. The average Zn (1.19 μg/L) concentrations were much higher than those of the global average (Table 1), indicating that the widespread use of phosphate fertilizers in planting activities in the upper reaches of the Chaobai River is through surface runoff entering the river, as phosphate fertilizers contain an appropriate amount of Zn [23]. In addition, previous studies have indicated that the Zn concentration in the Chaobai River, which drains through the Miyun District, is primarily derived from agricultural production residues [53,61]. Therefore, PC2 is attributed to agricultural cultivation processes as the main source of Zn in the river.

The third PC contributed 17.2% of the total variance with the significant loading of Cr and Cu. Urban wastewater may also contain heavy metals such as Cr and Cu [62]. If these wastes are not strictly managed or treated, they may be directly discharged into rivers. Therefore, here, we attributed PC3 to the impact of urban activities.

The sources of anthropogenic DHMs could be revealed by land use distribution and proportions through the implementation of Spearman correlation [57]. The relationships between land use proportion and metal concentration in the mainstream of the Chaobai River were shown in Table 4. The finding demonstrated significant correlation between Cu content and water proportion variation (R = 0.73, $p < 0.05$), implying the possible Cu contribution of natural process, which is consistent with PCA analysis above. Meanwhile, Zn content was significantly correlated with urban land proportion variation (R = 0.82, $p < 0.01$), indicating the potential Zn sources of urban discharge in river water. These results underscore the importance of monitoring industrial wastewater discharges, especially considering the rapid industrialization in the Chaobai River basin.

**Table 4.** Spearman correlations between land use proportion and DHM variations in the Chaobai River.

| | Cr | Ni | Cu | Zn | As | Mo | Pb | V |
|---|---|---|---|---|---|---|---|---|
| Cultivated land | −0.20 | −0.35 | −0.23 | −0.19 | −0.15 | 0.35 | −0.26 | −0.03 |
| Woodland | 0.19 | 0.06 | −0.11 | 0.30 | −0.38 | −0.09 | −0.51 | −0.32 |
| Grassland | 0.14 | −0.32 | −0.08 | −0.57 | −0.17 | −0.46 | 0.03 | −0.29 |
| Water | 0.46 | −0.20 | 0.73 * | −0.62 | 0.04 | −0.48 | 0.61 | 0.03 |
| Urban land | −0.39 | 0.57 | −0.45 | 0.82 ** | 0.29 | 0.34 | −0.18 | 0.21 |

Notes: ** Significance level at $p < 0.01$. * Significance level at $p < 0.05$.

### 3.3. Water Quality in Megacity River

In this study, 10 parameters of water quality (pH, EC, TDS, Ni, Cr, Cu, Zn, Pb, As, and Mo) were selected for the WQI calculation in Table 5, and the WQI results are exhibited in Figure 4. The average WQI value of the Chaobai River was 23.35, indicating excellent water quality. WQI gradually declined from downstream to upstream in the Chaobai River, with the highest in site CB-2 and the lowest in site CB-7. The sampling site CB5-11 had excellent water quality (WQI < 25), and site CB1-4 had good water quality (25 < WQI < 50). Therefore, according to the WQI index, the natural water of the Chaobai River was acceptable for drinking. It is worth noting that this study only calculated WQI from the perspective of heavy metals. However, according to previous discussions about metal contents and sources, the potential risks of these ions still exist, although the WQI results of the Chaobai River are quite good. Frequent anthropogenic activities including agriculture and industry in these urban districts could lead to the increase in DHM content and the alteration of physicochemical parameters in river water [63]. According to recent governmental policies, the improved water quality in the upper reaches of the river is attributed not only to effective water resource management, but also to the dilution effects from high-flow tributaries, which help reduce impurities. Moreover, as shown in Figure 4, the primary contributions of WQI were the physicochemical parameters of EC, TDS, and pH, whereas the metal contribution is relatively lower, except for As and Mo, indicating potential health risks of As and Mo to human health.

**Table 5.** The selected parameters with weights in WQI analysis.

| Parameters | Drinking Water Guidelines [1] | Weight (wi) | Relative Weight (Wi) |
|---|---|---|---|
| pH | 8 | 5 | 0.116 |
| EC (μS/cm) | 1500 | 4 | 0.093 |
| TDS (mg/L) | 600 | 4 | 0.093 |
| Cr (μg/L) | 50 | 5 | 0.116 |
| Ni (μg/L) | 70 | 5 | 0.116 |
| Cu (μg/L) | 2000 | 2 | 0.047 |
| Zn (μg/L) | 3000 | 3 | 0.070 |
| As (μg/L) | 10 | 5 | 0.116 |
| Mo (μg/L) | 70 | 5 | 0.116 |
| Pb (μg/L) | 10 | 5 | 0.116 |

Notes: [1] Chinese drinking water standards (GB 5749-2006) [45].

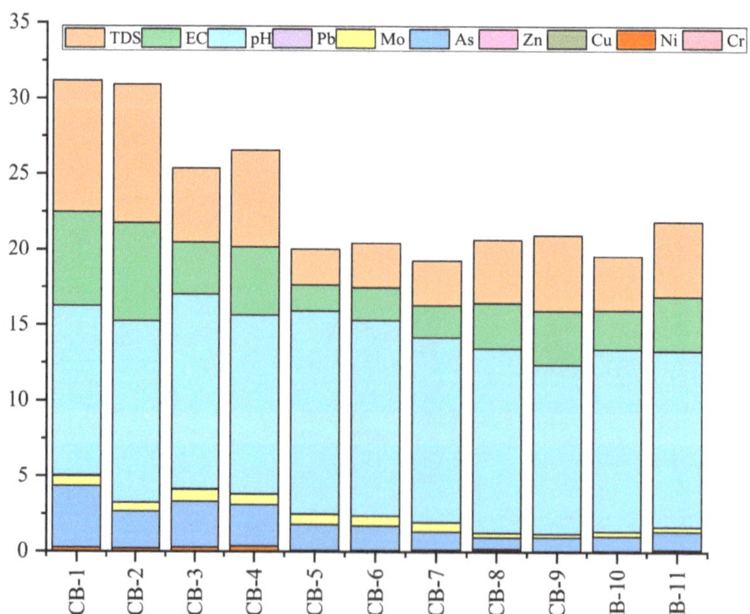


**Figure 4.** The spatial variations of WQI in the Chaobai River.

### 3.4. Non-Carcinogenic Risks of Metals in Urban Water

Water Ingestion

As a vital ecological resource in the BTH metropolitan region, the Chaobai River basin is not only the national key area for water conservation [44], but also a critical surface water source for drinking, particularly in Beijing [28,58,64]. Therefore, the risk evaluation of metals in this region is rather crucial to ensure the safety of drinking water [65–69]. According to the HQ calculation, Table S5 exhibited the HQ values via water ingestion. The calculation results indicated that the HQ values of Mo, V, Zn, Ni, Fe, and Cr via water ingestion were far less than the limitation values. However, the HQ values of As via water ingestion displayed a relatively higher risk in the Chaobai River ($HQ_{ingestion} > 0.1$). Furthermore, comparing with the 5th, 90th, 95th, and 99th percentiles of the potential As risks via water ingestion between child (0.04, 1.39, 1.78, 2.85) and adult (0.03, 0.71, 0.87, 1.24), child exhibited more potential risks than adult.

Figure 5 presented the predicted $HQ_{ingestion}$ value of As via MCS the method in the Chaobai River. As for the adult, the $HQ_{ingestion}$ value was less than 1, and 1.32% of the simulated value was higher than the USEPA standard. However, as for the child group, 19.48% of the simulation value was out of the limitation by the USEPA. Although the As value was less than the upper limitation of the WHO standard (10 μg/L), it was still much higher than the global mean value of urban rivers (0.62 μg/L), indicating potential health risks in the Chaobai River. These findings highlight the need for targeted water management policies and treatment strategies, particularly to control As contamination, in this typical megacity river. In previous studies, we used the same Monte Carlo simulation to assess the exposure risk of the Yongding River in the BTH region. Compared to the Yongding River, the average $HQ_{ingestion}$ of the study area is much lower than that of the Yongding River for both adult (0.63) and child (1.13) [26,70].

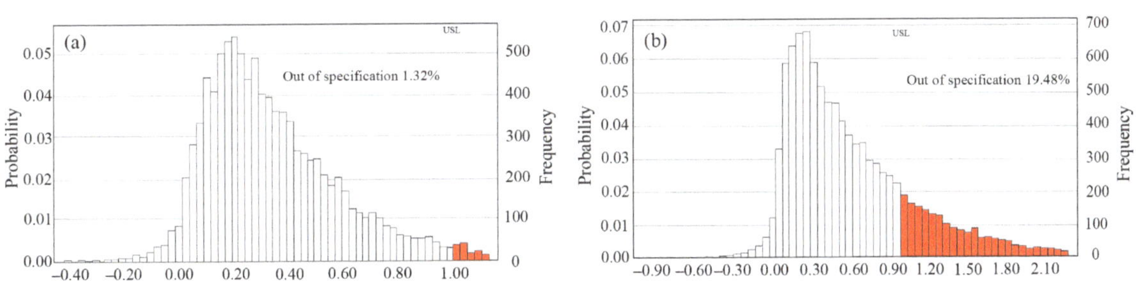


**Figure 5.** Predicted $HQ_{ingestion}$ value of As via Monte Carlo simulation in the Chaobai River on (**a**) adult and (**b**) child.

### 3.5. Carcinogenic Risks of Metals in Urban Water

Water Ingestion

Due to the limited non-carcinogenic dermal risk of metals in the Chaobai River, the analysis excluded carcinogenic risks via skin absorption. The urban residents were divided into four age groups to reveal a more accurate carcinogenic risk evaluation: infant (0~2 years old), child (2~6), teenager (7~16), and adult (≥17). According to the calculation, the limitation range of the carcinogenic risk of As was $10^{-6}$~$10^{-4}$. The predicted results are shown in Figure 6, which indicated that the average carcinogenic risks in these age groups were $1.21 \times 10^{-4}$ for infant group, $6.89 \times 10^{-5}$ for child group, $6.28 \times 10^{-5}$ for teenager group, and $1.72 \times 10^{-5}$ for adult group, respectively. Meanwhile, the 5% and 95% simulated risks were calculated as $1.50 \times 10^{-5}$ and $2.61 \times 10^{-4}$ for the infant group, 8.10 × $10^{-6}$ and 1.56 × $10^{-4}$ for the child group, 6.83 × $10^{-6}$ and 1.43 × 10−4 for the teenager group, and 2.38 × $10^{-6}$ and 3.70 × $10^{-5}$ for the adult group, respectively (Table S6). Furthermore, around 19.26% of As in the child group and 16.02% in the teenager group fell out of the allowable values, whereas only 1.27% of As in the adult group exceeded the acceptable ranges, implying a much lower carcinogenic risk of As in the adult group. On the contrary, 43.28% of As in the infant group far exceeded the acceptable standards, revealing a greater health threat for young ages, especially for the infant group.

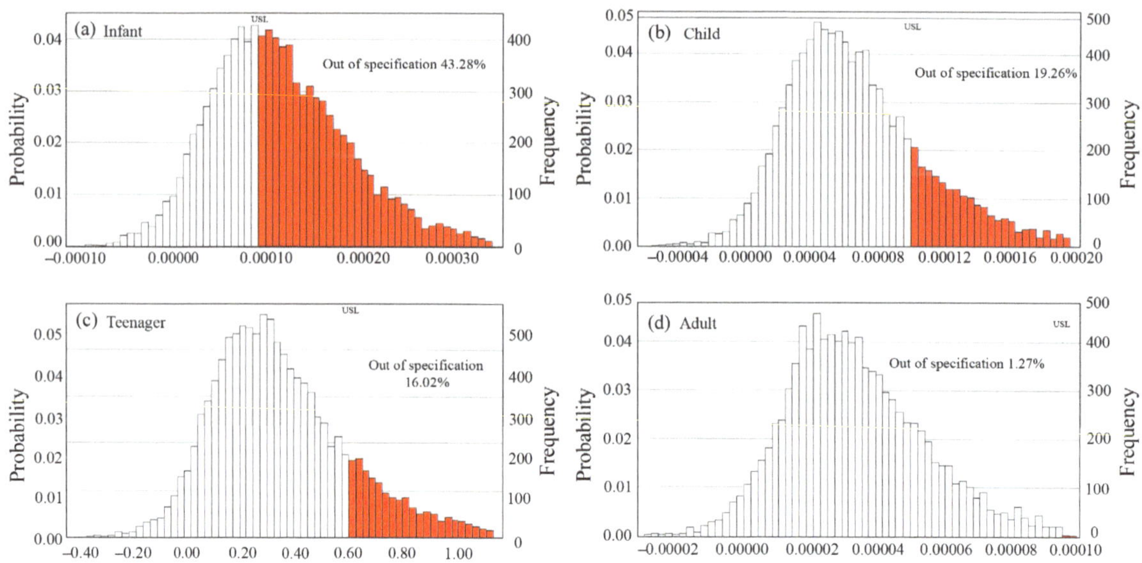


**Figure 6.** Predicted carcinogenic risk of As in (**a**) infant, (**b**) child, (**c**) teenager, and (**d**) adult groups in the Chaobai River.

Meanwhile, in order to investigate the parameters that could affect the calculated carcinogenic risks, we performed a sensitivity test in this study. Figure 7 illustrates the parameter contributions of risk uncertainty on As, and the most significant parameters in the infant and teenager groups followed the order of As content > ingestion rate > exposure frequency > body weight. However, body weight is relatively more influential in the child and adult groups than exposure frequency.

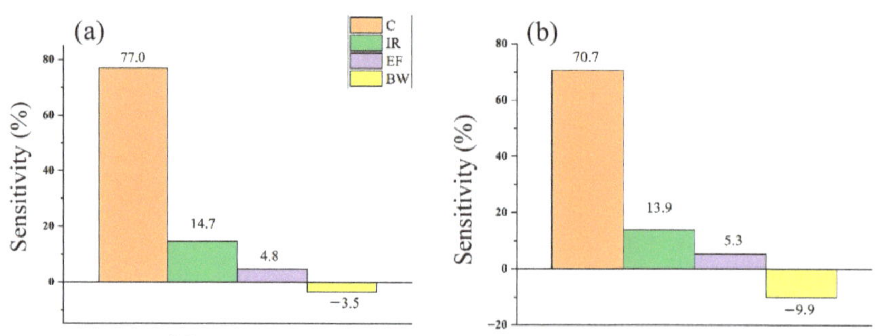


**Figure 7.** *Cont.*

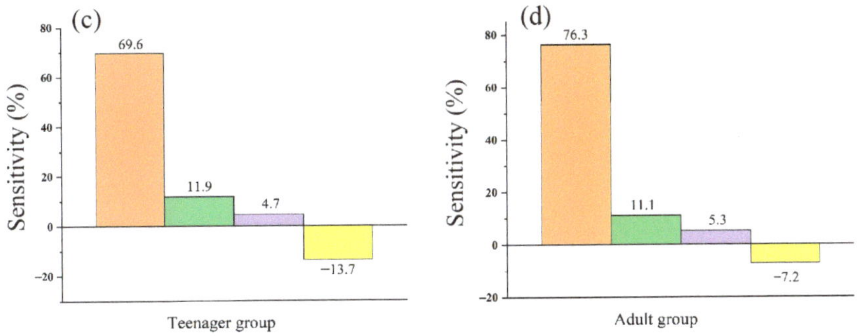


**Figure 7.** Contribution of input parameters on the uncertainty of the carcinogenic risk of As in (**a**) infant, (**b**) child, (**c**) teenager, and (**d**) adult age groups (C: concentration, IR: ingestion rate, EF: exposure frequency, and BW: body weight).

## 4. Conclusions

The spatial distribution, origins, and potential risks of urban water collected in the typical megacity river were fully investigated in this work. The concentration of DHMs in the Chaobai River showed obvious spatial variations, and V, As, and Mo exhibited the highest concentrations with the average value exceeding 1.60 μg/L. PCA analysis indicated three major PCs, among which V, Ni, As, Mo, and Pb originated from mixed sources directly impacted by industrial pollution and natural reactions; Zn was derived from the contribution of agriculture processes, whereas Cr and Cu were constrained by urban activities. The strong relationships between DHMs and land use proportions remained consistent with the PCA results. The Chaobai River water exhibited relatively good water quality and low risk level. The Monte Carlo simulation revealed that 1.32% and 19.48% of the simulated risk value of As fell out of the range of the USEPA standards for adults and children, respectively, indicating potential risks of As through water ingestion. Even though the carcinogenic risk of As in adults, children, and teenagers is weak, the risk in infants should not be neglected, which was significantly higher than other three groups. To ensure a safe water supply, it is essential to implement effective water resource management policies and monitoring programs in the future.

**Supplementary Materials:** The following supporting information can be downloaded at: https://www.mdpi.com/article/10.3390/w17020224/s1. Table S1: The parameter values and units of DHMs for hazard quotient (HQ) and hazard index (HI) analysis; Table S2: Parameters and values used for the deterministic and probabilistic calculations of non-carcinogenic risks [53,54,59–61]; Table S3: Parameters and values used for the deterministic and probabilistic calculations of cancer risks [56,60,71,72]; Table S4: The land use proportions of each sample site in the Chaobai River (unit as %); Table S5: Probabilistic risk of DHMs on adults and children ($HQ_{ingestion}$) in the Chaobai River; Table S6: Carcinogenic risk of As in the Chaobai River.

**Author Contributions:** Conceptualization, G.H. and X.G.; methodology, X.G.; software, X.G.; validation, X.G. and G.H.; formal analysis, X.G.; investigation, X.G. and J.Z.; resources, G.H.; data curation, X.G.; writing—original draft preparation, X.G.; writing—review and editing, X.G., G.H., S.Z. and J.Z.; visualization, X.G.; funding acquisition, G.H. All authors have read and agreed to the published version of the manuscript.

**Funding:** This work was supported by the "Deep-time Digital Earth" Science and Technology Leading Talents Team Funds for the Central Universities for the Frontiers Science Center for Deep-time Digital Earth, China University of Geosciences (Beijing) (Fundamental Research Funds for the Central Universities; grant number: 2652023001).

**Data Availability Statement:** Data will be made available on request.

**Acknowledgments:** The authors gratefully acknowledge Qian Zhang from the Institute of Geographic Sciences and Natural Resources Research, Chinese Academy of Sciences, for their assistance in the laboratory analysis.

**Conflicts of Interest:** The authors declare no conflicts of interest.

## References

1. Arslan, H.; Ayyildiz Turan, N. Estimation of spatial distribution of heavy metals in groundwater using interpolation methods and multivariate statistical techniques; its suitability for drinking and irrigation purposes in the Middle Black Sea Region of Turkey. *Environ. Monit. Assess.* **2015**, *187*, 516. [CrossRef] [PubMed]
2. Kumar, S.; Islam, A.R.M.T.; Hasanuzzaman, M.; Salam, R.; Khan, R.; Islam, M.S. Preliminary assessment of heavy metals in surface water and sediment in Nakuvadra-Rakiraki River, Fiji using indexical and chemometric approaches. *J. Environ. Manag.* **2021**, *298*, 113517. [CrossRef]
3. Zhuang, Q.; Li, G.; Liu, Z. Distribution, source and pollution level of heavy metals in river sediments from South China. *CATENA* **2018**, *170*, 386–396. [CrossRef]
4. Custodio, M.; Peñaloza, R.; Chanamé, F.; Hinostroza-Martínez, J.L.; De la Cruz, H. Water quality dynamics of the Cunas River in rural and urban areas in the central region of Peru. *Egypt. J. Aquat. Res.* **2021**, *47*, 253–259. [CrossRef]
5. Han, R.; Zhang, Q.; Xu, Z. Tracing Fe cycle isotopically in soils based on different land uses: Insight from a typical karst catchment, Southwest China. *Sci. Total Environ.* **2023**, *856*, 158929. [CrossRef] [PubMed]
6. Jaffar, S.T.A.; Luo, F.; Ye, R.; Younas, H.; Hu, X.-f.; Chen, L.-z. The Extent of Heavy Metal Pollution and Their Potential Health Risk in Topsoils of the Massively Urbanized District of Shanghai. *Arch. Environ. Contam. Toxicol.* **2017**, *73*, 362–376. [CrossRef] [PubMed]
7. Kumar, V.; Sharma, A.; Kumar, R.; Bhardwaj, R.; Kumar Thukral, A.; Rodrigo-Comino, J. Assessment of heavy-metal pollution in three different Indian water bodies by combination of multivariate analysis and water pollution indices. *Hum. Ecol. Risk Assess. Int. J.* **2020**, *26*, 1–16. [CrossRef]
8. Liu, J.; Han, G. Tracing Riverine Particulate Black Carbon Sources in Xijiang River Basin: Insight from Stable Isotopic Composition and Bayesian Mixing Model. *Water Res.* **2021**, *194*, 116932. [CrossRef] [PubMed]
9. Pekel, J.-F.; Cottam, A.; Gorelick, N.; Belward, A.S. High-resolution mapping of global surface water and its long-term changes. *Nature* **2016**, *540*, 418–422. [CrossRef] [PubMed]
10. Zhang, J.; Li, X.; Guo, L.; Deng, Z.; Wang, D.; Liu, L. Assessment of heavy metal pollution and water quality characteristics of the reservoir control reaches in the middle Han River, China. *Sci. Total Environ.* **2021**, *799*, 149472. [CrossRef]
11. Auerbach, D.A.; Deisenroth, D.B.; McShane, R.R.; McCluney, K.E.; Poff, N.L. Beyond the concrete: Accounting for ecosystem services from free-flowing rivers. *Ecosyst. Serv.* **2014**, *10*, 1–5. [CrossRef]
12. Chen, W.Y.; Li, X.; Hua, J. Environmental amenities of urban rivers and residential property values: A global meta-analysis. *Sci. Total Environ.* **2019**, *693*, 133628. [CrossRef] [PubMed]
13. Chen, D.; Chen, Y.-p.; Lin, Y. Heavy Rainfall Events Following the Dry Season Elevate Metal Contamination in Mining-Impacted Rivers: A Case Study of Wenyu River, Qinling, China. *Arch. Environ. Contam. Toxicol.* **2021**, *81*, 335–345. [CrossRef] [PubMed]
14. Wan, L.; Wang, H. Control of urban river water pollution is studied based on SMS. *Environ. Technol. Innov.* **2021**, *22*, 101468. [CrossRef]
15. Ali, M.M.; Rahman, S.; Islam, M.S.; Rakib, M.R.J.; Hossen, S.; Rahman, M.Z.; Kormoker, T.; Idris, A.M.; Phoungthong, K. Distribution of heavy metals in water and sediment of an urban river in a developing country: A probabilistic risk assessment. *Int. J. Sediment Res.* **2022**, *37*, 173–187. [CrossRef]
16. Hossain, M.S.; Ahmed, M.K.; Sarker, S.; Rahman, M.S. Seasonal variations of trace metals from water and sediment samples in the northern Bay of Bengal. *Ecotoxicol. Environ. Saf.* **2020**, *193*, 110347. [CrossRef] [PubMed]
17. Cai, L.M.; Wang, Q.S.; Wen, H.H.; Luo, J.; Wang, S. Heavy metals in agricultural soils from a typical township in Guangdong Province, China: Occurrences and spatial distribution. *Ecotoxicol Environ. Saf* **2019**, *168*, 184–191. [CrossRef] [PubMed]
18. Miao, F.; Zhang, Y.; Li, Y.; Fang, Q.; Zhou, Y. Implementation of an integrated health risk assessment coupled with spatial interpolation and source contribution: A case study of soil heavy metals from an abandoned industrial area in Suzhou, China. *Stoch. Environ. Res. Risk Assess.* **2022**, *36*, 2633–2647. [CrossRef]
19. Shen, L.; Zeng, J.; Guo, E.; Cheng, Y.; Li, L.; Niu, C.; Yuan, Y.; Ma, L. Source apportionment and risk assessment of heavy metals in urban soils from a central China city by using positive matrix factorization model coupled with Monte Carlo simulation. *Stoch. Environ. Res. Risk Assess.* **2023**, *37*, 291–304. [CrossRef]

20. Vardhan, K.H.; Kumar, P.S.; Panda, R.C. A review on heavy metal pollution, toxicity and remedial measures: Current trends and future perspectives. *J. Mol. Liq.* **2019**, *290*, 111197. [CrossRef]
21. Aghadadashi, V.; Neyestani, M.R.; Mehdinia, A.; Riyahi Bakhtiari, A.; Molaei, S.; Farhangi, M.; Esmaili, M.; Rezai Marnani, H.; Gerivani, H. Spatial distribution and vertical profile of heavy metals in marine sediments around Iran's special economic energy zone; Arsenic as an enriched contaminant. *Mar. Pollut. Bull.* **2019**, *138*, 437–450. [CrossRef] [PubMed]
22. Han, R.; Liu, W.; Zhang, J.; Zhao, T.; Sun, H.; Xu, Z. Hydrogeochemical characteristics and recharge sources identification based on isotopic tracing of alpine rivers in the Tibetan Plateau. *Environ. Res.* **2023**, *229*, 115981. [CrossRef]
23. Zhang, S.; Han, G.; Gao, X. The urbanization impacts on potentially toxic metals: The distribution, sources and contamination risks in river situated in typical megacity, China. *Sustain. Cities Soc.* **2023**, *97*, 104784. [CrossRef]
24. Zhi, X.; Chen, L.; Shen, Z. Impacts of urbanization on regional nonpoint source pollution: Case study for Beijing, China. *Environ. Sci. Pollut. Res.* **2018**, *25*, 9849–9860. [CrossRef]
25. World Health Organization. *Manganese in Drinking-Water: Background Document for Development of WHO Guidelines for Drinking-Water Quality*; World Health Organization: Geneva, Switzerland, 2004.
26. Gao, X.; Han, G.; Liu, J.; Zhang, S.; Zeng, J. Heavy Metal Accumulation and Source Apportionment in Urban River under Ecological Restoration: Relationships with Land Use and Risk Assessment Based on Monte Carlo Simulation. *ACS Earth Space Chem.* **2023**, *7*, 642–652. [CrossRef]
27. Duan, J.; Tan, J. Atmospheric heavy metals and Arsenic in China: Situation, sources and control policies. *Atmos. Environ.* **2013**, *74*, 93–101. [CrossRef]
28. Gao, X.; Han, G.; Zhang, S.; Liu, J.; Wang, D.; Ma, S.; Zhao, Y. Anthropogenic gadolinium sources and remediation in highly urbanized river in the Beijing-Tianjin-Hebei (BTH) region, China: Insights from buffer zone and three-dimensional tracer system. *Process Saf. Environ. Prot.* **2024**, *186*, 37–48. [CrossRef]
29. Bitton, G.; Jung, K.; Koopman, B. Evaluation of a microplate assay specific for heavy metal toxicity. *Arch. Environ. Contam. Toxicol.* **1994**, *27*, 25–28. [CrossRef]
30. De Miguel, E.; Iribarren, I.; Chacón, E.; Ordoñez, A.; Charlesworth, S. Risk-based evaluation of the exposure of children to trace elements in playgrounds in Madrid (Spain). *Chemosphere* **2007**, *66*, 505–513. [CrossRef] [PubMed]
31. Qu, R.; Han, G. Potassium isotopes of fertilizers as potential markers of anthropogenic input in ecosystems. *Environ. Chem. Lett.* **2023**, *21*, 41–45. [CrossRef]
32. Xia, F.; Qu, L.; Wang, T.; Luo, L.; Chen, H.; Dahlgren, R.A.; Zhang, M.; Mei, K.; Huang, H. Distribution and source analysis of heavy metal pollutants in sediments of a rapid developing urban river system. *Chemosphere* **2018**, *207*, 218–228. [CrossRef]
33. Chiu, W.A.; Barton, H.A.; DeWoskin, R.S.; Schlosser, P.; Thompson, C.M.; Sonawane, B.; Lipscomb, J.C.; Krishnan, K. Evaluation of physiologically based pharmacokinetic models for use in risk assessment. *J. Appl. Toxicol.* **2007**, *27*, 218–237. [CrossRef] [PubMed]
34. Jiang, C.; Zhao, Q.; Zheng, L.; Chen, X.; Li, C.; Ren, M. Distribution, source and health risk assessment based on the Monte Carlo method of heavy metals in shallow groundwater in an area affected by mining activities, China. *Ecotoxicol. Environ. Saf.* **2021**, *224*, 112679. [CrossRef] [PubMed]
35. Pang, K.; Li, M.; Liu, L.; Yang, J.L.; Zhao, H.J. [Evaluation and Source Analysis of Heavy Metal Pollution in Sediments of the Yellow River Basin Based on Monte Carlo Simulation and PMF Model]. *Huan Jing Ke Xue* **2022**, *43*, 4008–4017. [CrossRef]
36. Fu, J.; Zhao, C.; Luo, Y.; Liu, C.; Kyzas, G.Z.; Luo, Y.; Zhao, D.; An, S.; Zhu, H. Heavy metals in surface sediments of the Jialu River, China: Their relations to environmental factors. *J. Hazard. Mater.* **2014**, *270*, 102–109. [CrossRef]
37. Li, Y.; Chen, H.; Song, L.; Wu, J.; Sun, W.; Teng, Y. Effects on microbiomes and resistomes and the source-specific ecological risks of heavy metals in the sediments of an urban river. *J. Hazard. Mater.* **2021**, *409*, 124472. [CrossRef] [PubMed]
38. Paul, D. Research on heavy metal pollution of river Ganga: A review. *Ann. Agrar. Sci.* **2017**, *15*, 278–286. [CrossRef]
39. Xiao, J.; Wang, L.; Deng, L.; Jin, Z. Characteristics, sources, water quality and health risk assessment of trace elements in river water and well water in the Chinese Loess Plateau. *Sci. Total Environ.* **2019**, *650*, 2004–2012. [CrossRef] [PubMed]
40. Lu, W.; Zhang, G. Green development efficiency of urban agglomerations in a developing country: Evidence from Beijing-Tianjin-Hebei in China. *Environ. Dev. Sustain.* **2023**, *25*, 6939–6962. [CrossRef]
41. Yang, L.; He, J.; Liu, Y.; Wang, J.; Jiang, L.; Wang, G. Characteristics of change in water quality along reclaimed water intake area of the Chaobai River in Beijing, China. *J. Environ. Sci.* **2016**, *50*, 93–102. [CrossRef] [PubMed]
42. Zhang, Y.; Chen, H.; Jing, L.; Teng, Y. Ecotoxicological risk assessment and source apportionment of antibiotics in the waters and sediments of a peri-urban river. *Sci. Total Environ.* **2020**, *731*, 139128. [CrossRef] [PubMed]
43. He, B.; He, J.; Wang, J.; Li, J.; Wang, F. Characteristics of GHG flux from water-air interface along a reclaimed water intake area of the Chaobai River in Shunyi, Beijing. *Atmos. Environ.* **2018**, *172*, 102–108. [CrossRef]
44. Li, X.; Yu, X.; Wu, K.; Feng, Z.; Liu, Y.; Li, X. Land-use zoning management to protecting the Regional Key Ecosystem Services: A case study in the city belt along the Chaobai River, China. *Sci. Total Environ.* **2021**, *762*, 143167. [CrossRef]

45. *GB5749-2006;* Standard for Drinking Water Quality. Standard Press of China: Beijing, China, 2006.
46. Namieśnik, J.; Rabajczyk, A. The speciation and physico-chemical forms of metals in surface waters and sediments. *Chem. Speciat. Bioavailab.* **2010**, *22*, 1–24. [CrossRef]
47. Neal, C.; Williams, R.J.; Neal, M.; Bhardwaj, L.C.; Wickham, H.; Harrow, M.; Hill, L.K. The water quality of the River Thames at a rural site downstream of Oxford. *Sci. Total Environ.* **2000**, *251–252*, 441–457. [CrossRef]
48. Ouyang, T.P.; Zhu, Z.Y.; Kuang, Y.Q.; Huang, N.S.; Tan, J.J.; Guo, G.Z.; Gu, L.S.; Sun, B. Dissolved trace elements in river water: Spatial distribution and the influencing factor, a study for the Pearl River Delta Economic Zone, China. *Environ. Geol.* **2006**, *49*, 733–742. [CrossRef]
49. Xu, X.-Y.; Pu, X.; Liu, X.-L.; Zhang, R.-N.; Zhang, Y.-H.; Gao, J.; Dong, X.; Wang, Z.-K. Tempo-spatial variations of dissolved organic carbon and heavy metals in Miyun section of Chaobai River. *J. Ecol. Rural Environ.* **2020**, *36*, 1177–1184.
50. Gaillardet, J.; Viers, J.; Dupré, B. Trace Elements in River Waters. *Treatise Geochem.* **2003**, *5*, 605. [CrossRef]
51. *GB 3838-2002;* Environmental Quality Standard for Surface Water. State Environmental Protection Administration: Beijing, China, 2006.
52. Liu, S.; Pan, G.; Zhang, Y.; Xu, J.; Ma, R.; Shen, Z.; Dong, S. Risk assessment of soil heavy metals associated with land use variations in the riparian zones of a typical urban river gradient. *Ecotoxicol. Environ. Saf.* **2019**, *181*, 435–444. [CrossRef] [PubMed]
53. Wang, J.; Liu, G.; Liu, H.; Lam, P.K.S. Multivariate statistical evaluation of dissolved trace elements and a water quality assessment in the middle reaches of Huaihe River, Anhui, China. *Sci. Total Environ.* **2017**, *583*, 421–431. [CrossRef] [PubMed]
54. Emenike, P.C.; Tenebe, I.; Ogarekpe, N.; Omole, D.; Nnaji, C. Probabilistic risk assessment and spatial distribution of potentially toxic elements in groundwater sources in Southwestern Nigeria. *Sci. Rep.* **2019**, *9*, 1–15. [CrossRef] [PubMed]
55. Zheng, F.; Guo, X.; Tang, M.; Zhu, D.; Wang, H.; Yang, X.; Chen, B. Variation in pollution status, sources, and risks of soil heavy metals in regions with different levels of urbanization. *Sci. Total Environ.* **2023**, *866*, 161355. [CrossRef]
56. Fallahzadeh, R.A.; Khosravi, R.; Dehdashti, B.; Ghahramani, E.; Omidi, F.; Adli, A.; Miri, M. Spatial distribution variation and probabilistic risk assessment of exposure to chromium in ground water supplies; a case study in the east of Iran. *Food Chem Toxicol.* **2018**, *115*, 260–266. [CrossRef] [PubMed]
57. Tian, H.Z.; Lu, L.; Cheng, K.; Hao, J.M.; Zhao, D.; Wang, Y.; Jia, W.X.; Qiu, P.P. Anthropogenic atmospheric nickel emissions and its distribution characteristics in China. *Sci. Total Environ.* **2012**, *417–418*, 148–157. [CrossRef]
58. Wang, X.; Lu, C.; Cao, Y.; Chen, L.; Abedin, M.Z. Decomposition, decoupling, and future trends of environmental effects in the Beijing-Tianjin-Hebei region: A regional heterogeneity-based analysis. *J. Environ. Manag.* **2023**, *331*, 117124. [CrossRef] [PubMed]
59. Binkowitz, B.S.; Wartenberg, D. Disparity in quantitative risk assessment: A review of input distributions. *Risk Anal.* **2001**, *21*, 75–90. [CrossRef]
60. USDOE. *The Risk Assessment Information System (RAIS);* U.S. Department of Energy's Oak Ridge Operations Office (ORO): Washington, DC, USA, 2011.
61. Wang, B.; Yu, G.; Huang, J.; Hu, H. Development of species sensitivity distributions and estimation of HC5 of organochlorine pesticides with five statistical approaches. *Ecotoxicology* **2008**, *17*, 716–724. [CrossRef] [PubMed]
62. Zhao, B.; Bi, B.; Lu, S.; Liu, X.; Ni, B. Occurrence and risk assessment of heavy metals in urban water systems of Beijing, China. Desalin. *Water Treat* **2020**, *208*, 312–321. [CrossRef]
63. Wang, J.; Pu, X.; Zhang, R.; Liu, X.; Zhang, Y.; Wang, J.; Wang, Y.; Meng, M.; Li, S. Spatial-temporal distribution and source apportionment of riverine dissolved organic carbon in the Miyun section of the Chaobai River in Beijing, China. *Water-Energy Nexus* **2023**, *6*, 160–166. [CrossRef]
64. Zheng, F.; Liu, L.; Li, B.; Yang, Y.; Guo, M. Effects of reclaimed water use for scenic water on groundwater environment in a multilayered aquifer system beneath the Chaobai River, Beijing, China: Case study. *J. Hydrol. Eng.* **2015**, *20*, B5014003. [CrossRef]
65. Gupta, S.; Gupta, S.K. Application of Monte Carlo simulation for carcinogenic and non-carcinogenic risks assessment through multi-exposure pathways of heavy metals of river water and sediment, India. *Environ. Geochem. Health* **2023**, *45*, 3465–3486. [CrossRef] [PubMed]
66. Kattel, G.; Reeves, J.; Western, A.; Zhang, W.; Jing, W.; McGowan, S.; Cuo, L.; Scales, P.; Dowling, K.; He, Q. Healthy waterways and ecologically sustainable cities in Beijing-Tianjin-Hebei urban agglomeration (northern China): Challenges and future directions. *Wiley Interdiscip. Rev. Water* **2021**, *8*, e1500. [CrossRef]
67. Li, X.; Yan, C.; Wang, C.; Ma, J.; Li, W.; Liu, J.; Liu, Y. PM2.5-bound elements in Hebei Province, China: Pollution levels, source apportionment and health risks. *Sci. Total Environ.* **2022**, *806*, 150440. [CrossRef]
68. Liu, X.; Guo, C.; Wu, Y.; Huang, C.; Lu, K.; Zhang, Y.; Duan, L.; Cheng, M.; Chai, F.; Mei, F.; et al. Evaluating cost and benefit of air pollution control policies in China: A systematic review. *J. Environ. Sci.* **2023**, *123*, 140–155. [CrossRef]
69. Ren, J.; Han, G.; Liu, J.; Gao, X. Assessing wet season water quality and health risks of Chaobai River: Impacts of natural weathering and anthropogenic activities in a small urban watershed. *Urban Water J.* **2024**, *21*, 393–405. [CrossRef]
70. Yushin, N.; Jakhu, R.; Chaligava, O.; Grozdov, D.; Zinicovscaia, I. Evaluation of the potentially toxic elements and radionuclides in the soil sample of Novaya Zemlya in the Arctic Circle. *Environ. Pollut.* **2024**, *361*, 124871. [CrossRef] [PubMed]

71. Fitzpatrick, J.; Schoeny, R.; Gallagher, K.; Deener, K.; Dockins, C.; Firestone, M.; Jordan, W.; McDonough, M.; Murphy, D.; Olsen, M. US environmental protection Agency's framework for human health risk assessment to inform decision making. *Int. J. Risk Assess. Manag.* **2017**, *20*, 3–20. [CrossRef]
72. USEPA. Supplemental guidance for developing soil screening levels for superfund sites. US Environmental Protection Agency. 2002. Available online: https://nepis.epa.gov/ (accessed on 30 June 2022).

water

MDPI

Article

# Assessment of Aquatic Ecosystem Health in the Irtysh River Basin Using eDNA Metabarcoding

Tianjian Song [1,2,†], Fangze Zi [3,4,†], Yuxin Huang [2,5], Lei Fang [2], Yuna Zhang [2,5], Yu Liu [2], Jiang Chang [2,*] and Junsheng Li [6,*]

1 College of Water Sciences, Beijing Normal University, Beijing 100875, China; 202131470020@mail.bnu.edu.cn
2 State Key Laboratory of Environmental Criteria and Risk Assessment, Chinese Research Academy of Environmental Sciences, Beijing 100012, China; hyx573542801@126.com (Y.H.); 13955910967@163.com (L.F.); 220220932060@lzu.edu.cn (Y.Z.); liuyu.qmul@gmail.com (Y.L.)
3 College of Life Sciences and Technology, Tarim University, Alar 843300, China; 2024400189@buct.edu.cn
4 College of Material Science and Engineering, Beijing University of Chemical Technology, Beijing 100029, China
5 College of Ecology, Lanzhou University, Lanzhou 730000, China
6 Command Center for Comprehensive Survey of Natural Resources, China Geological Survey Bureau, Beijing 100055, China
* Correspondence: conservation1@126.com (J.C.); lijunsheng001@mail.cgs.gov.cn (J.L.)
† These authors contributed equally to this work.

**Abstract:** Environmental DNA (eDNA) metabarcoding is a powerful method for monitoring aquatic biodiversity and evaluating ecosystem health. In this study, we applied eDNA metabarcoding and a multi-species biotic integrity index (Mt-IBI), constructed by selecting and screening core metrics to capture key ecological responses, to assess the Irtysh River Basin in Xinjiang, China, analyzing samples from 52 sites. The community structure showed high sensitivity to environmental stressors, with dissolved oxygen (DO), total nitrogen (TN), and elevation being key factors, while alien fish richness negatively impacted ecosystem health. These findings highlight the importance of maintaining environmental parameters and controlling invasive species, demonstrating the potential of the Mt-IBI for early detection of ecological degradation and guiding freshwater conservation.

**Keywords:** index of biotic integrity; environmental DNA; health assessment; alien species

## 1. Introduction

Environmental DNA (eDNA) metabarcoding is an innovative technique for monitoring biodiversity and assessing ecosystem conditions in freshwater environments [1,2]. By analyzing genetic material in water samples, eDNA metabarcoding enables the detection of organisms, from microscopic to macroscopic [3,4]. eDNA metabarcoding is less labor-intensive than traditional survey methods, is non-destructive, and provides higher taxonomic resolution across multiple groups simultaneously [5,6]. In recent years, eDNA metabarcoding has been widely used to assess aquatic biodiversity and effectively indicate the health of river ecosystems [7–9]. By integrating eDNA metabarcoding with a multi-species biotic integrity index, we can comprehensively assess ecosystem health and monitor changes in community composition [10,11].

Microbial communities in aquatic ecosystems are sensitive to changes in ecological health, often more so than macroinvertebrates [12], particularly in severely disturbed regions. Microorganisms play a crucial role in biogeochemical processes, and their abundance responds rapidly to environmental changes [13,14]. Consequently, microbial indices are increasingly used in the assessment of aquatic ecosystems, helping to develop new concepts of ecosystem functioning under various stressors [15,16]. Microbial indices have become reliable tools for biodiversity conservation, and the monitoring and management of aquatic ecosystems [17], especially with advances in eDNA high-throughput sequencing methods [18]. Recent advancements, such as a bacteria-based index of biotic integrity developed by Liu et al. [19,20], demonstrate the effectiveness of microbial indicators in assessing aquatic ecosystem health. Phytoplankton and zooplankton play pivotal roles in aquatic ecosystems, and environmental changes can quickly disrupt their roles, causing notable alterations in community structure and diversity [21–24]. As foundational producers, phytoplankton form the base of food networks, while zooplankton control their growth, enabling energy transfer between trophic levels and supporting ecosystem function [25].

Previous research has highlighted that index of biotic integrity (IBI) assessments, when based on distinct biological groups, often yield inconsistent results within the same region [26]. Typically, IBIs focus on individual groups, which can overlook shifts in other ecosystem elements and the environmental pressures impacting them. Studies combining biotic indices from multiple biological groups have proven useful in providing a more comprehensive assessment of aquatic ecosystem health and guiding the selection of appropriate methods [8,27–29]. This suggests that integrating multiple taxa into assessment frameworks provides a more holistic view of ecosystem health, especially in environments impacted by varying levels of human activity.

Therefore, the primary goal of this study is to develop a multi-species biotic integrity index (Mt-IBI) for the Irtysh River Basin, utilizing eDNA metabarcoding data from bacterial, algal, fungal, and protozoan communities. This study aims to provide a quantitative and comprehensive framework for assessing ecosystem health in this region, offering insights that can guide conservation and management efforts in a river system under varied environmental pressures.

## 2. Materials and Methods

### 2.1. Study Area

This study took place in the Altay region (Figure 1) in northern Xinjiang Uygur Autonomous Region, China, near Russia, Kazakhstan, and Mongolia. The region has two major rivers: the Irtysh River and the Ulungur River. The Irtysh River, the second-largest in Xinjiang, is uniquely China's only river flowing toward the Arctic Ocean [30]. Within China, it spans 546 km and drains a basin covering 57,000 square kilometers, creating a distinctive comb-like water network [31]. The Ulungur River, the second-largest in the Altay region, stretches 821 km and has a basin area of 61,400 square kilometers [32]. Since 1987, a water diversion project has linked the Irtysh River to Ulungur Lake through a 3 km canal, making the Ulungur River an auxiliary water body of the Irtysh River [33].

The Irtysh River Basin is characterized by snow cover from November to March and snow-free conditions in July and August, with no significant glacier presence. Notably, discharge from May to July accounts for approximately 70% of annual runoff in the Kayiertesi River Basin, which is a major tributary [34]. Given these hydrological patterns, summer sampling provides a representative snapshot of the basin's ecological state.

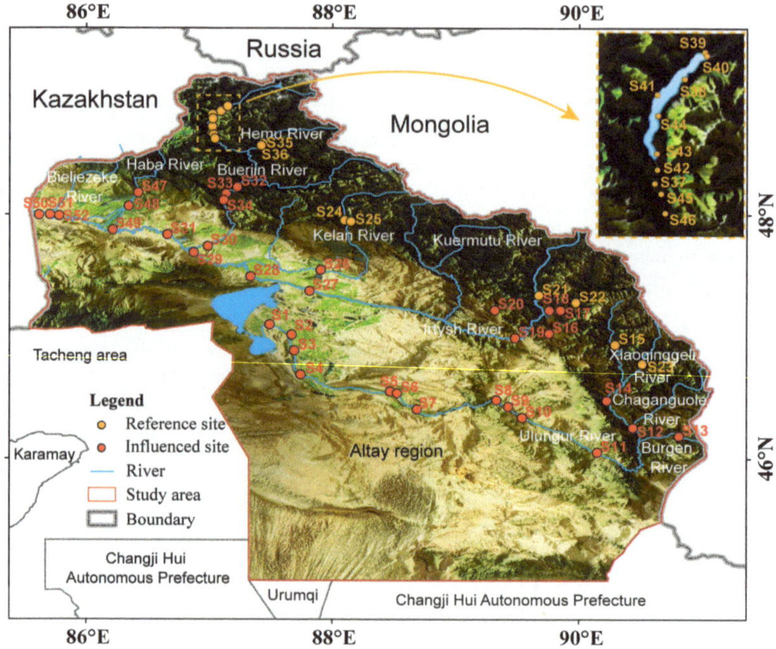


**Figure 1.** Location of 52 sampling sites in the Irtysh River basin.

## *2.2. eDNA Sample Collection and Environmental Variable Measurements*

Water sampling took place in July 2023 (Table S1). At each site, we collected 1 L of surface water in a sterile bottle, which was stored on ice. The water samples were vacuum filtered with polycarbonate filters (diameter: 50 mm, pore size: 0.22 μm) within 8 h of collection. Filtration units and tools were treated with a 10% bleach solution (~1% hypochlorous acid) and rinsed twice with $ddH_2O$ between samples to remove potential eDNA contamination. Filters were stored in sterile centrifuge tubes at −20 °C until eDNA extraction.

Geographical coordinates were recorded with a GPS device. Water physicochemical properties, such as water temperature (WT), pH, conductivity (Cond), oxidation–reduction potential (ORP), dissolved oxygen (DO), and total dissolved solids (TDS), were measured using a YSI Pro Plus multiparameter instrument (YSI Incorporated, Yellow Springs, OH, USA). Water samples were taken 0.5 m below the surface and analyzed in a lab for chemical oxygen demand (COD), total nitrogen (TN), and total phosphorus (TP). All measurements followed the Environmental Quality Standards for Surface Water in China (fourth edition).

## *2.3. eDNA Extraction, Sequencing, and Bioinformatic*

Genomic DNA was extracted from polycarbonate filters using a DNeasy Blood and Tissue Kit (QIAGEN GmbH, Hilden, Germany) following the manufacturer's protocol. DNA quality was verified through agarose gel electrophoresis, and the extracts were preserved at −20 °C for subsequent processing. The bacterial 16S rRNA gene and the eukaryotic 18S rRNA gene were amplified with primers 515F and 806R for 16S (515F: GTGYCAGCMGCCGCGGTAA; 806R: GGACTACNVGGGTWTCTAAT) [35], and primers 1380F and 1510R for 18S (1380F: TCCCTGCCHTTTGTACACAC; 1510R: CCTTCYGCAGGTTCACCTAC) [36], respectively. After PCR, the products were purified, quantified, and evaluated prior to high-throughput sequencing on an Illumina PE250

Platform (BIOZERON Biotechnology Co., Ltd., Shanghai, China) employing a 2 × 250 bp paired-end approach.

Bioinformatics analyses followed the pipeline described in He et al. [37]. Paired sequences were imported into QIIME 2 separately [38]. Quality control of the sequences was conducted, followed by the creation of amplicon sequence variant (ASV) tables using the DADA2 plugin [39]. Based on the resulting ASV count table and sequences, de novo OTU clustering at a 97% identity threshold was performed using the vsearch plugin [40]. Taxonomy assignment was carried out via the q2-feature-classifier plugin [41] and the Greengenes2 database for bacteria [42]. Following the CRABS tool protocol [43], we constructed a non-redundant comprehensive database by extracting 18S rRNA gene sequences for algae, fungi, and protozoa from both the Nucleotide Sequence database (NT) of the National Center for Biotechnology Information (NCBI-nt, https://www.ncbi.nlm.nih.gov/, accessed on 21 August 2024) and the Barcode of Life Data System (BOLD, https://www.boldsystems.org/, accessed on 16 August 2024). This approach ensures extensive coverage and minimizes redundancy, thereby enhancing the accuracy of our taxonomic assignments.

### 2.4. Development of the Mt-IBI

The Mt-IBI was established following Zhu et al. [10] and adjusted for the environmental conditions of the Irtysh River Basin. The index was created in four steps: (1) categorization of reference and influenced sites, (2) selection of candidate metrics, (3) screening of core metrics, and (4) calculation of the final index.

#### 2.4.1. Categorization of Reference and Influenced Sites

The initial phase of creating the biotic integrity index involves identifying the reference and influenced sites. Reference sites are usually chosen from river segments without human disturbances [31]. In this study, reference sites were selected based on three primary criteria to minimize anthropogenic impacts and ensure high water quality. Firstly, sites had to meet the COD Class I standard as defined by China's National Environmental Quality Standards for Surface Water (GB 3838-2002) [44], indicating excellent water quality with minimal organic pollutants. Secondly, extensive fish sampling at these sites revealed no presence of non-native or invasive species, thereby maintaining the biological integrity of the ecosystem. Lastly, reference sites were strategically located in the upstream regions of the main river and its tributaries, which are areas presumed to experience lower human disturbances. While this study did not consider certain environmental factors, such as heavy metals and soil physicochemical properties, the upstream location of these sites was used as a proxy for reduced anthropogenic influence (Table S2).

#### 2.4.2. Selection of Candidate Metrics

Candidate metrics were chosen to represent various aspects of community structure and function, such as diversity (e.g., Shannon and Simpson indices), composition (e.g., dominant taxa at different taxonomic levels), functional processes, and keystone species. To capture the ecological response of each species comprehensively, 235 candidate metrics were selected, covering bacteria, fungi, algae, and protozoan communities (Table S3). Diversity indices were computed in R using the vegan package, based on OTU tables [45]. Composition metrics were based on the dominant taxa at various taxonomic levels, determined by the relative abundance of OTUs. Function metrics were derived from microbial metabolic activities, phenotypic categories, and functional pathways using FAPROTAX v1.2.10 [46]. Keystone species were pinpointed by identifying high-degree nodes within the bipartite network [47].

2.4.3. Screening of the Metrics

Core metrics were screened through a three-step approach involving range, response, and redundancy tests. Metrics were excluded if they were absent in more than 30% of the sites, according to the range test. Next, Mann–Whitney U tests and boxplot comparisons were applied to identify metrics that exhibited notable differences between reference and influenced sites. Metrics that exhibited clear ecological responses to environmental stressors were retained [48]. Finally, metrics with high correlation coefficients ($\geq$0.70) were excluded to ensure that the remaining core metrics were independent and non-redundant (Table S4).

2.4.4. Calculation of the Mt-IBI

The values for each core metric were standardized and then combined to calculate the Mt-IBI (Table S5). Metrics that increased in response to disturbance were standardized using Equation (1), while those that decreased were standardized with Equation (2) [10]. The final Mt-IBI score for each site was calculated by summing the standardized values of the core metrics.

$$\text{Metric value} = (\text{maximum value} - \text{site value})/(\text{maximum value} - \text{5th percentile}) \quad (1)$$

$$\text{Metric value} = (\text{site value}/\text{95th percentile}) \quad (2)$$

We adopted the 25th-percentile IBI value from the reference site as a baseline for evaluating ecosystem health. Locations exceeding this threshold were categorized as healthy. Those with lower scores were classified into four categories, ranging from best to worst ecosystem health: sub-healthy, common, poor, and bad (Table S6).

*2.5. Statistical Analysis*

We used a constrained ordination method to examine the relationship between environmental variables and community structures. Detrended correspondence analysis (DCA) of bacterial, algal, fungal, and protozoan communities revealed axis lengths below three, suggesting that redundancy analysis (RDA) was appropriate for these groups. Total dissolved solids (TDS) were excluded from the analysis due to strong collinearity as shown by the variance inflation factor (VIF). Forward selection combined with Monte Carlo permutation tests were used to identify environmental variables strongly linked to community structures. All statistical procedures were conducted in R version 4.3.2, utilizing the stats and vegan packages [49].

We then identified five potential explanatory variables that could influence IBI: alien fish richness (unpublished data), human footprint index [50], population density [51], elevation, and river width (Table S1). These variables were selected to represent both anthropogenic impacts and natural environmental conditions. A simple linear regression analysis was conducted using the lm function in R to determine the effects of these five variables on the calculated Mt-IBI values. Regression diagnostics, including checks for multicollinearity and residual analysis, were performed to ensure the validity of the model [52]. Regression coefficients were visualized using the ggplot2 package [53]. By employing the plspm package, partial least squares path modeling (PLS-PM) was used to uncover both the direct and indirect variables influencing the Mt-IBI. Principal component analysis (PCA) was used to reduce the dimensionality of the observed variables related to water nutrients [54,55]. The first principal component was extracted to construct the path analysis (Table S7).

# 3. Results

## 3.1. Analysis of Community Composition

A total of 5,434,975 high-quality sequences were detected, with the number of sequences per sample ranging from 57,724 to 183,015 (Table S1). These eDNA data were classified into 2002 valid OTUs, including 1670 bacterial OTUs, 96 algal OTUs, 164 fungal OTUs, and 73 protozoan OTUs. These OTUs were assigned to 56 phyla, 152 classes, 370 orders, 693 families, and 1283 genera.

High-throughput 16S rRNA gene sequencing was used to analyze the species composition of prokaryotic microorganisms at both the phylum and genus levels. Pseudomonadota was the most abundant (49.5%), followed by Actinomycetota (17.3%), Bacillota_I (12.8%), and Deinococcota (8.5%), with the top eight phyla accounting for 99.2% (Figure 2a). At the family level, Moraxellaceae was the most abundant (12.8%), followed by Exiguobacteraceae (10.9%), Deinococcaceae (8.5%), and Sphingomonadaceae_486827 (8.5%), with the top ten families accounting for 70.0% (Figure 2b).

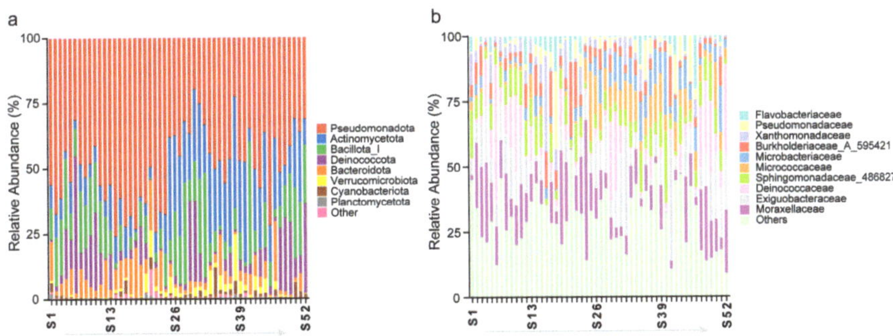


**Figure 2.** Relative abundance distribution of the microbial communities at the (**a**) phylum and (**b**) family levels across 52 sampling sites. The light-colored arrow from S1 to S52 indicates that the sample points are numbered incrementally in the direction of the arrow.

Using 18S high-throughput sequencing, we analyzed the distribution of algae, fungi, and protozoa at the class and genus levels. The most abundant class was Chlorophyceae (32.5%), followed by Dothideomycetes (26.1%), Spirotrichea (8.5%), and Leotiomycetes (7.7%), with the top 10 classes accounting for 93.7% (Figure 3a). At the genus level, Mychonastes (20.9%) was the most abundant, followed by Cladosporium (13.3%), Scenedesmus (6.4%), and Halteria (5.2%), with the 10 dominant genera accounting for 61.7% (Figure 3b).

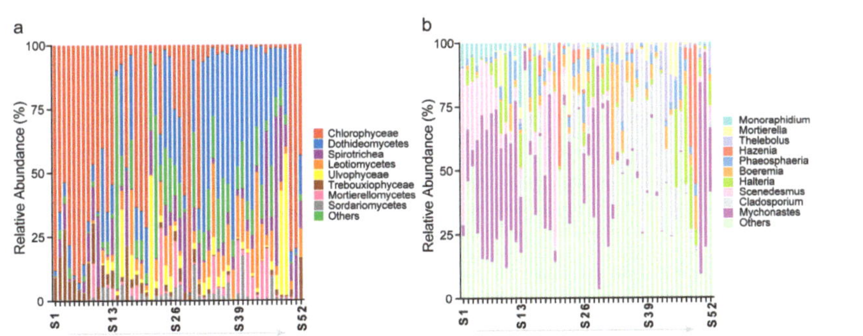


**Figure 3.** Relative abundance distribution of the algal, fungal, and protozoan communities at the (**a**) class and (**b**) genus levels across 52 sampling sites. The light-colored arrow from S1 to S52 indicates that the sample points are numbered incrementally in the direction of the arrow.

### 3.2. Associations Between Environmental Factors and Communities in the Irtysh River Basin

RDA analyses (Figure 4) showed that certain environmental variables were significantly correlated with microbial community composition. For the bacterial community, dissolved oxygen (DO) exhibited the strongest correlation ($R^2 = 0.5257$, $p < 0.001$), followed by total nitrogen (TN), conductivity (Cond), elevation (Elev), oxidation–reduction potential (ORP), pH, total phosphorus (TP), chemical oxygen demand (COD), and water temperature (WT) (Figure 4a). For algal, fungal, and protozoan communities, DO also had the highest correlation ($R^2 = 0.2962$, $p < 0.001$), whereas TN, TP, ORP, pH, Elev, and Cond were weaker but still significant (Figure 4b). COD and WT did not show significant correlations in this group. These findings highlight links between environmental factors and community composition.

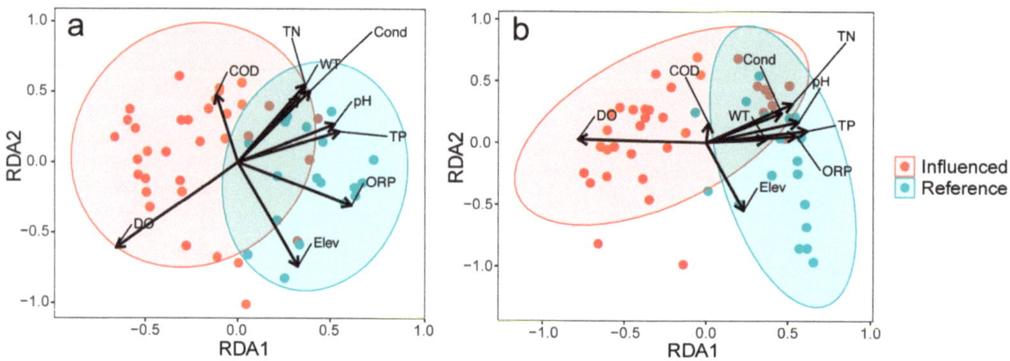


**Figure 4.** RDA ordination plot illustrating the relationship between environmental factors and communities of (**a**) microorganisms and (**b**) algae, fungi, and protozoa. Abbreviations: COD, chemical oxygen demand; Cond, conductivity; DO, dissolved oxygen; Elev, elevation; TN, total nitrogen; TP, total phosphorus; WT, water temperature.

### 3.3. Development of the Mt-IBI Using Taxonomy-Assigned eDNA Metabarcoding Data

Four screening tests were applied to identify the core metrics. Initially, 65 candidate metrics were removed after the range test. Next, 72 metrics demonstrating significant difference between the reference and influenced sites (Mann–Whitney U test, $p < 0.05$) were eliminated. The third step retained 63 metrics with strong sensitivity between the two site types, confirmed by box plot analysis. Finally, 24 metrics that showed high correlation with others (Pearson's correlation test, $p < 0.01$ and $r > 0.70$) were considered redundant and excluded (Table S4).

We identified eleven indicators (Table S3): M60 (order richness of fungi), M68 (order richness of bacteria), M78 (relative abundance of Trebouxiophyceae), M82 (relative abundance of Sphaeropleales), M93 (relative abundance of Agaricomycetes), M111 (relative abundance of Cladosporium), M120 (relative abundance of Oligohymenophorea), M147 (relative abundance of Verrucomicrobiae), M155 (relative abundance of Bacillales), M194 (relative abundance of Sphingomonas), and M224 (relative abundance of keystone species with a degree of five). Table S5 contains the formulas used to calculate these core metrics.

A consistent approach was applied to compute Mt-IBI values for each sampling site, forming the basis for an aquatic health assessment standard in our study region (Table S6). The results indicated that 30.8%, 7.8%, 13.7%, and 25.0% of the sites were classified as healthy, sub-healthy, common, and poor health categories, respectively.

Linear model results (Figure 5b) showed that the richness of alien fish species had a significant negative effect on the Mt-IBI ($p < 0.001$). In contrast, the human footprint index, population density, river width, and elevation had no statistically significant effect on the Mt-IBI. The estimates for these variables were −0.3614, 0.3240, 0.2611, and 0.3977, with $p$-values greater than 0.05. To further explore these interactions, we conducted a PLS path analysis encompassing alien fish richness, human footprint index, population density, and geographical factors (elevation and river width), as well as key water quality parameters (water organics and water nutrients). The resulting path diagram (Figure 5c) indicates that alien fish richness has a significant positive effect on water organic load (path coefficient = 0.595, $p < 0.001$). Notably, water organics and water nutrients both exhibit significant negative relationships with the Mt-IBI (path coefficients = $-0.427$, $p = 0.001$ and $-0.257$, $p = 0.041$, respectively).

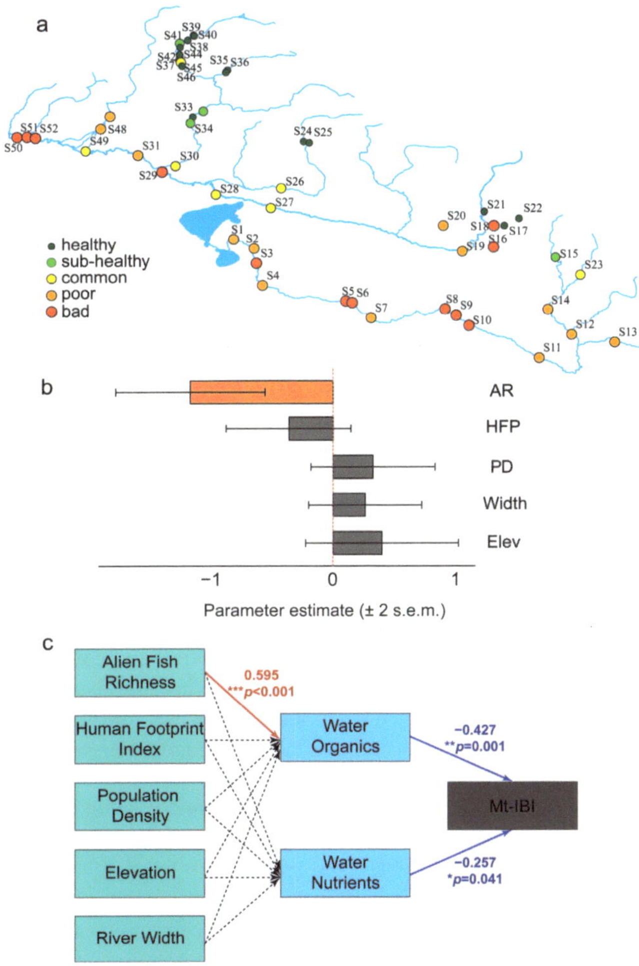


**Figure 5.** (**a**) A health assessment result using the multi-species biotic integrity index (Mt-IBI), (**b**) response of the Mt-IBI to human activities determined using a linear regression model, and (**c**) path diagrams indicating direct and indirect pathways affecting the Mt-IBI. Solid lines represent statistically significant ($p < 0.05$) effects, while dashed lines denote nonsignificant relationships. Abbreviations: AR, alien fish richness; Elev, elevation; HFP, human footprint index; PD, population density; Width, river width.

## 4. Discussion

The development and application of the Mt-IBI in this study provide an innovative approach to assess ecosystem health based on the integration of multiple biological groups. By focusing on microbial, protozoan, fungal, and algal communities, the Mt-IBI offers a more holistic perspective compared to traditional single-taxa indices. The higher sensitivity of the Mt-IBI to changes in ecological conditions, compared to conventional indices, demonstrates its potential as a valuable tool for early detection of ecosystem degradation. The inclusion of metrics related to keystone species and functional groups further enhances the index's ability to reflect changes in ecosystem dynamics and resilience [56,57]. Through analysis of the community composition across 52 sampling sites in the Irtysh River Basin, this research highlights the role of both environmental variables and anthropogenic factors in shaping microbial as well as algal, fungal, and protozoan communities. This approach provides a comprehensive understanding of ecosystem health, revealing new insights into how external pressures influence the structure and function of riverine biota.

The ordination analysis revealed that dissolved oxygen (DO), total nitrogen (TN), and elevation exhibited strong correlations with microbial, as well as algal, fungal, and protozoan communities. DO showed the highest correlation for both groups, suggesting its potential importance in shaping community composition and structure. This finding is consistent with earlier studies that demonstrated the sensitivity of microbial communities to oxygen availability, which directly influences metabolic activity and nutrient cycling processes [58,59]. Similarly, TN, conductivity, and elevation were significantly correlated with microbial community composition, reflecting how nutrient levels and geographic factors may shape the microbial diversity and distribution [60].

A possible explanation for the lack of significant correlation between COD and algae, fungi, and protozoa is the relatively low level of organic pollution in the watershed [61]. According to the National Environmental Quality Standards for Surface Water of China [44], COD values at 81% (42/52) of sites measured in this study were above Class I, and the total phosphorus (TP) values at all sites exceeded Class II (Table S1). Similarly, water temperature (WT) may not have had a significant effect because the temperature variations across sampling sites were relatively small, potentially falling within a tolerance range that does not substantially affect community composition. Seasonal variations in nutrient cycling and resource availability could further influence the sensitivity of various communities to COD. To fully understand these interrelationships and their effects on community composition, future studies should incorporate multi-seasonal sampling. Additionally, the environmental factors considered in this study were limited, potentially influencing the screening outcomes. Future research should incorporate factors like land use, soil quality indices, and biological interactions.

The separation between the reference and influenced sites was distinct in bacterial, algal, fungal, and protozoan RDA plots, with influenced sites closely associated with higher levels of nutrient enrichment (TN and TP) and lower levels of DO. In contrast, reference sites were generally characterized by a higher pH, ORP, and conductivity, which are factors that also contributed to the differentiation of bacterial community structure. These findings suggest that nutrient loading and changes in physical-chemical parameters associated with anthropogenic impacts are key drivers of variation in aquatic microbial communities [62]. Elevation also played a role in differentiating between reference and influenced sites, potentially due to its effect on water flow, temperature, and overall habitat characteristics, which in turn may be related to bacterial, algal, fungal, and protozoan community structures [63].

In response to fishery poverty alleviation policies, the Altay region of Xinjiang province has vigorously developed cold-water fish farming industries in the attached waters around

Ulungur Lake and the Irtysh River Basin. However, certain intentional introductions (such as stocked *Hucho taimen* and *Brachymystax lenok* from the Heilongjiang River Basin) and unintentional introductions (such as *Pseudorasbora parva* and *Abbottina rivularis*) have posed threats to the survival of native fish species and the stability of ecosystems [64]. Our results indicate that the richness of alien fish was the only variable with a significant negative estimate, underscoring its dominant role in influencing river health. Although previous research has shown that non-native species can reduce biodiversity and alter community composition and ecosystem functioning [65,66], our main purpose in using alien fish richness was to represent anthropogenic factors (e.g., fish stocking and aquaculture) in the Irtysh River Basin [31]. Still, the significant relationship between alien fish richness and the Mt-IBI underscores the need to understand the specific mechanisms by which non-native fish impact native communities. Alien species can exert significant pressure on native communities through competition, predation, and habitat alteration, leading to cascading effects that disrupt ecological balance [67].

The lack of significant relationships between the Mt-IBI and other explanatory variables, such as river width, and elevation, suggests that these factors may exert subtle or indirect influence on ecosystem health. The human footprint index and population density were assessed at a regional scale, which may not capture localized impacts on aquatic ecosystems. Fine-scale variations in land use and altered hydrology might have more direct effects on ecosystem health but were not fully resolved in our dataset [68,69]. Meanwhile, industrial activities, such as the discharge of pollutants and habitat alteration, as well as agricultural practices, including runoff of fertilizers and pesticides, can directly influence water quality and ecosystem health. To clarify these impacts, future studies will incorporate detailed industrial and agricultural metrics to better assess their relationships with biotic integrity [54]. It is also important to note that the single-season design of our survey is a limitation that may influence the detection of significant relationships. Aquatic ecosystems are subject to seasonal variations that affect both biotic and abiotic components. Conducting multi-seasonal surveys would provide a more comprehensive understanding of temporal dynamics and potentially reveal significant relationships that are not apparent in a single-season snapshot.

However, there are several limitations to the present study that should be addressed in future research. One limitation is the dependence on data from a single sampling event, which may not fully capture temporal variations in community composition and environmental conditions. Seasonal fluctuations can have significant effects on microbial as well as algal, fungal, and protozoan communities, and repeated sampling across different seasons would provide a more comprehensive assessment of ecosystem health [70]. Another limitation is the potential bias introduced by the eDNA metabarcoding approach, particularly in terms of primer specificity and amplification efficiency, which could influence the detection of certain taxa [71]. Budgetary constraints have led most current studies [9,55,72,73], including ours, to use single primer sets for DNA amplification. This approach facilitates the efficient processing of large sample numbers within existing financial constraints. Future studies should consider using multiple primer sets and improving the coverage of reference databases to enhance the accuracy and reliability of biodiversity assessments. Additionally, emerging technologies, such as digital droplet PCR and quantitative PCR methods, could be incorporated in future studies to enhance the accuracy of eDNA metabarcoding [74,75]. Primer selection bias represents just one factor contributing to the uncertainty of eDNA detection. Further research is required to determine how eDNA can reliably assess biodiversity. To assess species relative abundance or biomass, eDNA metabarcoding techniques should be integrated with traditional monitoring methods to mitigate biases in eDNA-based quantitative evaluations [76].

## 5. Conclusions

In conclusion, the results of this study highlight the practicality of combining environmental DNA (eDNA) metabarcoding with the multi-species biotic integrity index (Mt-IBI) as an effective method for assessing the health of aquatic ecosystems. The Mt-IBI developed here is highly sensitive to changes in community composition and environmental conditions, providing a robust framework for monitoring and managing freshwater ecosystems. Our study highlights the distinct responses of bacterial as well as algal, fungal, and protozoan communities to environmental factors, with dissolved oxygen being the most significant factor for both. Other important factors include nutrient concentrations (TN and TP), redox conditions (ORP), and physical parameters such as elevation and conductivity. Our findings also emphasize the significant adverse impacts of human-induced disturbances, such as the introduction of alien fish species, on the biological integrity of river ecosystems. By integrating data from different biological groups, the Mt-IBI offers a powerful tool for understanding and conserving the ecological health of complex aquatic systems. Although the Irtysh River Basin serves as an ideal test case for this study, broader applicability should be verified through future research conducted in varied ecological contexts.

**Supplementary Materials:** The following supporting information can be downloaded at https://www.mdpi.com/article/10.3390/w17020246/s1: Table S1: Summary of environmental variables and human activity information of 52 sampling sites; Table S2: Mean values (±SE) and the differences between influenced sites and reference sites; Table S3: The list of candidate metrics for developing the Mt-IBI; Table S4: Correlation matrix among metrics selection to compose the Mt-IBI of the Irtysh River; Table S5: The extreme values, anchors, and formulas calculation for core metrics in Mt-IBI development; Table S6: Mt-IBI health assessment at 52 sampling sites of the study area; Table S7: Results of principal component analysis of variables group.

**Author Contributions:** Conceptualization, T.S., F.Z., J.C. and J.L.; methodology, T.S. and F.Z.; software, T.S.; validation, J.C. and J.L.; formal analysis, T.S., F.Z. and Y.H.; investigation, T.S., F.Z., Y.H., L.F., Y.Z. and Y.L.; data curation, T.S. and F.Z.; writing—original draft preparation, T.S. and F.Z.; writing—review and editing, J.C. and J.L.; visualization, T.S., F.Z. and Y.H.; supervision, J.C. and J.L.; project administration, T.S., J.L. and J.C.; funding acquisition, J.C. All authors have read and agreed to the published version of the manuscript.

**Funding:** This research was funded by the State Key Development Programs of Basic Research of China [grant number 2024YFF1307200] and the Third Xinjiang Scientific Expedition Program [grant number 2021xjkk0600].

**Data Availability Statement:** The data supporting this study's findings are available from the corresponding authors upon reasonable request.

**Acknowledgments:** We would like to thank three anonymous reviewers who gave constructive comments on the manuscript.

**Conflicts of Interest:** The authors declare no conflicts of interest.

## References

1. Thomsen, P.F.; Kielgast, J.O.S.; Iversen, L.L.; Wiuf, C.; Rasmussen, M.; Gilbert, M.T.P.; Orlando, L.; Willerslev, E. Monitoring endangered freshwater biodiversity using environmental DNA. *Mol. Ecol.* **2012**, *21*, 2565–2573. [CrossRef] [PubMed]
2. Taberlet, P.; Bonin, A.; Zinger, L.; Coissac, E. *Environmental DNA: For Biodiversity Research and Monitoring*; Oxford University Press: Oxford, UK, 2018. [CrossRef]
3. Deiner, K.; Fronhofer, E.A.; Mächler, E.; Walser, J.-C.; Altermatt, F. Environmental DNA reveals that rivers are conveyer belts of biodiversity information. *Nat. Commun.* **2016**, *7*, 12544. [CrossRef] [PubMed]

4. Valentini, A.; Taberlet, P.; Miaud, C.; Civade, R.; Herder, J.; Thomsen, P.F.; Bellemain, E.; Besnard, A.; Coissac, E.; Boyer, F.; et al. Next-generation monitoring of aquatic biodiversity using environmental DNA metabarcoding. *Mol. Ecol.* **2016**, *25*, 929–942. [CrossRef] [PubMed]
5. Deiner, K.; Bik, H.M.; Mächler, E.; Seymour, M.; Lacoursière-Roussel, A.; Altermatt, F.; Creer, S.; Bista, I.; Lodge, D.M.; de Vere, N.; et al. Environmental DNA metabarcoding: Transforming how we survey animal and plant communities. *Mol. Ecol.* **2017**, *26*, 5872–5895. [CrossRef]
6. Eichmiller, J.J.; Miller, L.M.; Sorensen, P.W. Optimizing techniques to capture and extract environmental DNA for detection and quantification of fish. *Mol. Ecol. Resour.* **2016**, *16*, 56–68. [CrossRef]
7. Elbrecht, V.; Vamos, E.E.; Meissner, K.; Aroviita, J.; Leese, F.; Yu, D. Assessing strengths and weaknesses of DNA metabarcoding-based macroinvertebrate identification for routine stream monitoring. *Methods Ecol. Evol.* **2017**, *8*, 1265–1275. [CrossRef]
8. Huang, X.; Xu, J.; Liu, B.; Guan, X.; Li, J. Assessment of Aquatic Ecosystem Health with Indices of Biotic Integrity (IBIs) in the Ganjiang River System, China. *Water* **2022**, *14*, 278. [CrossRef]
9. Wang, S.; Zhang, P.; Zhang, D.; Chang, J. Evaluation and comparison of the benthic and microbial indices of biotic integrity for urban lakes based on environmental DNA and its management implications. *J. Environ. Manag.* **2023**, *341*, 118026. [CrossRef]
10. Zhu, M.; Yang, N.; Li, Y.; Zhang, W.; Wang, L.; Niu, L.; Wang, L.; Zhang, H. Assessing the effects of cascade dams on river ecological status using multi-species interaction-based index of biotic integrity (Mt-IBI). *J. Environ. Manag.* **2021**, *299*, 113585. [CrossRef]
11. Hu, H.; Wei, X.Y.; Liu, L.; Wang, Y.B.; Jia, H.J.; Bu, L.K.; Pei, D.S. Supervised machine learning improves general applicability of eDNA metabarcoding for reservoir health monitoring. *Water Res.* **2023**, *246*, 120686. [CrossRef]
12. Lear, G.; Boothroyd, I.K.G.; Turner, S.J.; Roberts, K.; Lewis, G.D. A comparison of bacteria and benthic invertebrates as indicators of ecological health in streams. *Freshw. Biol.* **2009**, *54*, 1532–1543. [CrossRef]
13. Sims, A.; Zhang, Y.; Gajaraj, S.; Brown, P.B.; Hu, Z. Toward the development of microbial indicators for wetland assessment. *Water Res.* **2013**, *47*, 1711–1725. [CrossRef] [PubMed]
14. David, G.M.; López-García, P.; Moreira, D.; Alric, B.; Deschamps, P.; Bertolino, P.; Restoux, G.; Rochelle-Newall, E.; Thébault, E.; Simon, M.; et al. Small freshwater ecosystems with dissimilar microbial communities exhibit similar temporal patterns. *Mol. Ecol.* **2021**, *30*, 2162–2177. [CrossRef]
15. Zhang, L.; Fang, W.; Li, X.; Lu, W.; Li, J. Strong linkages between dissolved organic matter and the aquatic bacterial community in an urban river. *Water Res.* **2020**, *184*, 116089. [CrossRef]
16. Li, Y.; Wu, H.; Shen, Y.; Wang, C.; Wang, P.; Zhang, W.; Gao, Y.; Niu, L. Statistical determination of crucial taxa indicative of pollution gradients in sediments of Lake Taihu, China. *Environ. Pollut.* **2019**, *246*, 753–762. [CrossRef]
17. Tamames, J.; Abellán, J.J.; Pignatelli, M.; Camacho, A.; Moya, A. Environmental distribution of prokaryotic taxa. *BMC Microbiol.* **2010**, *10*, 85. [CrossRef]
18. Huang, S.; Yoshitake, K.; Watabe, S.; Asakawa, S. Environmental DNA study on aquatic ecosystem monitoring and management: Recent advances and prospects. *J. Environ. Manag.* **2022**, *323*, 116310. [CrossRef]
19. Liu, Q.; Yin, S.; Yi, Y. A bacteria-based index of biotic integrity indicates aquatic ecosystem restoration. *Environ. Sci. Ecotechnol.* **2024**, *22*, 100451. [CrossRef]
20. Liu, Q.; Yin, S.; Luo, Q.; Yi, Y. A bacteria-based index of biotic integrity assesses aquatic ecosystems effectively in rewetted long-term dry river channel after water replenishment. *Ambio* **2024**. [CrossRef]
21. Wang, Z.; Huang, W.; Zhu, D.; Huang, Q.; Wu, L.; Liu, X. Determining Critical Thresholds of Environmental Flow Restoration Based on Planktonic Index of Biotic Integrity (P-IBI): A Case Study in the Typical Tributaries of Poyang Lake. *Int. J. Environ. Res. Public Health* **2023**, *20*, 169. [CrossRef]
22. Zhao, K.; Wang, L.; You, Q.; Zhang, J.; Pang, W.; Wang, Q. Impact of cyanobacterial bloom intensity on plankton ecosystem functioning measured by eukaryotic phytoplankton and zooplankton indicators. *Ecol. Indic.* **2022**, *140*, 109028. [CrossRef]
23. Zhang, Y.; Ban, X.; Li, E.; Wang, Z.; Xiao, F. Evaluating ecological health in the middle-lower reaches of the Hanjiang River with cascade reservoirs using the Planktonic index of biotic integrity (P-IBI). *Ecol. Indic.* **2020**, *114*, 106282. [CrossRef]
24. Qin, M.; Fan, P.; Li, Y.; Wang, H.; Wang, W.; Liu, H.; Messyasz, B.; Goldyn, R.; Li, B. Assessing the Ecosystem Health of Large Drinking-Water Reservoirs Based on the Phytoplankton Index of Biotic Integrity (P-IBI): A Case Study of Danjiangkou Reservoir. *Sustainability* **2023**, *15*, 5282. [CrossRef]
25. Yang, J.R.; Lv, H.; Isabwe, A.; Liu, L.; Yu, X.; Chen, H.; Yang, J. Disturbance-induced phytoplankton regime shifts and recovery of cyanobacteria dominance in two subtropical reservoirs. *Water Res.* **2017**, *120*, 52–63. [CrossRef]
26. Bae, M.-J.; Li, F.; Kwon, Y.-S.; Chung, N.; Choi, H.; Hwang, S.-J.; Park, Y.-S. Concordance of diatom, macroinvertebrate and fish assemblages in streams at nested spatial scales: Implications for ecological integrity. *Ecol. Indic.* **2014**, *47*, 89–101. [CrossRef]

27. Young, S.S.; Yang, H.N.; Huang, D.J.; Liu, S.M.; Huang, Y.H.; Chiang, C.T.; Liu, J.W. Using Benthic Macroinvertebrate and Fish Communities as Bioindicators of the Tanshui River Basin Around the Greater Taipei Area—Multivariate Analysis of Spatial Variation Related to Levels of Water Pollution. *Int. J. Environ. Res. Public Health* **2014**, *11*, 7116–7143. [CrossRef]
28. Wu, J.; Mao, R.; Li, M.; Xia, J.; Song, J.; Cheng, D.; Sun, H. Assessment of aquatic ecological health based on determination of biological community variability of fish and macroinvertebrates in the Weihe River Basin, China. *J. Environ. Manag.* **2020**, *267*, 110651. [CrossRef]
29. Yang, J.; Zhang, L.; Mu, Y.; Wang, J.; Yu, H.; Zhang, X. Unsupervised biological integrity assessment by eDNA biomonitoring of multi-trophic aquatic taxa. *Environ. Int.* **2023**, *175*, 107950. [CrossRef]
30. Duman, M.; Nadila, M. Analysis of the spatial and temporal variation characteristics of hydrometeorological elements in irtysh river basin. *Energy Energy Conserv.* **2017**, *126*, 119–120, (In Chinese with English abstract).
31. Li, X.; Jia, P.; Niu, C.; Xing, Y.; Li, H.; Liu, H.; Tang, W.; Zhao, Y. Species diversity of freshwater fish and assessment on watershed health in the Irtysh River and Ulungur River basins in Xinjiang, China. *Biodivers. Sci.* **2020**, *28*, 422–434. [CrossRef]
32. Nurlan, H. Hydrological characteristics in the ulungur river basin. *Arid Zone Res.* **2014**, *31*, 798–802. (In Chinese with English abstract).
33. Liu, J. Analysis on the water amount flowing into ulungur lake. *Energy Energy Conserv.* **2015**, *20*, 103–105. (In Chinese with English abstract).
34. Zhang, W.; Kang, S.-C.; Shen, Y.-P.; He, J.-Q.; Chen, A.-A. Response of snow hydrological processes to a changing climate during 1961 to 2016 in the headwater of Irtysh River Basin, Chinese Altai Mountains. *J. Mt. Sci.* **2017**, *14*, 2295–2310. [CrossRef]
35. Caporaso, J.G.; Lauber, C.L.; Walters, W.A.; Berg-Lyons, D.; Lozupone, C.A.; Turnbaugh, P.J.; Fierer, N.; Knight, R. Global patterns of 16S rRNA diversity at a depth of millions of sequences per sample. *Proc. Natl. Acad. Sci. USA* **2011**, *108*, 4516–4522. [CrossRef] [PubMed]
36. Amaral-Zettler, L.A.; McCliment, E.A.; Ducklow, H.W.; Huse, S.M. A Method for Studying Protistan Diversity Using Massively Parallel Sequencing of V9 Hypervariable Regions of Small-Subunit Ribosomal RNA Genes. *PLoS ONE* **2009**, *4*, e6372. [CrossRef]
37. He, X.; Gilmore, S.R.; Sutherland, T.F.; Hajibabaei, M.; Miller, K.M.; Westfall, K.M.; Pawlowski, J.; Abbott, C.L. Biotic signals associated with benthic impacts of salmon farms from eDNA metabarcoding of sediments. *Mol. Ecol.* **2021**, *30*, 3158–3174. [CrossRef]
38. Bolyen, E.; Rideout, J.R.; Dillon, M.R.; Bokulich, N.A.; Abnet, C.C.; Al-Ghalith, G.A.; Alexander, H.; Alm, E.J.; Arumugam, M.; Asnicar, F.; et al. Reproducible, interactive, scalable and extensible microbiome data science using QIIME 2. *Nat. Biotechnol.* **2019**, *37*, 852–857. [CrossRef]
39. Callahan, B.J.; McMurdie, P.J.; Rosen, M.J.; Han, A.W.; Johnson, A.J.A.; Holmes, S.P. DADA2: High-resolution sample inference from Illumina amplicon data. *Nat. Methods* **2016**, *13*, 581–583. [CrossRef]
40. Rognes, T.; Flouri, T.; Nichols, B.; Quince, C.; Mahé, F. VSEARCH: A versatile open source tool for metagenomics. *PeerJ* **2016**, *4*, e2584. [CrossRef]
41. Bokulich, N.; Dillon, M.; Bolyen, E.; Kaehler, B.; Huttley, G.; Caporaso, J. q2-sample-classifier: Machine-learning tools for microbiome classification and regression. *J. Open Source Softw.* **2018**, *3*, 934. [CrossRef]
42. McDonald, D.; Jiang, Y.; Balaban, M.; Cantrell, K.; Zhu, Q.; Gonzalez, A.; Morton, J.T.; Nicolaou, G.; Parks, D.H.; Karst, S.M.; et al. Greengenes2 unifies microbial data in a single reference tree. *Nat. Biotechnol.* **2024**, *42*, 715–718. [CrossRef] [PubMed]
43. Jeunen, G.-J.; Dowle, E.; Edgecombe, J.; von Ammon, U.; Gemmell, N.J.; Cross, H. crabs—A software program to generate curated reference databases for metabarcoding sequencing data. *Mol. Ecol. Resour.* **2023**, *23*, 725–738. [CrossRef] [PubMed]
44. *GB3838-2002;* State Environmental Protection Administration (SEPA) of China, Beijing of Regula. Environmental Quality Standard for Surface Water. Standards Press: Beijing, China, 2002; pp. 1–8.
45. Niu, L.; Li, Y.; Wang, P.; Zhang, W.; Wang, C.; Wang, Q. Understanding the Linkage between Elevation and the Activated-Sludge Bacterial Community along a 3,600-Meter Elevation Gradient in China. *Appl. Environ. Microbiol.* **2015**, *81*, 6567–6576. [CrossRef] [PubMed]
46. Louca, S.; Parfrey, L.W.; Doebeli, M. Decoupling function and taxonomy in the global ocean microbiome. *Science* **2016**, *353*, 1272–1277. [CrossRef]
47. Berry, D.; Widder, S. Deciphering microbial interactions and detecting keystone species with co-occurrence networks. *Front. Microbiol.* **2014**, *5*, 219. [CrossRef]
48. Yang, J.; Zhang, X. eDNA metabarcoding in zooplankton improves the ecological status assessment of aquatic ecosystems. *Environ. Int.* **2020**, *134*, 105230. [CrossRef]
49. R Core Team. *R: A Language and Environment for Statistical Computing;* R Foundation for Statistical Computing: Vienna, Austria, 2022. Available online: www.R-project.org/ (accessed on 14 November 2024).

50. Venter, O.; Sanderson, E.W.; Magrach, A.; Allan, J.R.; Beher, J.; Jones, K.R.; Possingham, H.P.; Laurance, W.F.; Wood, P.; Fekete, B.M.; et al. Global terrestrial Human Footprint maps for 1993 and 2009. *Sci. Data* **2016**, *3*, 160067. [CrossRef]
51. WorldPop; Center for International Earth Science Information Network (CIESIN), Columbia University. Global High Resolution Population Denominators Project-Funded by the Bill and Melinda Gates Foundation (OPP1134076). 2018. Available online: https://hub.worldpop.org/doi/10.5258/SOTON/WP00675 (accessed on 24 November 2024).
52. Zuur, A.; Leno, E.; Walker, N.; Saveliev, A.; Smith, G. *Mixed Effects Models and Extensions in Ecology with R*; Springer: New York, NY, USA, 2009.
53. Wickham, H. *ggplot2: Elegant Graphics for Data Analysis*; Springer: New York, NY, USA, 2016. Available online: https://ggplot2.tidyverse.org/ (accessed on 14 November 2024).
54. Niu, L.; Zou, G.; Guo, Y.; Li, Y.; Wang, C.; Hu, Q.; Zhang, W.; Wang, L. Eutrophication dangers the ecological status of coastal wetlands: A quantitative assessment by composite microbial index of biotic integrity. *Sci. Total Environ.* **2022**, *816*, 151620. [CrossRef]
55. Zhang, W.; Zhu, M.; Li, Y.; Wang, C.; Qian, B.; Niu, L.; Wang, P.; Gu, J.; Yang, N. How fluvial inputs directly and indirectly affect the ecological status of different lake regions: A bio-assessment framework. *J. Hydrol.* **2020**, *582*, 124502. [CrossRef]
56. Jordán, F.; Liu, W.; Davis, A.J. Topological keystone species: Measures of positional importance in food webs. *Oikos* **2006**, *112*, 535–546. [CrossRef]
57. Banerjee, S.; Schlaeppi, K.; van der Heijden, M.G.A. Keystone taxa as drivers of microbiome structure and functioning. *Nat. Rev. Microbiol.* **2018**, *16*, 567–576. [CrossRef] [PubMed]
58. Park, Y.; Cha, J.; Song, B.; Huang, Y.; Kim, S.; Kim, S.; Jo, E.; Fortin, S.; An, S. Total Microbial Activity and Sulfur Cycling Microbe Changes in Response to the Development of Hypoxia in a Shallow Estuary. *Ocean Sci. J.* **2020**, *55*, 165–181. [CrossRef]
59. Chen, J.; Hanke, A.; Tegetmeyer, H.E.; Kattelmann, I.; Sharma, R.; Hamann, E.; Hargesheimer, T.; Kraft, B.; Lenk, S.; Geelhoed, J.S.; et al. Impacts of chemical gradients on microbial community structure. *ISME J.* **2017**, *11*, 920–931. [CrossRef] [PubMed]
60. Wang, J.; Soininen, J.; Zhang, Y.; Wang, B.; Yang, X.; Shen, J. Patterns of elevational beta diversity in micro- and macroorganisms. *Glob. Ecol. Biogeogr.* **2012**, *21*, 743–750. [CrossRef]
61. Jiang, Y.; He, W.; Liu, W.; Qin, N.; Ouyang, H.; Wang, Q.; Kong, X.; He, Q.; Yang, C.; Yang, B.; et al. The seasonal and spatial variations of phytoplankton community and their correlation with environmental factors in a large eutrophic Chinese lake (Lake Chaohu). *Ecol. Indic.* **2014**, *40*, 58–67. [CrossRef]
62. Cao, J.; Liang, H.; Zhang, Y.; Du, S.; Zhang, J.; Tao, Y. Development and Evaluation of the Plankton Biological Integrity Index (P-IBI) in Dry and Wet Seasons for Dianchi Lake. *Ecologies* **2024**, *5*, 68–82. [CrossRef]
63. Yang, Q.; Zhang, P.; Li, X.; Yang, S.; Chao, X.; Liu, H.; Ba, S. Distribution patterns and community assembly processes of eukaryotic microorganisms along an altitudinal gradient in the middle reaches of the Yarlung Zangbo River. *Water Res.* **2023**, *239*, 120047. [CrossRef]
64. Xie, C.; Guo, Y.; Li, Y.; Zhou, Q.; Xie, P.; Huo, B. *Investigation on Water Ecological Environment and Fishery Resources of Xinjiang Transboundary Rivers: Irtysh River*; Science Press: Beijing, China, 2021.
65. Strayer, D.L. Alien species in fresh waters: Ecological effects, interactions with other stressors, and prospects for the future. *Freshw. Biol.* **2010**, *55*, 152–174. [CrossRef]
66. Xiong, W.; Sui, X.; Liang, S.; Chen, Y. Non-native freshwater fish species in China. *Rev. Fish Biol. Fish.* **2015**, *25*, 651–687. [CrossRef]
67. Dudgeon, D. Multiple threats imperil freshwater biodiversity in the Anthropocene. *Curr. Biol.* **2019**, *29*, R960–R967. [CrossRef]
68. Allan, J.D. Landscapes and Riverscapes: The Influence of Land Use on Stream Ecosystems. *Annu. Rev. Ecol. Evol. Syst.* **2004**, *35*, 257–284. [CrossRef]
69. Dudgeon, D.; Arthington, A.H.; Gessner, M.O.; Kawabata, Z.-I.; Knowler, D.J.; Lévêque, C.; Naiman, R.J.; Prieur-Richard, A.-H.; Soto, D.; Stiassny, M.L.J.; et al. Freshwater biodiversity: Importance, threats, status and conservation challenges. *Biol. Rev.* **2006**, *81*, 163–182. [CrossRef] [PubMed]
70. Shade, A.; Gregory Caporaso, J.; Handelsman, J.; Knight, R.; Fierer, N. A meta-analysis of changes in bacterial and archaeal communities with time. *ISME J.* **2013**, *7*, 1493–1506. [CrossRef] [PubMed]
71. Elbrecht, V.; Leese, F. Can DNA-Based Ecosystem Assessments Quantify Species Abundance? Testing Primer Bias and Biomass—Sequence Relationships with an Innovative Metabarcoding Protocol. *PLoS ONE* **2015**, *10*, e0130324. [CrossRef] [PubMed]
72. Zhang, S.; Zhao, J.; Yao, M. A comprehensive and comparative evaluation of primers for metabarcoding eDNA from fish. *Methods Ecol. Evol.* **2020**, *11*, 1609–1625. [CrossRef]
73. Xiong, W.; Huang, X.; Chen, Y.; Fu, R.; Du, X.; Chen, X.; Zhan, A. Zooplankton biodiversity monitoring in polluted freshwater ecosystems: A technical review. *Environ. Sci. Ecotechnol.* **2020**, *1*, 100008. [CrossRef]
74. Doi, H.; Uchii, K.; Takahara, T.; Matsuhashi, S.; Yamanaka, H.; Minamoto, T. Use of Droplet Digital PCR for Estimation of Fish Abundance and Biomass in Environmental DNA Surveys. *PLoS ONE* **2015**, *10*, e0122763. [CrossRef]

75. Takahashi, M.; Saccò, M.; Kestel, J.H.; Nester, G.; Campbell, M.A.; van der Heyde, M.; Heydenrych, M.J.; Juszkiewicz, D.J.; Nevill, P.; Dawkins, K.L.; et al. Aquatic environmental DNA: A review of the macro-organismal biomonitoring revolution. *Sci. Total Environ.* **2023**, *873*, 162322. [CrossRef]
76. Shu, L.; Ludwig, A.; Peng, Z. Environmental DNA metabarcoding primers for freshwater fish detection and quantification: In silico and in tanks. *Ecol. Evol.* **2021**, *11*, 8281–8294. [CrossRef]

*water*

MDPI

*Article*

# AFAR-WQS: A Quick and Simple Toolbox for Water Quality Simulation

**Carlos A. Rogéliz-Prada *,† and Jonathan Nogales *,†**

Nature for Water Facility, The Nature Conservancy, Arlington, VA 22203, USA
* Correspondence: carlos.rogeliz@tnc.org (C.A.R.-P.); jonathan.nogales@tnc.org (J.N.)
† These authors contributed equally to this work.

**Abstract:** Water quality management in large basins demands tools that balance scientific rigor with computational efficiency to avoid paralysis by analysis. While traditional models offer detailed insights, their complexity and resource intensity hinder timely decision-making. To address this gap, we present AFAR-WQS, an open-source MATLAB™ toolbox that introduces a novel integration of assimilation factors with graph theory and a Depth-First Search (DFS) algorithm to rapidly simulate 13 water quality determinants across complex topological networks. AFAR-WQS resolves cumulative processes in networks of up to 30,000 segments in just 163 s on standard hardware, enabling real-time scenario evaluations. Its object-oriented architecture ensures scalability, allowing customization for urban drainage systems or macro-basin studies while maintaining computational efficiency. Case studies demonstrate its utility in prioritizing sanitation investments, assessing water quality at the national scale and fostering stakeholder collaboration through participatory workshops. By bridging the gap between simplified and complex models, AFAR-WQS supports adaptive management in contexts of hydrological uncertainty, regulatory compliance, and climate change. The toolbox is freely available at GitHub, offering a transformative approach for integrated water resource management.

**Keywords:** water quality modeling; river basin management; Matlab toolbox; assimilation factor

## 1. Introduction

The analysis and understanding of water systems in large basins present significant challenges in the development of tools that enable a multidimensional understanding of the current state of natural resources and the impacts that may arise from specific actions within a basin. A comprehensive view of water systems requires not only an understanding of water quantity but also water quality [1–4], which is a crucial parameter for assessing the health of freshwater ecosystems and their potential for sustainable use [5].

According to [6], approximately 80% of global industrial and municipal wastewater is discharged into the environment without prior treatment. This percentage is even higher in less developed countries, where the lack of infrastructure for sanitation and wastewater treatment is more severe. This situation has significant negative repercussions, not only for human health but also for the functioning of aquatic ecosystems and socio-economic development [6,7]. Moreover, the continuous growth of the global population, along with increased freshwater consumption for human use, agriculture, and industry, amplifies these challenges [8].

One of the most useful tools for understanding water quality is mathematical modeling [9]. Currently, the range of available models that allow us to understand water quality dynamics in watersheds is quite diverse. Some of the models cited in the literature include Agricultural Non-Point Source, AGWA, ANSWERS-2000, APEX, AQUATOX, BASINS, EFDC, EPD-RIV1, GLEAMS/CREAMS, HSPF, InVEST, KINEROS2, LSPC, MIKE SHE, NLEAP, PRMS, QUAL2E, QUAL2K, SIGA-CAL, SWAT, SWMM, WAM, WARMF, WASP7, WCS, AquaChem, TOMCAT, SisBaHIA, SIMCAT, QUAL2KW, MOHID, MIKE HYDRO, ECETOC, and CE-QUAL-W2 [4,10,11]. Some of these models rely on physical equations to provide a detailed description of water quality determinant fate and transport processes, whereas others employ simpler conceptual approaches. Although each approach has its advantages and limitations, numerous studies have focused on comparing these models to guide selection according to specific objectives [5,12].

However, many of the aforementioned models exhibit significant computational complexity and extensive data requirements, which pose challenges for their implementation and calibration in large basins [13]. In most cases, prolonged execution times represent a major constraint, particularly in contexts requiring rapid decision-making and exploration of multiple scenarios [14,15]. For instance, [14] reported that simulating a single hydrological year at an hourly time step for a 139 km river with cross sections spaced every 50 m required approximately four days of computation. This computational burden complicates the execution of uncertainty and risk analyses, as well as the evaluation of multiple scenarios within limited timeframes—both of which are fundamental for water resources planning [15]. Consequently, simplified modeling approaches have gained traction as viable alternatives for integrated water quality assessments in large basins as they facilitate large-scale modeling with lower computational and data demands. Furthermore, these models can be incorporated into decision-support systems that integrate optimization techniques and risk-and-uncertainty analyses [14]. In such contexts, methodologies capable of producing preliminary results with reduced computational effort are critical for informing resource allocation strategies and assessing the feasibility of management interventions [16].

In this context, simplified models offer a compelling alternative, enabling water quality simulations with lower computational demand and more accessible input data. Their efficiency makes them particularly valuable for preliminary assessments and large-scale applications. Due to their speed and reduced calibration complexity, these models can be seamlessly integrated into decision-support platforms. One example of such tools is the InVEST models [17], which are designed to support natural resource management, particularly in implementing nature-based solutions. There are cases, such as SWAT, which have been integrated into the Hydrologic and Water Quality System (HAWQS), a national-scale decision support system (DSS) that allows for the simple and intuitive modeling of large watersheds [18]. Similarly, InVEST has been incorporated into WaterProof, a web-based decision support tool that provides quick calculations of return on investment (ROI) and early indications of an optimal portfolio of nature-based solutions (NBS) for any watershed globally [19]. However, while the use of such platforms should be encouraged, they still present limitations for global-scale basin analysis. These limitations include longer calculation times, complex preliminary configurations before the model becomes operational on the platform, and default configurations based on global databases that do not adequately capture the local characteristics of the territory.

Given this situation, there is a significant gap in water quality modeling innovation, particularly in designing efficient algorithms and computational procedures that enhance efficiency while preserving scientific rigor. One of the key challenges in chemical risk assessments in the 21st century, as identified by [20], is the need to develop parsimonious tools that are both scientifically rigorous and practical for regulatory applications. These tools must strike a balance between capturing the complexity of environmental processes and providing rapid, actionable insights for decision-makers. This balance is crucial to avoid paralysis by analysis, where overly complex models hinder timely decision-making. AFAR-WQS directly addresses this challenge by offering a simplified yet robust modeling approach that enables efficient simulations of water quality determinants, making it a valuable tool for regulatory risk assessment and integrated water resources management.

Recently, ref. [21] presented a simplified water quality model based on the concept of assimilation factors [22,23] for 13 water quality determinants: temperature (T), organic nitrogen (NO), ammoniacal nitrogen ($NH_4$), nitrates ($NO_3$), organic phosphorus (Po), inorganic phosphorus (Pi), organic matter (OM), dissolved oxygen (DO), suspended solids (SS), pathogenic organisms (X), elemental mercury (Hg0), divalent mercury (Hg2), and methylmercury (MeHg). This parsimonious approach builds on the foundations of conceptual models such as ADZ-QUASAR [24], which focus on the transport and fate of water quality determinants in water bodies. However, unlike these models, which are primarily designed to assess the behavior of specific substances, AFAR-WQS is tailored for the integrated water quality management at the basin scale, enabling rapid simulations of multiple determinants and facilitating decision-making processes.

This approach has been successfully applied in various contexts to identify priorities in sanitation [25–28], support national-scale water resource management [27], and approximate the health and integrity of freshwater ecosystems [21,29]. For instance, it has been used to prioritize investments in sanitation infrastructure, evaluate the impacts of point and non-point source pollution, and assess the effectiveness of nature-based solutions in improving water quality. These applications demonstrate the versatility and practical relevance of the assimilation factor concept, which now serves as the foundation for AFAR-WQS.

In this research, we present a computational implementation of the model proposed by [21] incorporating a novel approach in the calculation scheme for the rapid simulation of water quality determinants. The implementation is carried out as an open-source MATLAB™ toolbox (1.0 tested on Matlab R2023b, https://github.com/N4W-Facility/AFAR-WQS_Toolbox) (accessed on 25 February 2025) called AFAR-WQS (Assimilation Factor Analysis of Rivers for Water Quality Simulation). This toolbox employs a Depth-First Search (DFS) algorithm to efficiently resolve cumulative processes in complex topological networks, maximizing computational performance. AFAR-WQS simulates the concentration, loads, and assimilation factors of 13 water quality determinants. Additionally, it allows for the integration of diffuse sources of pollution by basin, point discharges, and boundary conditions. The toolbox was developed using an object-oriented programming approach, which facilitates scalability and maintenance. Furthermore, it incorporates an interactive user interface that enables visual analysis of the model results, allowing users to explore variations in water quality conditions from the reach to the network scale, thus facilitating interpretation and decision-making in water resource management.

## 2. Materials and Methods

AFAR-WQS was developed using the concept of assimilation factors to model the 13 water quality parameters it evaluates. Conceptually, this factor is defined as the stream's capacity to absorb and mitigate the impact of a contamination event [22,23]. According to [22,23], this concept is intrinsically linked to the physical, chemical, and biological processes that influence the contaminant within the receiving water body. Mathematically, the assimilation factor is expressed as follows:

$$a = \frac{W_u}{C}$$

where $a$ represents the assimilation factor, which depends on the characteristics of the receiving water body and the nature of the contaminant; $W_u$ corresponds to the water quality determinant load originating from upstream segments (with the subscript $u$ indicating the concentration of the parameter at that point); and $C$ denotes the downstream concentration in the segment.

Operationally, AFAR-WQS represents the drainage network of a watershed as a directed graph, allowing each river segment to be modeled as a connection between two nodes, as illustrated in Figure 1a. Using this framework, AFAR-WQS resolves the topological connectivity of the network through a Depth-First Search (DFS) algorithm [30], which begins at the segment corresponding to the watershed outlet. From this point, the algorithm progresses upstream through the drainage network until it identifies a headwater segment, defined as one with no incoming connections. (Figure 1b). Upon identifying a headwater segment, the model begins aggregating loads for various water quality determinants, estimating both assimilation factors and concentrations for that specific reach (Figure 1c). The algorithm then proceeds downstream to the immediately adjacent reach. If this downstream reach has incoming connections, the model recursively moves upstream, repeating the process described until it returns to the original confluence point. Figure 1c illustrates the manner in which concentrations are estimated for intermediate and headwater reaches.

This modeling scheme assumes that the physicochemical conditions are uniform throughout the entire reach, as it represents the smallest modeling unit in AFAR-WQS. Therefore, the assimilation factors are calculated based on the average values of the physicochemical characteristics of the reach. Moreover, the recursive approach used by the tool to solve the topological network allows for the analysis of networks with a large number of nodes while using minimal storage resources, a capability that would not be possible with adjacency matrix-based schemes, which require significantly higher computational resources [31].

Reach segmentation can be performed automatically, either based on an area accumulation threshold or a maximum length defined by the user, employing tools such as Topotoolbox [32], ArcHydro [33], and ArcSWAT [34], which facilitate the generation of river networks from a digital elevation model (DEM). AFAR-WQS offers the advantage of resolving connections involving more than two reaches connected to a node, as well as one-to-one reach connections. However, it is important to note that the tool cannot represent cyclic connections or bifurcations within the network. Additionally, it assumes that the graph is unidirectional, with connections directed toward the reach representing the basin's outlet. This flexibility allows the topological network to be segmented at points of interest defined by the user. AFAR-WQS also supports the representation of topological networks with multiple outlets, which is particularly relevant in decision-making processes where boundaries are guided not strictly by watershed divides but by political or administrative limits.

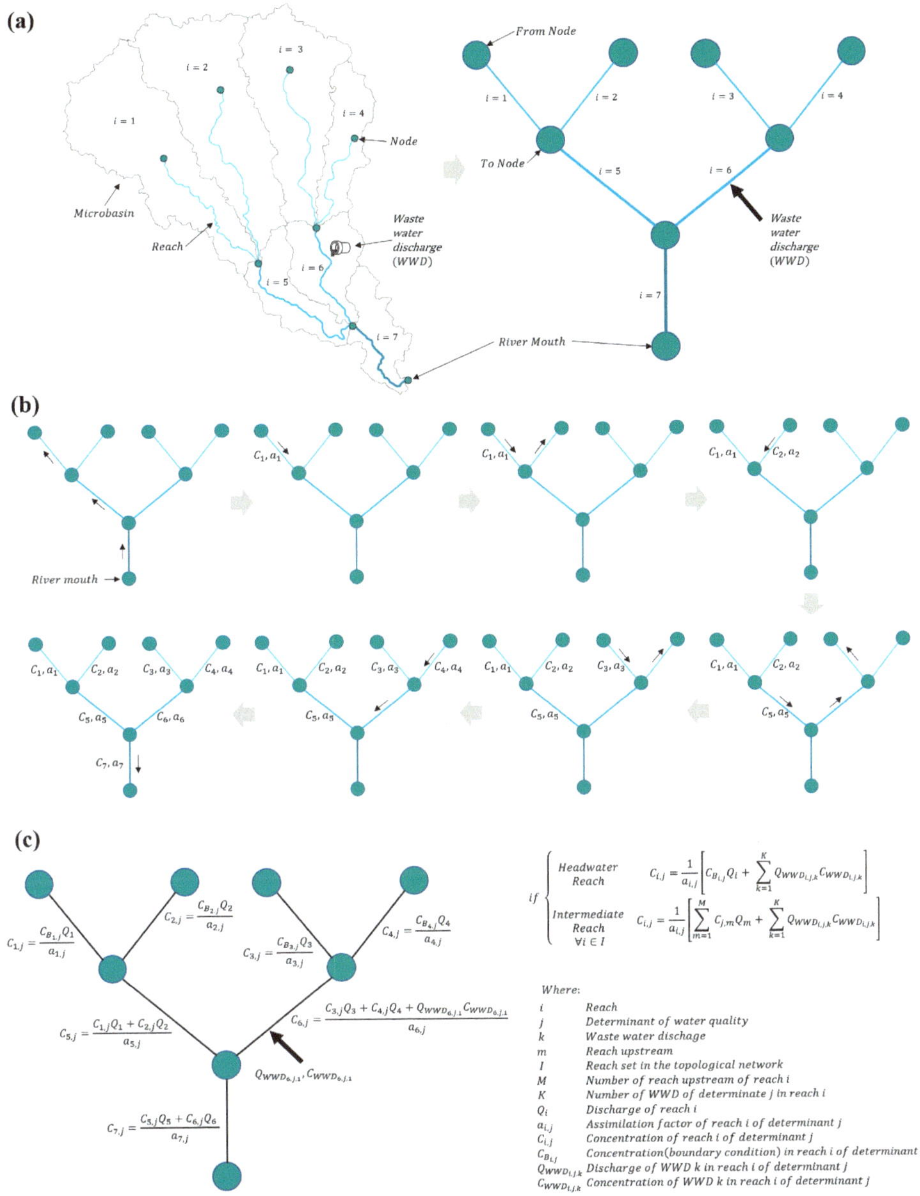


**Figure 1.** (**a**) Graph representation of a synthetic drainage network with seven reaches. (**b**) Schematic of the recursive solution framework used by AFAR-WQS to estimate assimilation factors and concentrations across the drainage network. (**c**) Example calculation of the concentration of a determinant *j* for a synthetic network with seven reaches.

*Toolbox Configuration*

The configuration of AFAR-WQS requires assigning each reach in the network a set of physical attributes related to the geomorphological characteristics of the river reaches to be modeled, which govern the transport conditions of the water quality determinants (see Table 1). All the necessary attributes for the tool can be derived from mathematical models (e.g., hydrological, hydrodynamic, etc.), prior studies, field data, or a combination of these sources. However, AFAR-WQS estimates certain key parameters by default, such as average travel time, dispersive fraction, advection time, and residence time. Given that the tool is developed using an object-oriented programming approach, users can customize it by incorporating additional methods tailored to their specific study area, allowing for the estimation of other attributes required by the tool.

**Table 1.** Attributes that are assigned to each topological network reach.

| Attributes | Unit | Description |
|---|---|---|
| ReachID | - | Unique positive integer numeric identifier of each reach in the topological network. |
| FromNode | - | Positive integer numeric identifier of the initial node of a reach of the topological network. |
| ToNode | - | Positive integer numerical identifier of the end node of a reach of the topological network. |
| ReachType | - | Identifies whether the reach represents a plain or mountain river. If false is specified, the tool will assume that the reach represents a plain river. To define whether a river is a plain or a mountain river, the first criterion may be to assume that the former is limited by capacity (slope $\leq$ 0.025 m/m) and the latter by supply (slope > 0.025 m/m), following the slope thresholds defined by [35]. A second criterion may be to use the slope threshold defined by [36] to define whether a river is mountain (slope > 0.002 m/m) or plain (slope < 0.002 m/m). |
| RiverMouth | - | Identifies the river reach that corresponds to the basin closure point. If the value is false, it is considered not to be a closure point reach. |
| $DF$ ** | dimensionless | The tool estimates the dispersive fraction following the criteria of [37]. For the sections of the topological network representing mountain rivers, an overall value of 0.27 is considered, while for plain rivers it is 0.40. |
| $\bar{t}$ | day | The tool estimates the average travel time as follows: $U_s = \frac{U}{1+\beta}$ $\bar{t} = \left(\frac{L}{U_s}\right) \times \left(\frac{1}{3600 * 24}\right)$ where $U_s$ is solute velocity (m/s), and $\beta$ is the effective delay coefficient. According to [37], the effective delay coefficient for mountain rivers has an overall magnitude of 1.10, while for plain rivers it is 2.0. |
| $\tau$ ** | day | The tool estimates the advection time as follows: $\tau = \bar{t} \times (1 - DF)$ |
| $TR$ ** | day | The tool estimates the residence time as follows: $TR = \bar{t} \times DF$ |
| L | m | River length representing the reach in the topological network. |
| Z | m.a.s.l | Average elevation of the river representing the reach in the topological network. |
| A | $m^2$ | Drainage area of the river representing the reach in the topological network, accumulated up to the ToNode of the reach. |
| Q | $m^3/s$ | Average discharge of the river representing the reach in the topological network, for a selected discharge scenario. |
| W | m | Average width of the river's cross section representing the reach in the topological network, for the selected discharge scenario. The width can be estimated from the DEM, satellite imagery [31,38], physically based relationships [39,40], field studies, or global datasets. |

**Table 1.** *Cont.*

| Attributes | Unit | Description |
|---|---|---|
| H | m | Average depth of the water column in the river representing the reach in the topological network, for the selected discharge scenario. The depth can be estimated from physically based relationships [39,40], field studies, or global datasets. |
| U | m/s | Average velocity of the water column in the river representing the reach in the topological network, for the selected discharge scenario. The velocity can be derived by continuity or through physically based relationships [39,40] as well as from field studies or global datasets. |
| S | m/m | Slope of the river representing the reach in the topological network. The slope can be estimated from the DEM, field studies, or global datasets. |
| T | °C | Average river water temperature representing the reach of the topological network. |
| Load_T | °C | Average temperature of the wastewater discharges entering the river representing the reach of the topological network. |
| Load_SS * | mg/d | The load of solids entering the river reach. |
| Load_X * | MPN/day | Total coliform load entering the river reach. |
| Load_NO * | mg/day | Organic nitrogen load entering the river reach. |
| Load_NH4 * | mg/day | Ammonia nitrogen load entering the river reach. |
| Load_NO3 * | mg/day | Nitrates load entering the river reach. |
| Load_PO * | mg/day | Organic phosphorus load entering the river reach. |
| Load_PI * | mg/day | Inorganic phosphorus load entering the river reach. |
| Load_OM * | mg/day | Organic matter load entering the river reach. |
| Load_DO * | mg/day | The load of dissolved oxygen entering the river reach. |
| Load_Hg0 * | mg/day | Elemental mercury load entering the river reach. |
| Load_Hg2 * | mg/day | Divalent mercury load entering the river reach. |
| | mg/day | Methylmercury load entering the river reach. |

Notes: * The loads of various water quality determinants entering the river, represented by a reach in the topological network, are estimated as the sum of diffuse loads (contributions from land covers, livestock, fertilization, etc.) within the sub-basin draining into the river reach, along with the loads contributed by point-source wastewater discharges within the same reach. ** The user can assign a unique value for reach section of the topological network.

Depending on the water quality determinant to be modeled, it is necessary to assign the corresponding reaction rates and decay parameters (see Table 2). By default, the tool incorporates the values used by [21] for all determinants. However, these parameters are fully editable by the user, allowing for greater flexibility to adapt the model to the specific conditions of the study area.

**Table 2.** Key physicochemical input parameters for each of the water quality determinants.

| Water Quality Determinants | Parameter | Unit | Description |
|---|---|---|---|
| Temperature | No parameters | - | - |
| Suspended Solids | $v_{SS}$ | m/day | Sedimentation velocity |
| | $k_{dX}$ | dimensionless | Constant decay of pathogenic organisms (mortality) |
| Pathogenic Organisms | $F_{px}$ | dimensionless | Fraction of pathogenic organisms adsorbed on solid particles |
| | $v_X$ | m/day | Sedimentation velocity of the adsorbed fraction of pathogens on solid particles |

**Table 2.** *Cont.*

| Water Quality Determinants | Parameter | Unit | Description |
|---|---|---|---|
| Organic Nitrogen | $k_{NO}$ | 1/day | Decay rate by hydrolysis of organic nitrogen |
| | $v_{NO}$ | m/day | Sedimentation velocity of organic nitrogen |
| Ammoniacal Nitrogen | $k_{NH4}$ | 1/day | Nitrification decay rate |
| Nitrates | $k_{dNO3}$ | 1/day | Denitrification rate |
| | $F_{oxdNO3}$ | dimensionless | Factor considering the effect of low oxygen on denitrification |
| Organic Phosphorus | $k_{Po}$ | 1/day | Organic phosphorus hydrolysis decay rate |
| | $v_{Po}$ | m/day | Organic phosphorus sedimentation velocity |
| Inorganic Phosphorus | $v_{Pi}$ | m/day | Inorganic phosphorus sedimentation velocity |
| Organic Matter | $k_{dOM}$ | 1/day | Organic matter oxidation decay rate |
| | $F_{oxdOM}$ | dimensionless | Factor considering the effect of low oxygen on organic matter |
| Oxygen Deficit | $k_a$ | 1/day | Reaeration rate |
| Elemental Mercury | $v_v$ | m/day | Elemental mercury volatilization velocity |
| | $k_{ox}$ | 1/day | Elemental mercury oxidation reaction rate |
| | $k_{rx}$ | 1/day | Oxidation decay rate of mercury |
| Divalent Mercury | $k_{me,a}$ | 1/day | Adsorbed divalent mercury methylation rate |
| | $k_{me,d}$ | 1/day | Dissolved divalent mercury methylation rate |
| | $F_{pHg2}$ | dimensionless | Fraction of divalent mercury adsorbed on solid particles |
| | $v_{Hg2}$ | m/day | Sedimentation velocity of divalent mercury |
| Methyl mercury | $F_{pMeHg}$ | dimensionless | Fraction of methyl mercury adsorbed on solid particles |
| | $v_{MeHg}$ | m/day | Sedimentation velocity of methyl mercury |

## 3. Results and Discussion

### 3.1. Toolbox Structure

The file and folder structure used by the toolbox is illustrated in Figure 2. The @ClassNetwork object represents the topological network and includes the recursive accumulation operations across this network. In turn, @AFAR_WQS groups the necessary methods for estimating the concentrations and assimilation factors of the various water quality determinants, inheriting the properties and methods of @ClassNetwork. Each determinant is modeled as a specific method within the @AFAR_WQS object, and these methods are executed independently. This allows the user to configure only the required inputs for the specific water quality determinant they intend to model.

AFAR-WQS takes the drainage network as input, organized in a data structure called ReachData. Each row in ReachData corresponds to an individual reach in the network, while the columns contain the geomorphological and physicochemical characteristics of the reach (detailed in Table 1). The ReachData structure can be exported as a shapefile, allowing for visualization and further analysis in any Geographic Information System (GIS) software.

As for the outputs, the toolbox calculates the concentration for each water quality determinant (expressed in mg/L for most determinants, except for coliforms, which is expressed in MPN/L), the assimilation factors in liters, and the loads in mg/day (except for coliforms, which is expressed in MPN/day). These results correspond to the flow scenario previously configured by the user. If different flow conditions need to be evaluated, the user can run multiple model executions to represent various discharge scenarios.

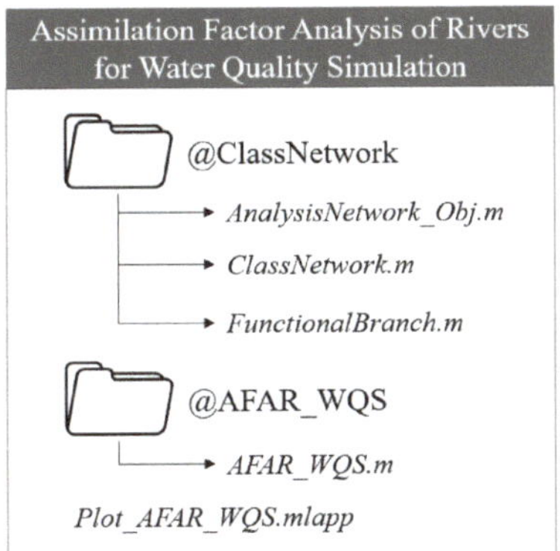


**Figure 2.** Folder structure and functions of the AFAR-WQS Toolbox.

### 3.2. Computational Performance

The DFS algorithm [30] employed by AFAR-WQS to resolve the topological connectivity of the network exhibits strong computational performance, delivering results within minutes. On a computer equipped with an Intel(R) Core(TM) i7-10750H CPU @ 2.60GHz and 64.0 GB of RAM, AFAR-WQS completes the modeling of a single water quality parameter for a network comprising 30,000 river segments in just 163 s (see Figure 3). A key advantage of AFAR-WQS is its low memory usage, as it does not generate vector data structures larger than the total number of segments in the network. Notably, due to the recursive nature of the DFS algorithm, the computational time scales following a power-law relationship as the number of segments increases. However, for networks containing up to 100,000 segments, the computation times remain manageable and aligned with the requirements of macro-basin scale modeling. The implementation of AFAR-WQS was validated using the results from [21], where we verified that the approach accurately reproduces the original outcomes.

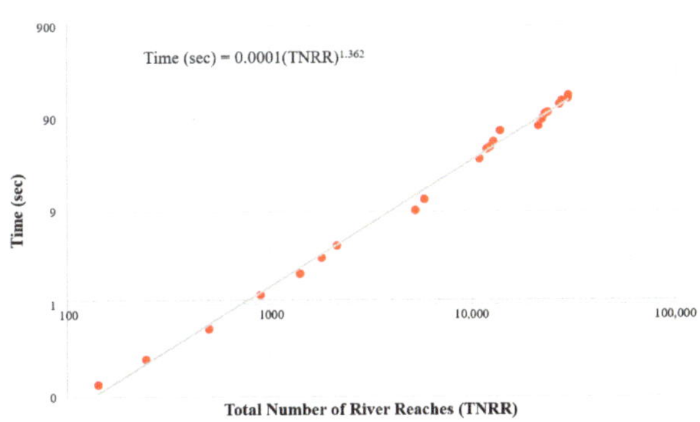


**Figure 3.** AFAR-WQS computational performance by number of drainage network segments.

### 3.3. Outputs and Visualization Options

When modeling a water quality determinant with AFAR-WQS, three key outputs are generated: (1) the concentration of the water quality determinant at the outlet of each reach (represented by object variables starting with "C_"), (2) the assimilation factor for the water quality determinant within the reach (represented by variables beginning with "AF_"), and (3) the load of the water quality determinant at the reach outlet (represented by variables beginning with "W_"). These outputs are generated for each reach in the analysis network and are stored as column vectors within the object representing the network, thereby facilitating efficient data management and analysis.

The toolbox also provides a Matlab application developed in App Designer, enabling users to visualize the results of each water quality determinant and support model interpretation. In the application interface, users can select both the variable and the water quality determinant they wish to visualize (see Figure 4a). The tool offers two visualization options: a network-level representation and a longitudinal profile. For the longitudinal profile, users must specify the Reach ID of the initial reach for the determinant profile. In this visualization, the graph displays the cumulative reach distance on the horizontal axis, spanning from the selected starting reach to the network outlet (river mouth), while the vertical axis illustrates changes in the analyzed determinant variable along the profile (see Figure 4b).

Additionally, the Matlab application allows users to customize the color palette used to visualize the water quality determinant variable, thereby improving visual comprehension and facilitating the analysis of model results.

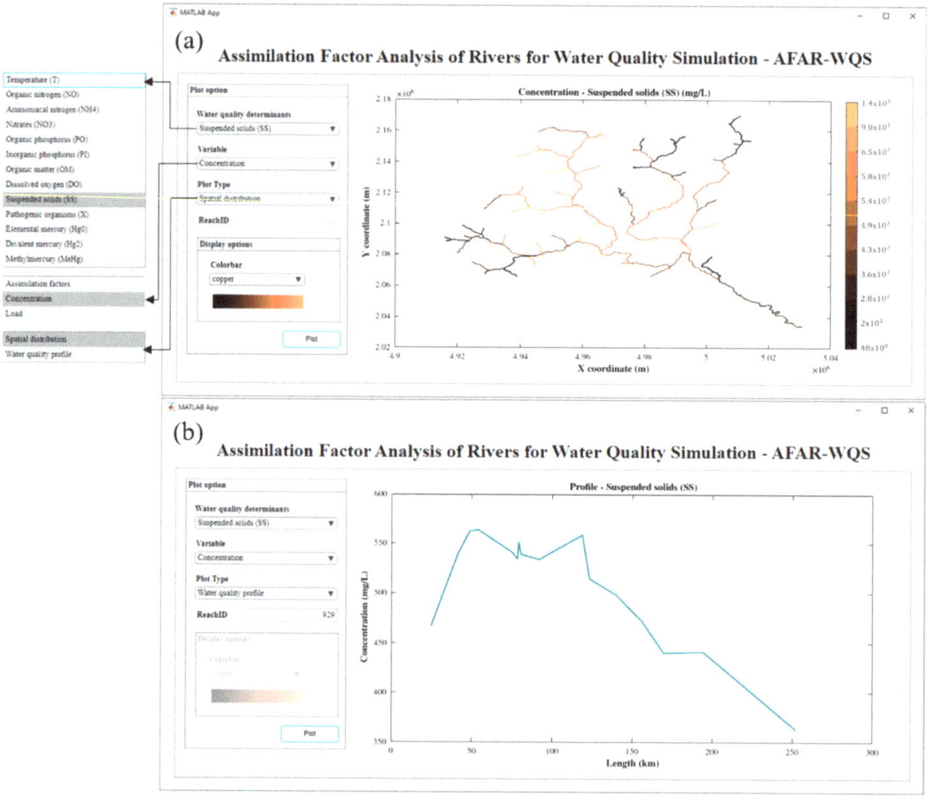


**Figure 4.** Visualization examples of suspended solids for a synthetic network. (**a**) llustrates the network-wide distribution of suspended solids concentration. (**b**) depicts the suspended solids profile extending from reach 929 to the network outlet.

### 3.4. Application in the Decision-Making Process

The concept of assimilation factors [22,23], applied to topological networks via graph theory and solved using a Depth-First Search algorithm, provides AFAR-WQS with the ability to run multiple simulations of the same water quality determinant in macro-basins within a matter of minutes. This speed not only optimizes computational times but also creates new opportunities for participatory management and real-time decision-making. For instance, the tool facilitates interactive workshops with key stakeholders, allowing for dynamic evaluation and comparison of scenarios, thereby fostering a deeper understanding and stronger community engagement in water resource management.

Furthermore, the rapid execution of simulations facilitates risk and uncertainty analyses, thereby enhancing support for decision-making in large-scale systems [14]. This capability is especially relevant in contexts where hydrological variability and anthropogenic pressures demand swift, well-founded responses. AFAR-WQS does not aim to replace more detailed studies that can be performed using robust models; rather, its goal is to serve as a preliminary analysis tool offering an integrated view of water quality at the macro-basin scale. This approach helps identify critical areas requiring more detailed study, thus optimizing the allocation of technical and financial resources.

AFAR-WQS's straightforward and quick configuration enables the evaluation of multiple scenarios in short timeframes, allowing decision-makers to analyze the potential effects of interventions such as constructing wastewater treatment plants or regulating discharges over several downstream kilometers. This capability is particularly useful in infrastructure planning and public policy evaluation, where multiple variables and scenarios must be considered within tight deadlines.

Likewise, the tool accommodates a broad spectrum of variants, enabling comprehensive risk and uncertainty analyses. This feature provides more robust support for selecting which interventions to implement, reducing the likelihood of suboptimal decisions. For instance, in climate change scenarios, where hydrological uncertainty is high, AFAR-WQS can be used to assess the resilience of different management strategies in the face of extreme events such as droughts or floods.

Another potential application of AFAR-WQS is in the context of freshwater ecosystem restoration. With the adoption of the Convention on Biological Diversity [41] by several countries, efforts to restore degraded ecosystems have intensified. According to [42], the success of these initiatives depends largely on the support of local communities, as their active participation contributes to the long-term effectiveness of restoration projects. However, in large areas or at the country scale, this process represents a significant challenge, particularly in communicating to communities the synergies between rivers and how the impacts of restoration projects propagate from upstream to downstream.

Modeling frameworks such as SIGA-CAL [4] allow for robust assessments of the impacts of nature-based solutions (NBS) on water quality. However, their use in workshops with local communities is limited due to the long computation times required; for example, a single simulation for a watershed of more than 10,000 km$^2$ can take more than 10 h. In contrast, AFAR-WQS is specifically designed for such contexts, where an exhaustive level of detail is not required, but rather an approximate and rapid response to the impacts that could be generated by interventions across the territory. AFAR-WQS can be configured for very large areas, including entire countries, enabling the evaluation of effects at the basin or national scale in a matter of minutes.

This capability is particularly useful in participatory workshops, where stakeholders often want to explore the potential effects of changes or measures and how they affect water quality. With AFAR-WQS, this is possible in just a few minutes, making it feasible to conduct these interactions in a dynamic and efficient manner. By allowing communities to quickly visualize how interventions propagate from upstream to downstream, AFAR-WQS fosters a deeper understanding of the impacts and benefits of restoration projects, enhancing community engagement and ensuring the sustainability of interventions in the long term.

The output generated by AFAR-WQS aids in identifying priority areas not only for infrastructure investment but also for research and the development of more detailed models focused on specific regions. Various applications reported in the literature corroborate the usefulness of this approach. For example, ref. [25] employed the estimation of assimilation factors to prioritize investments in water sanitation at the national level in Colombia, focusing on pathogens, nitrogen, and phosphorus by calculating the maximum permissible loads in rivers. Similarly, ref. [26] applied the same concept to create a tool aimed at generating and selecting optimal sanitation scenarios based on water quality standards and cost-efficiency using the Bogotá and Teusacá river basins as a case study. Ref. [27] also implemented assimilation factors at the national scale, including uncertainty analyses and designing an investment plan for the water sanitation sector across Peru. Ref. [21] used assimilation factors to evaluate the health and integrity of freshwater ecosystems in the Caquetá River basin, thus demonstrating the applicability of this strategy in freshwater ecosystem conservation. And, more recently, refs. [28,43] employed the same approach to model the transport and fate of mercury in surface waters at the national scale in Colombia.

The recursive algorithm implemented in AFAR-WQS confers scalability when configuring and applying the tool to different network topologies. The only requirement is that the network be a unidirectional graph (i.e., with no cycles), which does not preclude detailed subdivisions of specific reaches for more precise modeling when needed. This flexibility allows its use in urban environments with complex structures and extensive interconnections, as well as the option to combine simplified and detailed segments within a single study. For example, in urban areas with intricate drainage networks, AFAR-WQS can be employed to evaluate the impact of water quality determinant discharges at different points in the network, identifying critical areas that require immediate intervention.

The concept of assimilation factors has also been integrated into regulatory instruments within the Colombian context, such as the official methodology for determining the length of influence of discharges in surface watercourses [44], underscoring the relevance and currency of this approach in public management. This institutional adoption facilitates the uptake of AFAR-WQS by governmental agencies and non-governmental organizations, promoting its use in water resource planning and management at multiple scales.

Finally, by incorporating object-oriented programming, AFAR-WQS aligns with the component-based modeling paradigm [45]. This modular architecture simplifies the addition of new water quality determinants and the extension of conceptual representations to hydrological, sedimentological, and biological processes. Consequently, decision-making becomes more informed by enabling a comparative evaluation of alternative interventions within an integrated water resources management framework.

## 4. Conclusions

The AFAR-WQS toolbox represents a significant breakthrough in water quality modeling by transforming the theoretical concept of assimilation factors into a practical, scalable, and parsimonious decision-support tool. Developed as an open-source MATLAB™ toolbox, AFAR-WQS leverages graph theory and a Depth-First Search (DFS) algorithm to efficiently resolve cumulative processes in extensive, unidirectional river networks. This approach enables rapid simulations of 13 water quality determinants—including suspended solids, pathogens, nutrients, dissolved oxygen, biochemical oxygen demand, and mercury—across macro-basins, achieving computation times of 163 s for networks with 30,000 segments on standard hardware. Such efficiency addresses the critical challenge of analysis paralysis, empowering stakeholders to evaluate scenarios, conduct risk analyses, and prioritize interventions within feasible timeframes.

The novelty of this study lies in AFAR-WQS's capability to efficiently resolve cumulative processes in complex topological networks. This not only maximizes computational performance but also enables the rapid simulation of water quality determinants, making it a valuable tool for preliminary assessments and large-scale applications. Furthermore, AFAR-WQS is designed for high adaptability, allowing its application in diverse watershed contexts—from urban environments with intricate drainage networks to large basins. The toolbox's object-oriented programming approach ensures scalability, permitting users to tailor the model to the specific characteristics of each basin without compromising computational efficiency. Its interactive user interface facilitates result visualization and interpretation, making it accessible for communicating technical information to non-technical stakeholders. This feature is particularly relevant for fostering collaboration among scientists, decision-makers, and local communities, as it promotes participatory decision-making processes. For instance, AFAR-WQS can be employed in stakeholder workshops to simulate and compare various intervention scenarios very quickly, cultivating a shared understanding of potential impacts and trade-offs. This collaborative approach not only enhances transparency in decision-making but also increases the likelihood of successfully implementing water management strategies.

The flexibility of AFAR-WQS is also one of its most notable features. The toolbox accommodates various data sources, ranging from field measurements to outputs from hydrological models, and adapts to diverse geographic contexts. Its object-oriented architecture ensures scalability, allowing users to customize river segmentation, integrate region-specific reaction rates, or refine network topology without undermining performance. Moreover, seamless GIS compatibility facilitates data preparation, visualization, and sharing, bridging the gap between technical modeling and spatial decision-making.

The assimilation factor framework underpinning AFAR-WQS is already institutionally recognized, particularly in Colombia's regulatory methodology for determining the length of influence for discharges in surface water bodies. While this regulatory framework adopts the conceptual basis of assimilation factors, AFAR-WQS transforms the theory into a dynamic tool for real-world applications. Case studies demonstrate its versatility: from prioritizing sanitation investments in the Bogotá region's watersheds to modeling mercury transport at the national scale in Colombia and designing climate-resilient water management plans in Peru. These applications highlight AFAR-WQS's dual role as both a preliminary evaluation tool and a potential platform for stakeholder engagement through participatory workshops that enable real-time scenario comparisons, thus promoting transparency and community acceptance.

By democratizing access to rapid, large-scale water quality assessments, AFAR-WQS supports adaptive governance in the face of climate change, population growth, and regulatory complexity. Its open-source nature fosters collaboration, encouraging the research community to expand its library of determinants or integrate modules for sediment dynamics and ecosystem services. The toolbox, along with its documentation and user-friendly interface, is freely available on GitHub (https://github.com/N4W-Facility/AFAR-WQS_Toolbox) (accessed on 25 February 2025), ensuring that policymakers, scientists, and communities worldwide can harness its potential.

**Author Contributions:** Conceptualization, C.A.R.-P. and J.N.; methodology, C.A.R.-P. and J.N.; software, J.N.; validation, C.A.R.-P.; writing—original draft preparation, J.N.; writing—review and editing, C.A.R.-P.; visualization, C.A.R.-P. and J.N.; supervision, C.A.R.-P.; project administration, C.A.R.-P.; funding acquisition, C.A.R.-P. All authors have read and agreed to the published version of the manuscript.

**Funding:** This research was funded by the Nature for Water Facility (funding number P120989 A111088).

**Data Availability Statement:** The code associated with this paper is openly available at https://github.com/N4W-Facility/AFAR-WQS_Toolbox (accessed on 25 February 2025).

**Acknowledgments:** The authors extend their sincere gratitude to the Nature for Water Facility for its invaluable support in fostering and advancing this research, as well as for its unwavering commitment to the advancement of scientific knowledge and the implementation of nature-based solutions.

**Conflicts of Interest:** The authors declare no conflicts of interest.

## References

1. Tao, Y.; Tao, Q.; Qiu, J.; Pueppke, S.G.; Gao, G.; Ou, W. Integrating Water Quantity- and Quality-Related Ecosystem Services into Water Scarcity Assessment: A Multi-Scenario Analysis in the Taihu Basin of China. *Appl. Geogr.* **2023**, *160*, 103101. [CrossRef]
2. Hwang, J.H.; Park, S.H.; Song, C.M. A Study on an Integrated Water Quantity and Water Quality Evaluation Method for the Implementation of Integrated Water Resource Management Policies in the Republic of Korea. *Water* **2020**, *12*, 2346. [CrossRef]
3. Nowicki, S.; Koehler, J.; Charles, K.J. Including Water Quality Monitoring in Rural Water Services: Why Safe Water Requires Challenging the Quantity versus Quality Dichotomy. *NPJ Clean Water* **2020**, *3*, 14. [CrossRef]
4. Jiménez, M.; Usma, C.; Posada, D.; Ramirez, J.; Rogéliz, C.A.; Nogales, J.; Spiro-Larrea, E. Planning and Evaluating Nature-Based Solutions for Watershed Investment Programs with a SMART Perspective Using a Distributed Modeling Tool. *Water* **2023**, *15*, 3388. [CrossRef]
5. Talukdar, P.; Kumar, B.; Kulkarni, V.V. A Review of Water Quality Models and Monitoring Methods for Capabilities of Pollutant Source Identification, Classification, and Transport Simulation. *Rev. Environ. Sci. Biotechnol.* **2023**, *22*, 653–677. [CrossRef]
6. Krishan, A.; Yadav, S.; Srivastava, A. Water Pollution's Global Threat to Public Health: A Mini-Review. *Int. J. Sci. Res. Sci. Eng. Technol.* **2023**, *10*, 321–334. [CrossRef]
7. Du Plessis, A. Persistent Degradation: Global Water Quality Challenges and Required Actions. *One Earth* **2022**, *5*, 129–131. [CrossRef]
8. Rajak, P.; Ganguly, A.; Nanda, S.; Mandi, M.; Ghanty, S.; Das, K.; Biswas, G.; Sarkar, S. Toxic Contaminants and Their Impacts on Aquatic Ecology and Habitats. In *Spatial Modeling of Environmental Pollution and Ecological Risk*; Elsevier: Amsterdam, The Netherlands, 2024; pp. 255–273.
9. Benedini, M.; Tsakiris, G. *Water Quality Modelling for Rivers and Streams*; Springer: Dordrecht, The Netherlands, 2013; Volume 70, ISBN 978-94-007-5508-6.
10. Ejigu, M.T. Overview of Water Quality Modeling. *Cogent. Eng.* **2021**, *8*, 1891711. [CrossRef]
11. Ecetoc, H. *A Mathematical Model for Use in Risk Assessment of Substances*; Special Report No. 8, in Preparation; ECETOC: Brussels, Belgium, 1994.
12. Darji, J.; Lodha, P.; Tyagi, S. Assimilative Capacity and Water Quality Modeling of Rivers: A Review. *AQUA Water Infrastruct. Ecosyst. Soc.* **2022**, *71*, 1127–1147. [CrossRef]
13. Hofstra, N.; Kroeze, C.; Flörke, M.; van Vliet, M.T. Editorial Overview: Water Quality: A New Challenge for Global Scale Model Development and Application. *Curr. Opin. Environ. Sustain.* **2019**, *36*, A1–A5. [CrossRef]
14. Keupers, I.; Willems, P. Development and Testing of a Fast Conceptual River Water Quality Model. *Water Res.* **2017**, *113*, 62–71. [CrossRef] [PubMed]
15. Nguyen, T.T.; Keupers, I.; Willems, P. Conceptual River Water Quality Model with Flexible Model Structure. *Environ. Model. Softw.* **2018**, *104*, 102–117. [CrossRef]
16. Beven, K. How to Make Advances in Hydrological Modelling. *Hydrol. Res.* **2019**, *50*, 1481–1494. [CrossRef]
17. Natural Capital Project. InVEST (Version 3.13.0) [Software]. Stanford University, University of Minnesota, Chinese Academy of Sciences, The Nature Conservancy, World Wildlife Fund, Stockholm Resilience Centre, and the Royal Swedish Academy of Sciences. 2025. Available online: https://naturalcapitalproject.stanford.edu/software/invest (accessed on 25 February 2025).
18. Yen, H.; Daggupati, P.; White, M.; Srinivasan, R.; Gossel, A.; Wells, D.; Arnold, J. Application of Large-Scale, Multi-Resolution Watershed Modeling Framework Using the Hydrologic and Water Quality System (HAWQS). *Water* **2016**, *8*, 164. [CrossRef]
19. Rogéliz, C.A.; Vigerstol, K.; Galindo, P.; Nogales, J.; Raepple, J.; Delgado, J.; Piragauta, E.; González, L. WaterProof—A Web-Based System to Provide Rapid ROI Calculation and Early Indication of a Preferred Portfolio of Nature-Based Solutions in Watersheds. *Water* **2022**, *14*, 3447. [CrossRef]
20. Di Guardo, A.; Gouin, T.; MacLeod, M.; Scheringer, M. Environmental Fate and Exposure Models: Advances and Challenges in 21st Century Chemical Risk Assessment. *Environ. Sci. Process Impacts* **2018**, *20*, 58–71. [CrossRef] [PubMed]

21. Nogales, J.; Rogéliz-Prada, C.; Cañon, M.A.; Vargas-Luna, A. An Integrated Methodological Framework for the Durable Conservation of Freshwater Ecosystems: A Case Study in Colombia's Caquetá River Basin. *Front. Environ. Sci.* **2023**, *11*, 1264392. [CrossRef]
22. Chapra, S.C. *Surface Water-Quality Modeling*, 15th ed.; Waveland Press: Long Grove, IL, USA, 2008.
23. Chapra, S.C. *Surface Water Quality Modelling*; Mc Graw Hill: New York, NY, USA, 1997.
24. Lees, M.J.; Camacho, L.; Whitehead, P. Extension of the QUASAR River Water Quality Model to Incorporate Dead-Zone Mixing. *Hydrol. Earth Syst. Sci.* **1998**, *2*, 353–365. [CrossRef]
25. Rojas, A. Aplicación de Factores de Asimilación Para La Priorización de La Inversión En Sistemas de Saneamiento Hídrico En Colombia. Master's Thesis, Universidad Nacional de Colombia, Bogotá, Colombia, 2011.
26. Navas, A. *Factores de Asimilación de Carga Contaminante en Ríos—Una Herramienta Para la Identificación de Estrategias de Saneamiento Hídrico en Países en Desarrollo*; Universidad de los Andes: Bogotá, Colombia, 2016.
27. Mamani, J.A. Desarrollo de Un Modelo Numérico de Calidad Del Agua en Un Marco Probabilístico de Soporte a Las Decisiones a Escala Nacional. Master's Thesis, Univerdad de los Andes, Bogotá, Colombia, 2022.
28. Correa-Caselles, D. Metodología Para La Estimación Del Destino y Transporte de Mercurio Presente En Los Ríos de Colombia. Master's Thesis, Univerdad de los Andes, Bogotá, Colombia, 2022.
29. Rivera Gutiérrez, J.V. Evaluation of the Kinetics of Oxidation and Removal of Organic Matter in the Self-Purification of a Mountain River. *Dyna* **2015**, *82*, 183–193. [CrossRef]
30. Hopcroft, J.; Tarjan, R. Algorithm 447: Efficient Algorithms for Graph Manipulation. *Commun. ACM* **1973**, *16*, 372–378. [CrossRef]
31. Tangi, M.; Schmitt, R.; Bizzi, S.; Castelletti, A. The CASCADE Toolbox for Analyzing River Sediment Connectivity and Management. *Environ. Model. Softw.* **2019**, *119*, 400–406. [CrossRef]
32. Schwanghart, W.; Scherler, D. Short Communication: TopoToolbox 2—MATLAB-Based Software for Topographic Analysis and Modeling in Earth Surface Sciences. *Earth Surf. Dyn.* **2014**, *2*, 1–7. [CrossRef]
33. ESRI, M.D. *ArcHydro: GIS for Water Resources*; ESRI, Inc.: Redlands, CA, USA, 2013.
34. Winchell, M.; Srinivasan, R.; Di Luzio, M.; Arnold, J.G. *ArcSWAT 2.3. 4 Interface For SWAT2005*; Grassland, Soil and Research Service: Temple, TX, USA, 2009.
35. Flores, A.N.; Bledsoe, B.P.; Cuhaciyan, C.O.; Wohl, E.E. Channel-reach Morphology Dependence on Energy, Scale, and Hydroclimatic Processes with Implications for Prediction Using Geospatial Data. *Water Resour. Res.* **2006**, *42*, 6412. [CrossRef]
36. Wohl, E.; Merritt, D. Prediction of Mountain Stream Morphology. *Water Resour. Res.* **2005**, *41*, W08419. [CrossRef]
37. González, R. *Determinación Del Comportamiento de La Fracción Dispersiva En Ríos Característicos de Montaña*; UNAL: Bogotá, Colombia, 2008.
38. Schmitt, R.J.P.; Bizzi, S.; Castelletti, A. Tracking Multiple Sediment Cascades at the River Network Scale Identifies Controls and Emerging Patterns of Sediment Connectivity. *Water Resour. Res.* **2016**, *52*, 3941–3965. [CrossRef]
39. Wilkerson, G.V.; Parker, G. Physical Basis for Quasi-Universal Relationships Describing Bankfull Hydraulic Geometry of Sand-Bed Rivers. *J. Hydraul. Eng.* **2011**, *137*, 739–753. [CrossRef]
40. Parker, G.; Wilcock, P.R.; Paola, C.; Dietrich, W.E.; Pitlick, J. Physical Basis for Quasi-Universal Relations Describing Bankfull Hydraulic Geometry of Single-Thread Gravel Bed Rivers. *J. Geophys. Res. Earth Surf.* **2007**, *112*, 1–21. [CrossRef]
41. Chandra, A.; Idrisova, A. Convention on Biological Diversity: A Review of National Challenges and Opportunities for Implementation. *Biodivers. Conserv.* **2011**, *20*, 3295–3316. [CrossRef]
42. Higgins, J.; Zablocki, J.; Newsock, A.; Krolopp, A.; Tabas, P.; Salama, M. Durable Freshwater Protection: A Framework for Establishing and Maintaining Long-Term Protection for Freshwater Ecosystems and the Values They Sustain. *Sustainability* **2021**, *13*, 1950. [CrossRef]
43. Correa-Caselles, D.; Camacho, L.A.; Fernandez, N. A Modeling Framework to Estimate the Transport and Fate of Mercury in Nationwide Surface Waters. *Water* **2025**, *17*, 250. [CrossRef]
44. MADS; ANLA. *Metodología Para La Definición de La Longitud de Influencia de Vertimientos Sobre Corrientes de Agua Superficial*; MADS and ANLA: Bogotá, Colombia, 2013.
45. Buahin, C.A.; Horsburgh, J.S. Evaluating the Simulation Times and Mass Balance Errors of Component-Based Models: An Application of OpenMI 2.0 to an Urban Stormwater System. *Environ. Model. Softw.* **2015**, *72*, 92–109. [CrossRef]

water

MDPI

Article

# Spatial-Temporal Monitoring of Water Quality in Rural Property Enrolled in a Program for Payment for Environmental Water Services (PES-Water)—A Case Study in Brazil

**Joice Machado Garcia [1], Regina Márcia Longo [1], Adélia Nobre Nunes [2] and Raissa Caroline Gomes [1,*]**

[1] Postgraduate Program in Urban Infrastructure System, Polytechnic School, Pontifical Catholic University of Campinas, Campinas 13087-571, Brazil

[2] CEGOT—Centre for Studies in Geography and Spatial Planning, Department of Geography and Tourism, University of Coimbra, 3004-504 Coimbra, Portugal

* Correspondence: raissa.cgt@hotmail.com

**Abstract:** Payments for ecosystem (or environmental) services (PES) encourage land users to manage their land in ways that deliver environmental benefits. This study aimed to assess the water quality in a rural property located in the Protection and Recovery of Watersheds of Campinas, which has been participating in the payment for ecosystem services program since 2018. More specifically, seven points of interest regarding the physicochemical indicators of the water were raised, which were subjected to descriptive statistical and variance analysis. The results revealed significant spatio-temporal variability in the monitored water quality indicators for dissolved oxygen, biochemical oxygen demand, pH, total phosphorus, and total nitrogen. More significant fluctuations were observed in the spatial location of the sampling points for turbidity, temperature, and electrical conductivity. However, the greatest variability depends on the time of year when the samples were collected.

**Keywords:** payments for environmental services; water quality; land use; land cover; urban river basin

## 1. Introduction

Ecosystem services are benefits provided, directly or indirectly, by biophysical elements and ecological functions to human beings, such as the provision of food and drinking water, climate regulation, and protection from natural disasters, and their existence is associated with the integrity of ecosystems, biodiversity, and anthropic actions [1].

However, the provision of ecosystem services has become scarce, mainly in urban centers, given that urban expansion causes changes in land use and coverage, especially benefiting the industrial and construction sector [2]. In addition to their reduction, the fact that these services are not priced by the economic market is another risk element. This is based on the assumption that environmental services are free, although they are used in the production of goods and services [3]. Given this scenario, the identification, management, and valuation of ecosystem services constitute some of the greatest challenges for urban planning in the 21st century.

Payments for environmental or ecosystem services (PES) can be seen as (1) a tool to provide economic incentives or compensation for land users that can promote ecosystem services [4]; (2) an instrument or mechanism capable of correcting market failure through the recognition of practices that guarantee, recover, or improve the provision of ecosystem services [5]. It could also become a means to promote biodiversity conservation and rural development, particularly in tropical and subtropical regions [6]; or (3) a policy tool to fund nature conservation efforts [7].

Payments for ecosystem services (PES) encourage land users to manage land in a way that provides environmental benefits. For example, the focus of many PES programs in Latin America is to increase or maintain the supply of good-quality water [8,9]. It is important to note that the importance of payment to land users is not restricted to a single service provided by the area in question, but may also be linked to a change in perception regarding the environmental services provided, leading to a change in habits and a feeling of recognition by society (4). However, this is a reality that is not well understood, with 42% of the PES analyzed in a study showing a crowding-out effect [7]. PES schemes are defended when environmental externalities are observed, such as a deforestation process that leads to loss of water quality. These situations can be corrected by creating ad hoc markets based on the Coase postulate, where the aim is to achieve optimum social welfare through negotiation, which in this case can be through payments to farmers in exchange for change to good management practices, which makes the cost of giving up forest exploitation practices worthwhile [8].

A PES proposal assumes that conservationist practices generate benefits for society and, more directly, for agents who benefit from cost reduction or improvement in the quality of inputs needed for their production processes. PES suggest the internalization of external benefits by sharing with the beneficiaries of environmental services the costs of carrying out the actions taken by the providers [10]. PES present an economic perspective of conservation, since the service provider finds in payment an incentive to change their behavior and guarantee the ecological balance of their productive activities [11]. However, for their correct functioning, PES programs need the main components to be defined: which services are a priority, which practices lead to the generation of this service, pre-disposition to payment, and possible payers [12].

Latin America has been a pioneer in the implementation of PES [8,13] by addressing four main ecosystem services: biodiversity conservation, watershed services, carbon sequestration, and scenic beauty [14]. The Atlantic Forest biome has stood out, being present in most of the 200 PES projects according to Forest Trends [15], and the projects are related to payments for watershed services (PWS), with a focus on water resource conservation [12]. Therefore, river ecosystems are important providers of essential services, such as the provision of water for human consumption [16]

In the case of Campinas (Brazil), the motivation for implementing the PES program was given in the face of the water crisis that affected its metropolitan region in the period from 2013 to 2015. The package to combat water scarcity provided the drafting of Municipal Law no. 15.046 of 2015, which instituted the PES program [17], and Municipal Decree no. 19.441 of 2017, which regulated the PES-Water subprogram [18].

The PES-Water proposal in the municipality was drawn up based on successful national and international experiences, such as the Conservador de Águas program in Extrema, Minas Gerais. From its implementation in 2017 until the end of 2020, the PES-Water Campinas had already benefited 127 properties, which included the planting of more than 109 thousand seedlings and the installation of 230 sanitary sewage systems [19]. The reviews of activity found for this experience are mostly qualitative and theoretical despite the volume of schemes currently in force [8]. For territorial management to be validated as a generator of water ecosystem services and to prove the effectiveness of PWS, hydrological monitoring is essential [20]. There is a gap in understanding how to evaluate and improve PES monitoring plans [21].

The analysis is consolidated as an important tool that aims to convince water users (supply and hydroelectric energy generation companies) to support the project, ensuring its financial sustainability and financing for impact assessments as a whole [22,23]. There is a relationship between the quality of water in the watershed and society's environmental perception, where it is found that improving the quality of the resource maximizes the positive impact on social well-being [16]. A study that evaluated PES in Brazil presents data that the first PES program was initiated in 1997 in the state of Santa Catarina as part of a municipal policy to protect water resources [4]. Based on the above assumption that monitoring water quality on rural properties is part of PES programs and can contribute to the implementation of planned actions and help improve the quality and supply of water resources, the aim of this study was to assess the spatio-temporal variation in water quality on a rural property participating in a PES-Water program. To this end, seven sampling points were considered and evaluated over a period in order to gauge whether one year is sufficient to attribute changes in water quality as a result of the improvements implemented on the property under the PES program.

## 2. Materials and Methods

### *2.1. Study Area*

The property used for the study is located in the hydrographic basin of the Atibaia River in the municipality of Campinas, São Paulo. It has been involved in the PES program since February 2018 and has an area of 102.9 ha, of which 14.4 ha constitute a permanent preservation area (PPA), 16.8 ha are vegetation remnants, and 58.3 ha are a grassland area.

There are four blocks into which the PES actions and goals were divided by the Personalized Property Program, defined as (i) the one that covers obligations pre-defined in legislation for forest restoration of protected areas and protection of native vegetation, called the forest code; (ii) the one that deals with rainwater drainage and domestic effluent management, called rural sanitation; (iii) the one that concerns the implementation of soil conservation practices, as well as the maintenance of existing practices, called soil conservation; and finally (iv) the one that concerns the authorization for the use of water resources, called grant. To evaluate the scenario, seven areas for bimonthly monitoring were proposed, as can be seen in Figure 1. The objective was to verify whether the actions and improvements implemented and/or seasonality could bring noticeable changes to the parameters considered. Therefore, the seven areas considered were defined ass (1) downstream property (external point to assess the quality of water leaving the property); (2) forest remnant (external course, PES planting); (3) sewer amount (before the treatment system and external course; higher water flow); (4) erosive process (conservation practices application; low water quantity); (5) consumption—nascent (spring displacement; livestock amount fencing); (6) livestock interference (cattle trampling; PES surrounding; downstream planting); and (7) property amount (prior to improvements; siltation; drainage work) [24].

### *2.2. Methodology*

The water collection took place in the months of April 2019, July 2019, September 2019, November 2019, and January 2020, obtaining four samples at each point. The first stage of water quality analysis consisted of measuring the following parameters: dissolved oxygen—DO (oximeter); biochemical oxygen demand—BOD (oximeter and incubation); pH (potentiometer); temperature (oximeter); turbidity (turbidimeter); electrical conductivity (conductivity meter); total dissolved solids—TDS (gravimetry); total phosphorus (ascorbic acid method); total nitrogen (Kjeldahl digestion and distillation); and total coliforms (multiple tubes) [24]. All these were selected in view of their relevance in monitoring PES programs—water [25].

For the parameters DO, BOD, pH, turbidity, total phosphorus, and TDS, the contents of the elements were compared to the limits established by the Brazilian resolution of the Conselho Nacional de Meio Ambiente—CONAMA no. 357/05, which "provides for the classification of the bodies of water and environmental guidelines for its setting" [26]. This resolution presents threshold values for classifying water resources according to their quality and possible uses. Class 1 is considered the highest-quality water and can be used for more restricted uses (primary contact recreation, such as swimming, water skiing, and diving; irrigating vegetables that are eaten raw), while Class 4 can only be used for landscape harmony and navigation without human contact [26].

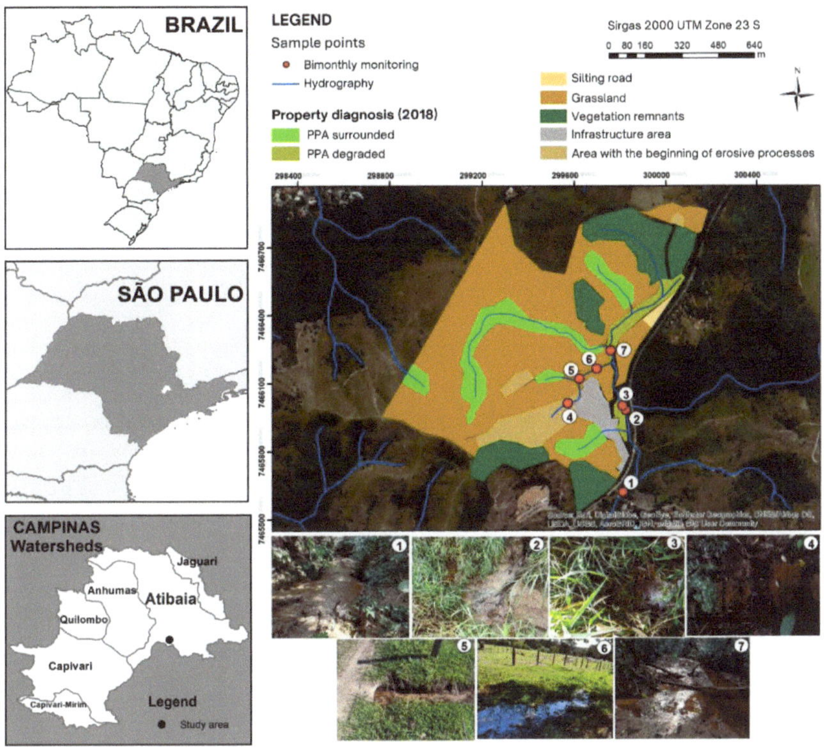


**Figure 1.** Sample points at the study property, Campinas, São Paulo [20].

As the temperature, electrical conductivity, total nitrogen, and total coliforms parameters do not have limits defined by CONAMA Resolution no. 357/05 [26], they were compared under Ministry of Health Ordinance no. 2.914/2011 [27], updated in 2017 by Annex XX of Consolidation Ordinance no. 5/2017 [28], as well as legislation and/or academic works.

Figure 2 shows the average monthly rate of precipitation and temperature in the study period and before the collection days, according to the Centro Integrado de Informações Agrometeorológicas—CIIAGRO [29].

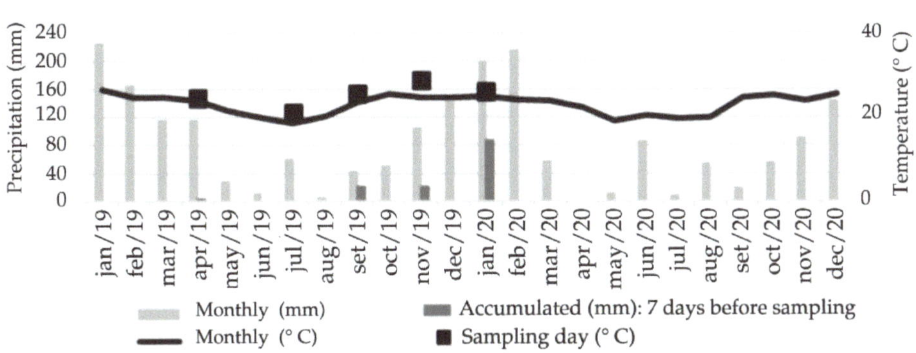


**Figure 2.** Climatic data in the municipality of Campinas during the study period. (Authors' own.)

The water data were initially processed in Microsoft Excel, in which the position measures were defined: minimum, first quartile, median, third quartile, and maximum, used for the boxplot composition, grouping, and classification technique that consists of a box diagram for viewing trends, dispersion, form of distribution or symmetry, and deviations within the analyzed data [30].

Then, to check the effects of spatial and seasonal variation on the water quality parameters, and using Statistica 10.0 software, a two-way ANOVA was applied at a 5% level of significance. The experimental design adopted was a split-plot design or subdivided plots, with the treatments arranged in a 7 × 5 factorial scheme, i.e., seven collection points and five months of sampling, with four repetitions per point. The parameters that showed a statistically significant difference ($p < 0.05$) were subjected to the Tukey test to compare means between groups [31].

## 3. Results and Discussion

### 3.1. Spatial-Temporal Variability on Water Oxygen, Temperature, and pH

Table 1 presents the descriptive statistics and Tukey test for the parameters of dissolved oxygen, biochemical oxygen demand, pH, and temperature. To assist in the discussion, Figures 3–6 show the boxplot distribution of the referred parameters, respectively.

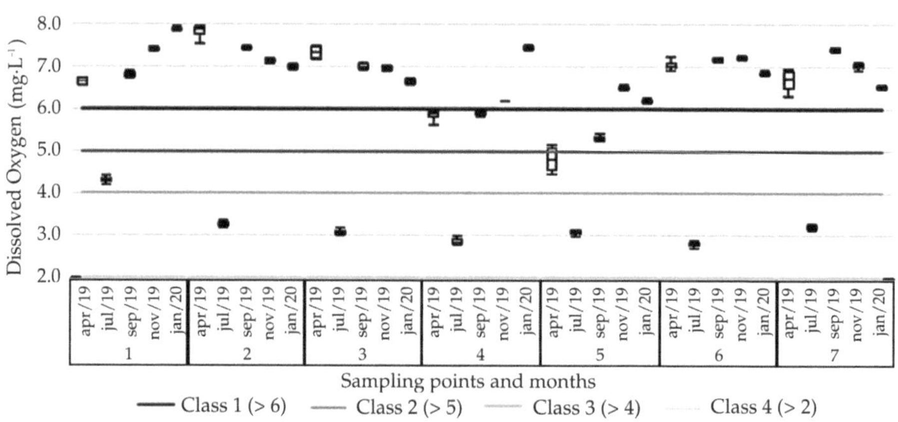


**Figure 3.** Descriptive analysis for the indicators dissolved oxygen—OD (mg·L$^{-1}$). (Authors' own).

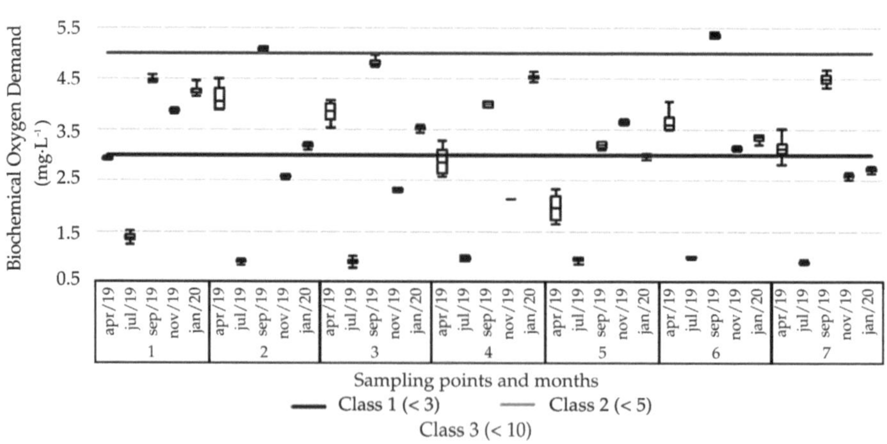


**Figure 4.** Descriptive analysis for the indicators biochemical oxygen demand—BOD (mg·L$^{-1}$). (Authors' own).

**Table 1.** Descriptive statistics and Tukey test for dissolved oxygen, biochemical oxygen demand, pH, and temperature parameters (authors' own).

| Point | Apr 2019 | Jul 2019 | Sep 2019 | Nov 2019 | Jan 2020 | Average | SD | CV (%) |
|---|---|---|---|---|---|---|---|---|
| | | | Dissolved Oxygen (mg·L$^{-1}$) | | | | | |
| 1 | 6.66 $^{cB}$ | 4.30 $^{bB}$ | 6.80 $^{dB}$ | 7.41 $^{aB}$ | 7.89 $^{aB}$ | 6.61 | 1.38 | 20.91 |
| 2 | 7.82 $^{cB}$ | 3.27 $^{bB}$ | 7.44 $^{dB}$ | 7.14 $^{aB}$ | 7.00 $^{aB}$ | 6.53 | 1.85 | 28.36 |
| 3 | 7.33 $^{cA}$ | 3.08 $^{bA}$ | 7.01 $^{dA}$ | 6.96 $^{aA}$ | 6.66 $^{aA}$ | 6.21 | 1.76 | 28.42 |
| 4 | 5.86 $^{cD}$ | 2.86 $^{bD}$ | 5.89 $^{dD}$ | 6.18 $^{aD}$ | 7.45 $^{aD}$ | 5.65 | 1.69 | 29.88 |
| 5 | 4.81 $^{cC}$ | 3.08 $^{bC}$ | 5.33 $^{dC}$ | 6.51 $^{aC}$ | 6.19 $^{aC}$ | 5.18 | 1.36 | 26.21 |
| 6 | 7.06 $^{cA}$ | 2.80 $^{bA}$ | 7.18 $^{dA}$ | 7.23 $^{aA}$ | 6.86 $^{aA}$ | 6.22 | 1.92 | 30.83 |
| 7 | 6.68 $^{cA}$ | 3.21 $^{bA}$ | 7.43 $^{dA}$ | 7.04 $^{aA}$ | 6.52 $^{aA}$ | 6.18 | 1.69 | 27.41 |
| Average | 6.60 | 3.23 | 6.73 | 6.92 | 6.94 | - | - | - |
| SD | 1.00 | 0.50 | 0.81 | 0.43 | 0.58 | - | - | - |
| CV (%) | 15.14 | 15.56 | 12.07 | 6.21 | 8.30 | - | - | - |
| | | | Biochemical Oxygen Demand (mg·L$^{-1}$) | | | | | |
| 1 | 2.93 $^{cD}$ | 1.38 $^{aD}$ | 4.49 $^{eD}$ | 3.88 $^{bD}$ | 4.28 $^{dD}$ | 3.39 | 1.27 | 37.50 |
| 2 | 4.14 $^{cBC}$ | 0.90 $^{aBC}$ | 5.08 $^{eBC}$ | 2.58 $^{bBC}$ | 3.19 $^{dBC}$ | 3.18 | 1.59 | 49.92 |
| 3 | 3.84 $^{cB}$ | 0.90 $^{aB}$ | 4.83 $^{eB}$ | 2.31 $^{bB}$ | 3.53 $^{dB}$ | 3.08 | 1.52 | 49.17 |
| 4 | 2.90 $^{cA}$ | 0.97 $^{aA}$ | 4.00 $^{eA}$ | 2.13 $^{bA}$ | 4.55 $^{dA}$ | 2.91 | 1.44 | 49.32 |
| 5 | 1.98 $^{cE}$ | 0.95 $^{aE}$ | 3.20 $^{eE}$ | 3.65 $^{bE}$ | 2.99 $^{dE}$ | 2.55 | 1.09 | 42.60 |
| 6 | 3.69 $^{cCD}$ | 0.99 $^{aCD}$ | 5.36 $^{eCD}$ | 3.14 $^{bCD}$ | 3.33 $^{dCD}$ | 3.30 | 1.56 | 47.39 |
| 7 | 3.15 $^{cA}$ | 0.90 $^{aA}$ | 4.51 $^{eA}$ | 2.6 $^{bA}$ | 2.72 $^{dA}$ | 2.78 | 1.30 | 46.65 |
| Average | 3.23 | 1.00 | 4.49 | 2.90 | 3.51 | - | - | - |
| SD | 0.73 | 0.17 | 0.72 | 0.67 | 0.67 | - | - | - |
| CV (%) | 22.51 | 17.37 | 16.08 | 23.16 | 19.11 | - | - | - |
| | | | pH | | | | | |
| 1 | 7.48 $^{abD}$ | 7.44 $^{aD}$ | 7.25 $^{bD}$ | 7.63 $^{dD}$ | 6.96 $^{cD}$ | 7.35 | 0.26 | 3.49 |
| 2 | 7.21 $^{abC}$ | 6.82 $^{aC}$ | 7.37 $^{bC}$ | 7.90 $^{dC}$ | 6.99 $^{cC}$ | 7.26 | 0.41 | 5.70 |
| 3 | 7.02 $^{abBC}$ | 7.10 $^{aBC}$ | 7.51 $^{bBC}$ | 7.57 $^{dBC}$ | 6.81 $^{cBC}$ | 7.20 | 0.33 | 4.54 |
| 4 | 6.85 $^{abA}$ | 6.92 $^{aA}$ | 7.06 $^{bA}$ | 7.44 $^{dA}$ | 6.95 $^{cA}$ | 7.04 | 0.24 | 3.35 |
| 5 | 7.14 $^{abA}$ | 6.91 $^{aA}$ | 6.82 $^{bA}$ | 7.49 $^{dA}$ | 7.12 $^{cA}$ | 7.09 | 0.26 | 3.66 |
| 6 | 7.06 $^{abB}$ | 7.01 $^{aB}$ | 7.35 $^{bB}$ | 7.48 $^{dB}$ | 7.02 $^{cB}$ | 7.18 | 0.22 | 3.02 |
| 7 | 6.95 $^{abA}$ | 7.21 $^{aA}$ | 6.71 $^{bA}$ | 7.32 $^{dA}$ | 7.16 $^{cA}$ | 7.07 | 0.24 | 3.44 |
| Average | 7.10 | 7.06 | 7.15 | 7.55 | 7.00 | - | - | - |
| SD | 0.21 | 0.21 | 0.30 | 0.18 | 0.11 | - | - | - |
| CV (%) | 2.91 | 3.04 | 4.20 | 2.41 | 1.64 | - | - | - |
| | | | Temperature (°C) | | | | | |
| 1 | 20.43 $^{dC}$ | 21.25 $^{bcC}$ | 20.08 $^{abC}$ | 21.55 $^{cC}$ | 20.08 $^{aC}$ | 20.68 | 0.69 | 3.31 |
| 2 | 18.58 $^{dA}$ | 19.03 $^{bcA}$ | 19.28 $^{abA}$ | 19.98 $^{cA}$ | 19.05 $^{aA}$ | 19.18 | 0.51 | 2.67 |
| 3 | 20.53 $^{dD}$ | 21.03 $^{bcD}$ | 21.23 $^{abD}$ | 21.70 $^{cD}$ | 20.80 $^{aD}$ | 21.06 | 0.44 | 2.11 |
| 4 | 19.50 $^{dB}$ | 20.15 $^{bcB}$ | 19.98 $^{abB}$ | 20.50 $^{cB}$ | 20.03 $^{aB}$ | 20.03 | 0.36 | 1.80 |
| 5 | 18.75 $^{dA}$ | 19.33 $^{bcA}$ | 18.78 $^{abA}$ | 18.98 $^{cA}$ | 18.93 $^{aA}$ | 18.95 | 0.23 | 1.22 |
| 6 | 21.25 $^{dE}$ | 21.75 $^{bcE}$ | 21.85 $^{abE}$ | 21.98 $^{cE}$ | 21.80 $^{aE}$ | 21.73 | 0.28 | 1.28 |
| 7 | 20.00 $^{dB}$ | 20.60 $^{bcB}$ | 20.85 $^{abB}$ | 19.63 $^{cB}$ | 20.28 $^{aB}$ | 20.27 | 0.48 | 2.38 |
| Average | 19.86 | 20.45 | 20.29 | 20.61 | 20.14 | - | - | - |
| SD | 0.98 | 1.01 | 1.09 | 1.15 | 0.99 | - | - | - |
| CV (%) | 4.92 | 4.92 | 5.36 | 5.60 | 4.92 | - | - | - |

1. Downstream property; 2. Native forest; 3. Sewer amount; 4. Erosive process; 5. Consumption—nascent; 6. Livestock interference; 7. Property amount. The data used were taken from the first author's master's dissertation. The results are expressed by an average of four repetitions. SD—standard deviation; CV—coefficient of variation. Means followed by the same lowercase letter do not differ significantly from each other by the Tukey test at a 5% probability in relation to the sampling time. Means followed by the same capital letter do not differ significantly by the Tukey test at a 5% probability in relation to the sampling location.

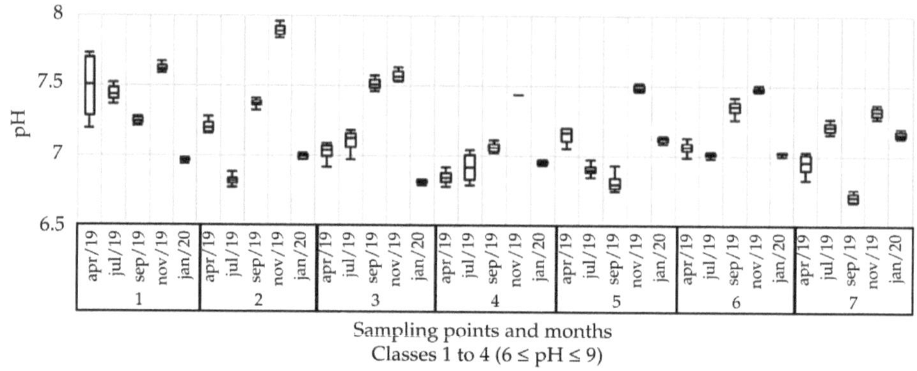


**Figure 5.** Descriptive analysis of the indicator pH, (Authors' own).

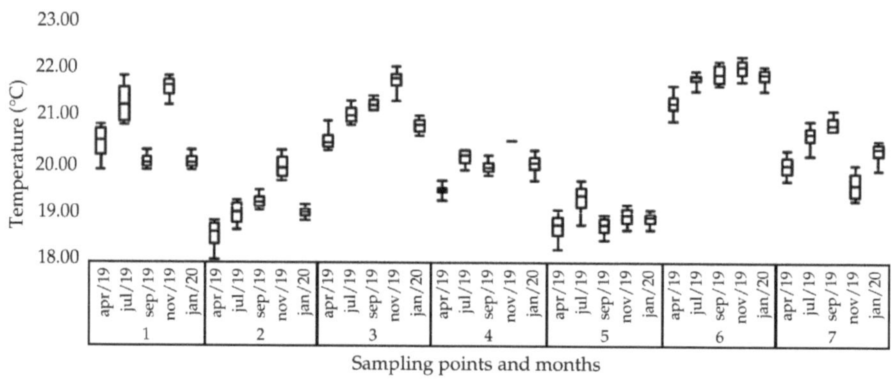


**Figure 6.** Descriptive analysis for the indicator temperature (°C), (Authors' own).

For the DO parameter, there is a predominance of classification by Class 1 according to CONAMA Resolution 357/05 [26]: 68.6% of the samples have a DO content greater than 6 mg·L$^{-1.}$ The levels classified as Class 4 (DO content > 2 mg·L$^{-1}$), which total 17.1% of the samples and refer to the July 2019 sampling, are explained by a technical failure in the measuring equipment, which postponed the measurement of the contents in one week and caused the lowering of the DO content.

There was no statistically significant difference between points 3, 6, and 7, whose averages are intermediate and around 6.20 mg·L$^{-1}$. Point 5 showed, on average, that the DO content (5.18 mg·L$^{-1}$) was lower than the other sample points. This finding was indicated by the Tukey test. This can probably be attributed to the thickness of the water layer from the point in question, since, visually, the water did not show any indications of contamination; at the same time, the BOD content obtained for it was lower than the other monitoring points, indicating a possible absence of organic contamination.

Point 1, which does not differ statistically from point 2, presented the highest DO content, considering the process of horizontal displacement of the water body and slope of the terrain, which allow greater swirling of the water and incorporation of oxygen in it. Of the points considered, this is the one that presented, in all samples, the largest volume of water.

From Figure 3, it is not possible to observe a pattern of increase or decrease in the DO content in the context of precipitation events. Even so, the highest DO average was observed in January 2020, the month in which the highest accumulated precipitation was observed prior to collection, and the lowest DO value was in July 2019, when precipitation prior to collection was zero.

The analysis of the DO content in terms of spatial variability indicated greater dispersion compared with temporal variability. The coefficient obtained thus ranged between 20.9% for point 1 and 30.8% for point 6 for samplings between 6.2% for the November 2019 sampling and 15.5% for the July 2019 sampling, implying that this parameter is influenced more by the sampling location than by the month. This configuration result was also reported in other research, in which the importance of addressing the spatial dimension in studies of the physical-chemical parameters of water is also emphasized [32].

As observed for DO, the BOD values of the July 2019 sampling, despite being classified as Class 1, were low due to the technical failure of reading equipment, which resulted in an already reduced DO content. Of the samples, 42.9% were Class 1 and 48.6% Class 2. Only 5.7% of the samples presented BOD greater than 5 $mg{\cdot}L^{-1}$, which characterizes them as being Class 3, referring to sampling points 2 and 6 in September 2019.

In agreement with the values raised for DO, Table 1 also shows greater variability of the data when analyzing the spatial disposition on the property, with CV oscillating between 37.5% for point 1, which presented the highest BOD value, and 49.9% for point 2. Point 5 had the lowest mean BOD value, 2.55 $mg{\cdot}L^{-1}$, since the values were already reduced for initial DO and, as previously reported, they do not characterize or evidence concentration of organic matter in the water body.

The temporal evaluation points to greater homogeneity of the data, since all the coefficients found are lower than the minimum spatial variation CV of 37.5%. Even so, there was no statistical similarity between the samples. Contrary to what is presented in the literature, BOD concentrations did not show a tendency to increase or decrease due to the occurrence of precipitation events, with the highest average observed for September 2019 and the lowest average for November 2019 when disregarded, due to technical failure, or values obtained for July.

In a study of a spring, the BOD values might be linked to situations typical of natural interferences in the water body, such as vegetation elements that have not yet been degraded [33]. The organic matter can reach the body of water through natural transport and is accentuated when the erosion rate is higher [34,35] in urban rivers. Therefore, the statistical similarity observed for points 4 and 7 is justified, considering the supply of material in these points by the terrain configurations and use and occupation described.

As for the pH, it can be seen from Figure 5 that all the data fell within the range of 6 to 9, considered ideal for the conservation of aquatic life and required by the resolution in all framing classes. The lowest homogeneity of the data was found for the samples whose coefficient of variation oscillated around 7.2, with the lowest value being in January 2020 (7.00) and the highest, 7.54, in November 2019. The pH tends to be lower in periods of lower precipitation [36].

On Morro do Céu in Niterói/RJ, the authors reported a correlation between the pH of soil and water, so that slightly acidic values in the waters, 5.6 to 6.6, were associated with the acid character of the soils of the region, around 5.0 [37]. A similar behavior is observed for the present study, although the variation reported between points was low (average CV = 3.88%), and the pH was neutral at point 5 (water pH = 7.04/soil pH = 4.5) and slightly alkaline at point 1 (water pH = 7.35/soil pH = 4.9).

The analysis of temperature (Figure 6) shows that the values obtained because of the spatial evaluation are in the range 18–22 °C, with point 5 being the lowest average in the samples (18.95 °C) and point 6 the highest average (21.73 °C), although they are in a pasture area. When analyzed individually, there is no significant variability in the data per sampling point, although the upstream and downstream characteristics of the data favor greater or lesser heating of the water.

With respect to temporal variability, it is observed that the heating of the waters is directly proportional to sunlight [38], having confirmed that the temperature of the waters was highest in November 2019 (20.6 °C), associated with the values of city air temperature (28.7 °C).

*3.2. Spatial-Temporal Variability of Water Electrical Conductivity, Turbidity, Total Dissolved Solids, Total Phosphorus, and Total Nitrogen*

Table 2 presents the descriptive statistics and Tukey test for electrical conductivity, turbidity, total dissolved solids, total phosphorus, and total nitrogen. To assist in the discussion, Figures 7–10 show the boxplot distribution of those parameters, respectively.

**Table 2.** Descriptive statistics and Tukey test for the parameters electrical conductivity, turbidity, total dissolved solids, total phosphorus, and total nitrogen. (Authors' own.)

| Point | Apr 2019 | Jul 2019 | Sep 2019 | Nov 2019 | Jan 2020 | Average | SD | CV (%) |
|---|---|---|---|---|---|---|---|---|
| | | | | Electrical Conductivity ($\mu S \cdot cm^{-1}$) | | | | |
| 1 | 97.03 $^{cA}$ | 93.68 $^{bA}$ | 103.40 $^{aA}$ | 113.30 $^{dA}$ | 101.63 $^{aA}$ | 101.81 | 7.48 | 7.35 |
| 2 | 89.40 $^{cA}$ | 73.13 $^{bA}$ | 112.38 $^{aA}$ | 141.83 $^{dA}$ | 91.08 $^{aA}$ | 101.56 | 26.48 | 26.07 |
| 3 | 114.63 $^{cF}$ | 96.88 $^{bF}$ | 102.28 $^{aF}$ | 117.53 $^{dF}$ | 111.43 $^{aF}$ | 108.55 | 8.68 | 8.00 |
| 4 | 97.75 $^{cE}$ | 95.55 $^{bE}$ | 88.33 $^{aE}$ | 107.30 $^{dE}$ | 103.58 $^{aE}$ | 98.50 | 7.35 | 7.46 |
| 5 | 101.20 $^{cC}$ | 87.03 $^{bC}$ | 91.20 $^{aC}$ | 97.95 $^{dC}$ | 97.70 $^{aC}$ | 95.02 | 5.75 | 6.06 |
| 6 | 105.18 $^{cD}$ | 85.80 $^{bD}$ | 94.00 $^{aD}$ | 101.48 $^{dD}$ | 94.98 $^{aD}$ | 96.29 | 7.46 | 7.75 |
| 7 | 77.85 $^{cB}$ | 50.98 $^{bB}$ | 77.73 $^{aB}$ | 102.33 $^{dB}$ | 66.95 $^{aB}$ | 75.17 | 18.74 | 24.93 |
| Average | 97.58 | 83.29 | 95.61 | 111.67 | 95.33 | - | - | - |
| SD | 11.68 | 16.38 | 11.41 | 14.97 | 14.12 | - | - | - |
| CV (%) | 11.97 | 19.67 | 11.93 | 13.41 | 14.81 | - | - | - |
| | | | | Turbidity (NTU) | | | | |
| 1 | 13.37 $^{eC}$ | 17.21 $^{aC}$ | 10.58 $^{dC}$ | 9.30 $^{bC}$ | 30.98 $^{cC}$ | 16.29 | 8.75 | 53.76 |
| 2 | 11.52 $^{eD}$ | 6.51 $^{aD}$ | 2.61 $^{dD}$ | 17.45 $^{bD}$ | 59.08 $^{cD}$ | 19.43 | 22.85 | 117.57 |
| 3 | 4.36 $^{eB}$ | 16.33 $^{aB}$ | 4.56 $^{dB}$ | 7.25 $^{bB}$ | 11.13 $^{cB}$ | 8.73 | 5.06 | 57.93 |
| 4 | 61.80 $^{eF}$ | 26.15 $^{aF}$ | 75.65 $^{dF}$ | 27.20 $^{bF}$ | 6.42 $^{cF}$ | 39.44 | 28.41 | 72.02 |
| 5 | 10.43 $^{eE}$ | 23.45 $^{aE}$ | 25.18 $^{dE}$ | 32.51 $^{bE}$ | 24.90 $^{cE}$ | 23.29 | 8.01 | 34.39 |
| 6 | 70.40 $^{eG}$ | 40.95 $^{aG}$ | 56.68 $^{dG}$ | 52.00 $^{bG}$ | 24.55 $^{cG}$ | 48.92 | 17.24 | 35.24 |
| 7 | 14.11 $^{eA}$ | 4.08 $^{aA}$ | 3.66 $^{dA}$ | 1.68 $^{bA}$ | 11.91 $^{cA}$ | 7.09 | 5.54 | 78.10 |
| Average | 3.23 | 1.00 | 4.49 | 2.90 | 3.51 | - | - | - |
| SD | 0.73 | 0.17 | 0.72 | 0.67 | 0.67 | - | - | - |
| CV (%) | 22.51 | 17.37 | 16.08 | 23.16 | 19.11 | - | - | - |
| | | | | Total dissolved solids ($mg \cdot L^{-1}$) | | | | |
| 1 | 116.0 $^{aC}$ | 148.0 $^{eC}$ | 198.5 $^{bC}$ | 159.5 $^{dC}$ | 84.5 $^{cC}$ | 141.3 | 43.3 | 30.7 |
| 2 | 95.5 $^{aD}$ | 126.0 $^{eD}$ | 187.5 $^{bD}$ | 240.5 $^{dD}$ | 154.5 $^{cD}$ | 160.8 | 56.1 | 34.9 |
| 3 | 134.5 $^{aB}$ | 268.5 $^{eB}$ | 218.0 $^{bB}$ | 243.5 $^{dB}$ | 185.5 $^{cB}$ | 210.0 | 52.2 | 24.9 |
| 4 | 141.50 $^{aA}$ | 282.5 $^{eA}$ | 171.0 $^{bA}$ | 326.0 $^{dA}$ | 211.0 $^{cA}$ | 226.4 | 76.8 | 33.9 |
| 5 | 253.0 $^{aA}$ | 309.5 $^{eA}$ | 162.0 $^{bA}$ | 210.5 $^{dA}$ | 244.5 $^{cA}$ | 235.9 | 54.5 | 23.1 |
| 6 | 241.0 $^{aE}$ | 376.0 $^{eE}$ | 215.5 $^{bE}$ | 238.0 $^{dE}$ | 277.0 $^{cE}$ | 269.5 | 63.5 | 23.6 |
| 7 | 170.5 $^{aAB}$ | 280.5 $^{eAB}$ | 145.5 $^{bAB}$ | 220.5 $^{dAB}$ | 322.0 $^{cAB}$ | 227.8 | 73.8 | 32.4 |
| Average | 164.6 | 255.9 | 185.4 | 234.1 | 211.3 | - | - | - |
| SD | 60.9 | 88.8 | 27.4 | 49.8 | 79.2 | - | - | - |
| CV (%) | 37.0 | 34.7 | 14.8 | 21.3 | 37.5 | - | - | - |

**Table 2.** *Cont.*

| Point | Apr 2019 | Jul 2019 | Sep 2019 | Nov 2019 | Jan 2020 | Average | SD | CV (%) |
|---|---|---|---|---|---|---|---|---|
| | | | | Total phosphorus (mg·L$^{-1}$) | | | | |
| 1 | 0.02 $^{aA}$ | 0.03 $^{aA}$ | 0.02 $^{aA}$ | 0.06 $^{cA}$ | 0.01 $^{bA}$ | 0.03 | 0.02 | 63.95 |
| 2 | 0.02 $^{aB}$ | 0.04 $^{aB}$ | 0.04 $^{aB}$ | 0.18 $^{cB}$ | 0.02 $^{bB}$ | 0.06 | 0.07 | 114.05 |
| 3 | 0.01 $^{aB}$ | 0.05 $^{aB}$ | 0.05 $^{aB}$ | 0.14 $^{cB}$ | 0.01 $^{bB}$ | 0.06 | 0.05 | 96.62 |
| 4 | 0.07 $^{aE}$ | 0.03 $^{aE}$ | 0.03 $^{aE}$ | 0.31 $^{cE}$ | 0.03 $^{bE}$ | 0.09 | 0.12 | 131.65 |
| 5 | 0.03 $^{aC}$ | 0.04 $^{aC}$ | 0.04 $^{aC}$ | 0.09 $^{cC}$ | 0.03 $^{bC}$ | 0.05 | 0.03 | 57.37 |
| 6 | 0.07 $^{aD}$ | 0.05 $^{aD}$ | 0.05 $^{aD}$ | 0.14 $^{cD}$ | 0.06 $^{bD}$ | 0.08 | 0.04 | 48.51 |
| 7 | 0.02 $^{aA}$ | 0.02 $^{aA}$ | 0.03 $^{aA}$ | 0.06 $^{cA}$ | 0.02 $^{bA}$ | 0.03 | 0.02 | 50.91 |
| Average | 0.03 | 0.04 | 0.04 | 0.14 | 0.03 | - | - | - |
| SD | 0.02 | 0.01 | 0.01 | 0.09 | 0.02 | - | - | - |
| CV (%) | 69.26 | 30.60 | 30.60 | 61.25 | 64.38 | - | - | - |
| | | | | Total nitrogen (mg·L$^{-1}$) | | | | |
| 1 | 55.33 $^{bA}$ | 12.00 $^{aA}$ | 18.79 $^{aA}$ | 30.69 $^{aA}$ | 11.52 $^{aA}$ | 25.67 | 18.30 | 71.30 |
| 2 | 47.71 $^{bA}$ | 3.43 $^{aA}$ | 3.43 $^{aA}$ | 12.12 $^{aA}$ | 0.93 $^{aA}$ | 13.52 | 19.58 | 144.78 |
| 3 | 84.74 $^{bA}$ | 8.79 $^{aA}$ | 8.79 $^{aA}$ | 18.43 $^{aA}$ | 4.86 $^{aA}$ | 25.12 | 33.70 | 134.17 |
| 4 | 47.60 $^{bA}$ | 18.43 $^{aA}$ | 18.43 $^{aA}$ | 24.14 $^{aA}$ | 3.07 $^{aA}$ | 22.33 | 16.15 | 72.30 |
| 5 | 40.10 $^{bA}$ | 2.95 $^{aA}$ | 2.95 $^{aA}$ | 8.79 $^{aA}$ | 10.69 $^{aA}$ | 13.10 | 15.48 | 118.25 |
| 6 | 50.21 $^{bA}$ | 27.24 $^{aA}$ | 27.24 $^{aA}$ | 13.31 $^{aA}$ | 7.83 $^{aA}$ | 25.17 | 16.41 | 65.20 |
| 7 | 110.33 $^{bA}$ | 18.79 $^{aA}$ | 12.00 $^{aA}$ | 11.05 $^{aA}$ | 13.43 $^{aA}$ | 33.12 | 43.27 | 130.64 |
| Average | 62.29 | 13.09 | 13.09 | 16.93 | 7.48 | - | - | - |
| SD | 25.58 | 8.92 | 8.92 | 7.97 | 4.68 | - | - | - |
| CV (%) | 41.07 | 68.12 | 68.12 | 47.05 | 62.60 | - | - | - |

1. Downstream property; 2. Native forest; 3. Sewer amount; 4. Erosive process; 5. Consumption—nascent; 6. Livestock interference; 7. Property amount. The data used were taken from the first author's master's dissertation. The results are expressed by an average of four repetitions. SD—standard deviation; CV—coefficient of variation. Means followed by the same lowercase letter do not differ significantly from each other by the Tukey test at a 5% probability in relation to the sampling time. Means followed by the same capital letter do not differ significantly by the Tukey test at a 5% probability in relation to the sampling location.

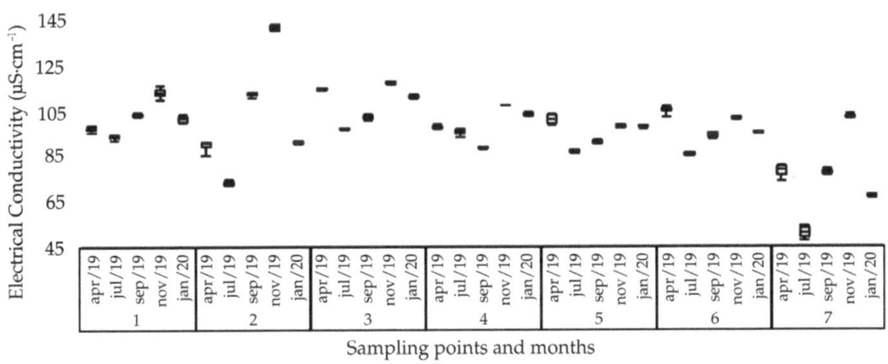


**Figure 7.** Descriptive analysis for the indicator electrical conductivity (μS·cm$^{-1}$), (Authors' own).

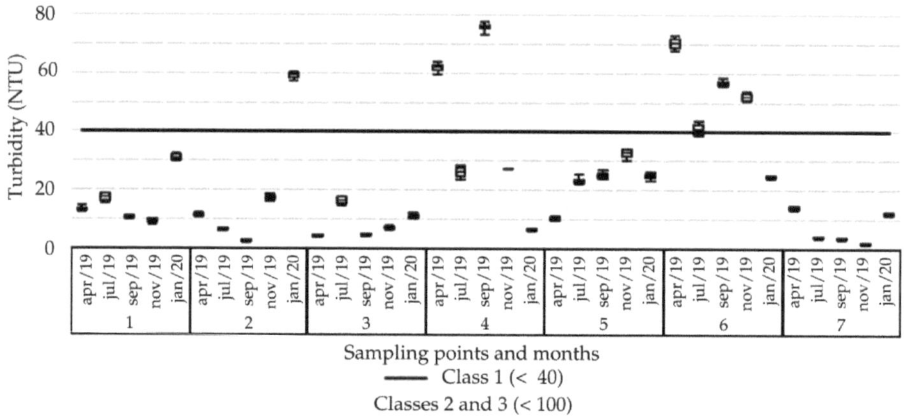


**Figure 8.** Descriptive analysis for the indicator turbidity (NTU). (Authors' own.)

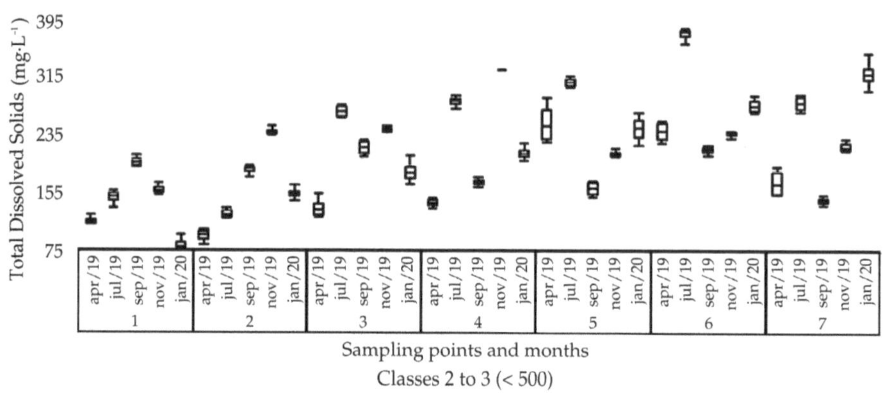


**Figure 9.** Descriptive analysis for the indicator total dissolved solids (mg·L$^{-1}$), (Authors' own).

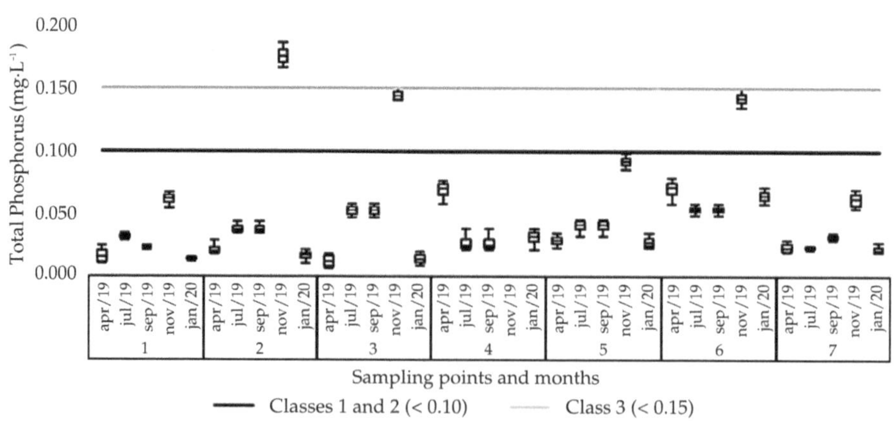


**Figure 10.** Descriptive analysis for the indicator total phosphorus (mg·L$^{-1}$), (Authors' own).

As for temperature, electrical conductivity does not have a limit defined by CONAMA Resolution 357/05 [26]; however, [39] reports that natural waters usually have conductivity levels between 10 and 100 $\mu S \cdot cm^{-1}$, and in environments polluted by domestic or industrial effluents, the values can reach up to 1000 $\mu S \cdot cm^{-1}$.

Figure 7 and Table 2 show that point 3 (Sewer amount) has the highest value for electrical conductivity, with an average of 108.5 $\mu S \cdot cm^{-1}$ and low variation between samples. Point 7, whose sample average was the lowest at 75.17 $\mu S \cdot cm^{-1}$, showed greater variability between the samples. Still, in the temporal analysis, the lowest conductivity value was found in the month when there was no precipitation, 83.29 $\mu S \cdot cm^{-1}$.

In a space-time assessment of the waters of the tributaries of the Barra dos Coqueiros hydroelectric plant reservoir in Goiás, the authors found conductivity levels between 10.5 $\mu S \cdot cm^{-1}$ and 135.8 $\mu S \cdot cm^{-1}$ and ascribed the values found to local lithological factors [40]. Twenty-five points were analyzed in the das Velhas' River in Minas Gerais and EC values above 100 $\mu S \cdot cm^{-1}$ in all data were obtained [41]. This result can be attributed to the fact that the points are in areas where land use is intensive, which favors the transport of dissolved solids to water bodies and a consequent increase in conductivity. In a study of water quality in watersheds with different land uses in the region of Cunha, São Paulo, the authors found that the average values of electrical conductivity in forested watersheds were less than 20 $\mu S \cdot cm^{-1}$ [42]. While the presence or absence of riparian vegetation is important, temperature, the total volume of water flow, and the geology of the area also influence this parameter, thus accounting for its lack of standardization [43].

For the turbidity parameter, Figure 8 shows the predominance of the classification of samples as Class 1 according to CONAMA Resolution 357/05 [26]: 80% of the samples have turbidity below 40 NTU. Only 20% of the sample data, of which 80% relate to samples collected in point 6, are categorized as Class 2, whose resolution stipulates a maximum limit of 100 NTU. The average values found for points 3 and 7 are slightly close, with a point average of less than 10 NTU. This configuration reflects that the role of the riparian forest in containing solids that could reach the body of water is fundamental to reducing turbidity [44]. In contrast, point 6 presented the highest observed value in the spatial average, at 48.92 NTU. It is a point with no tree/shrub vegetation cover, and is in a grassland area, which contributes to greater sediment transport to the water body. A study of the water quality of microbasins with different land uses in the Cunha region found that the values of temperature, turbidity, and apparent water color in agricultural recharge areas were higher than those recorded in forested recharge areas [42], which was also observed in this study.

In the temporal analysis, Table 2 shows the lowest turbidity rates in the July 2019 sample due to the absence of precipitation. However, the month with the highest average—April 2019, 26.5 NTU—does not correspond, as expected, to the month with the highest accumulated precipitation. This fact was also confirmed in the study of the Amazon River [45] and in the study of rivers in the south-central region of Paraná [46].

The TDS values obtained for the sampled points, as shown in Figure 9, are all below the limit of 500 $mg \cdot L^{-1}$ established by CONAMA Resolution 357/05 [26], and also the limit of 1000 $mg \cdot L^{-1}$ established by Consolidation Ordinance no. 5/2017 of the Ministry of Health [28]. Table 2 shows values close to data variability when analyzing spatial and temporal factors, so that the coefficient obtained for the points fluctuated between 23.1% for point 5 and 34.9% for point 2, and between 14.7% for the September sample and 37.49% for the July sample. The greatest variability of the variable occurred in the dry season—July 2019 [47].

In the average of samples, point 1 presented the lowest TDS value (141.3 $mg \cdot L^{-1}$), according to the Tukey test, which was attributed to the degree of preservation of its riparian forest that allows less sediment transport to the water body. In contrast, point 6 had the highest TDS content (269.5 $mg \cdot L^{-1}$). This point showed a higher value for turbidity, and these parameters are positively related [48].

The grassland areas, such as the location of point 6, although managed and with soil cover, are characterized by lower values of water infiltration in the soil, hydraulic conductivity, and greater resistance to penetration in relation to natural vegetation, all of which tend to delay runoff and consequently the entrainment of solids [49]. They point out that the increase in the concentration of solids in the water starts in the agricultural areas, which, in the absence of proper conservation management, become major potential sources of sediment [50]. There was no statistically significant difference between points 4 and 5, located in the pasture area and with TDS averages around 231 mg·L$^{-1}$.

Consolidation Ordinance No. 5/2017 of the Ministry of Health [28] establishes the absence of total coliforms in 100 mL of water (for human consumption or treated water), either at the treatment outlet or in the distribution system, to ensure its potability according to the microbiological standard; that is, in 100 mL samples collected, total coliforms must be absent. Therefore, for the sampling points, it was observed that they did not comply with the legislation, since it was noted that they showed behavior similar to one another and throughout the samplings, with a result greater than 1600 M.L.N. (Most Likely Number) per 100 mL of sample, with a 95% confidence limit [51]. Values above 1600 NMP are not detectable.

The presence of the total coliform group in natural waters does not necessarily imply contamination by fecal material, since, as established by the Water Supply Sanitary Inspection Manual [52], this group of coliforms includes bacteria that are found especially in tropical climates where their multiplication is more evident and naturally in soils, water, and plants; therefore, it is an analysis of limited sanitary value.

In a study of the Salgadinho River in the municipality of Juazeiro do Norte, 100% positivity was observed for total coliforms in the samples collected, with proportions greater than 1600 NMP·mL$^{-1}$ [53], as observed in a spring in the state of Rio de Janeiro, municipality of Varre-Sai, southeastern Brazil [54]. Another study in the extreme west of Santa Catarina, southern Brazil, showed that 64.1% of the water samples analyzed for human consumption were contaminated with total coliforms [55]. Such findings highlighted the urgency of adopting preventive measures, as well as water treatment, to minimize the chances of occurrence of waterborne diseases.

Regarding the total phosphorus content in the sample points, Figure 10 shows a predominance of classification by Class 1 and 2 according to CONAMA Resolution 357/05 [26]: 88.6% of the samples have a content of less than 0.10 mg·L$^{-1}$. A total of 5.7% of the samples categorized as Class 3 (phosphorus content <0.15 mg·L$^{-1}$), and only 2 samples (point 4, November 2019 and point 2, November 2019) had a content higher than 0.15 mg·L$^{-1}$, although the relevant legislation does not present a limit content for Class 4.

In the analysis of temporal variation, it can be seen from Table 2 that the greatest variation of the data occurred in the April 2019 sample, with a CV of 69.26%, whose accumulated precipitation in the seven days prior to collection was the lowest, being 3.8 mm, not considering July 2019, in which previous precipitation was not accounted for. Similarly, from the analysis of Figure 10, it is noticeable that the highest concentrations for phosphorus at the sampling points were obtained in the November 2019 sampling, in which the water density at the sampling points was low, although the accumulation of previous precipitation was 19.9 mm.

In January 2020, when precipitation amounted to 86.5 mm, the lowest value for total phosphorus was verified in the study points. Therefore, there was a tendency for the parameter of total phosphorus to increase in the dry season and decline in the rainy season. This finding corroborates the results obtained in a study of supply sources in Goiânia, in a study of Ribeirão Anhumas in Campinas/SP, and in an analysis of the Atibaia and Jaguari Rivers near the city of Paulínia/SP [56–58], respectively.

Regarding the variability of the data in the spatial evaluation, Table 2 also shows a variation of 50.9% for point 7 and 131.7% for point 4. There was no statistically significant difference between points 1 and 7, whose averages are the lowest among the analyzed points, at 0.029 and 0.033 mg·L$^{-1}$, respectively. In contrast, point 4 presented the highest total phosphorus content. Although the presence of phosphorus in the water body is commonly associated with the discharge of effluents, wastewater, or the leaching of regions where agriculture is practiced, the available phosphorus can result from the release of phosphates with the decomposition of rocks through natural weathering processes [59].

Figure 11 shows the results of the sampled points for total nitrogen. For this, the organic nitrogen content and the ammoniacal nitrogen present in water bodies are measured. In addition to this, the water body may contain nitrogen in the forms of nitrite and nitrate, which were not analyzed in this work. Only the ammoniacal form has concentration limits established according to CONAMA 357/05 [26].

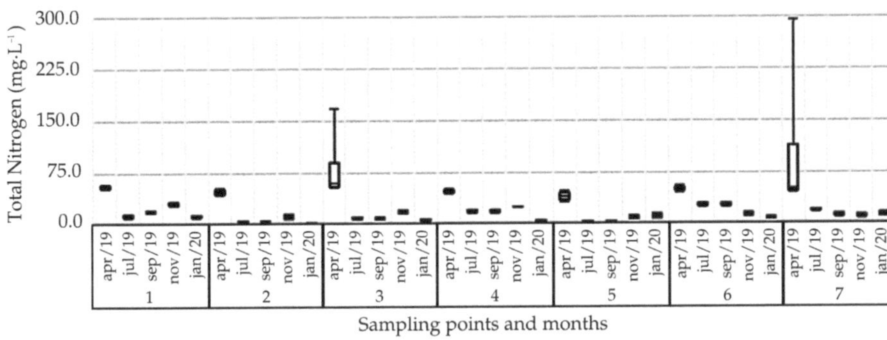


**Figure 11.** Descriptive analysis for the total nitrogen indicator (mg·L$^{-1}$), (Authors' own).

As shown in Table 2, it was found that 34.29% of the data had TN content of less than 10 mg·L$^{-1}$, 34.3% content between 10 and 20 mg·L$^{-1}$, 14.3% concentration of TN in the range 20 to 30 mg·L$^{-1}$, 11.4% between 30 and 40 mg·L$^{-1}$, and 5.7% greater than 50 mg·L$^{-1}$.

Although there is no limit for this parameter in CONAMA Resolution 357/05 [26], the levels obtained are higher than those reported in work carried out in natural waters, which found levels between 0.64 mg·L$^{-1}$ and 0.53 mg·L$^{-1}$ in rivers in the region of Cadeia do Espinhaço/MG [60]. Another example is the study in the Itamambuca River in Caraguatatuba, São Paulo, which observed a content of 0.08 mg·L$^{-1}$ [61].

Although this element does not have a minimum release pattern determined by the environmental agency, it should be considered since in high concentrations it can lead to excessive growth of algae, thus causing eutrophication [62].

The metrics normally used for hydrological monitoring of ecosystem services are not easily interpreted by researchers and decision makers [16], highlighting the importance of this study as a basis for replication in other locations where the PES-Water program exists and there is difficulty in monitoring to verify results. In addition, most PES programs do not clearly define which services are provided by the area in question and are linked to payment [63], which highlights the importance of this study that proposes seasonal monitoring of water quality based on previously established parameters.

## 4. Conclusions

The water quality indicators obtained for the sampling points implied good quality, especially at points 1 (Fazenda Santa Margarida exit), 5 (collection for consumption), and 7 (property amount), so that the values obtained for DO, BOD, pH, turbidity, TDS, and total phosphorus met the limits for Class 1 and 2 of CONAMA Resolution 357/05.

Spatial-temporal heterogeneity of the water quality indicators was observed at the sampling points. This is borne out by the analysis of variance performed, so that for the parameters DO, BOD, pH, phosphorus, and total nitrogen, greater fluctuation was noticed in relation to the spatial location of the samples. Sampling points for the turbidity, temperature, and EC parameters also show greater variability between samples. The TDS and total coliform parameters were influenced by the same magnitude when the two evaluation criteria were analyzed.

The monitoring carried out on the property over the course of a year helped check the compliance and implementation of actions foreseen by the PES project. However, it is understood that long-term monitoring would make it possible to differentiate between the improvements found in precipitation seasonality and those arising from the actions taken on the property under the program. Therefore, this local study consolidates the significant scientific contribution regarding the monitoring time of areas used for PES in tropical and subtropical regions.

For future work, it is suggested that a spatial-temporal evaluation be adopted, as the method has proven to be efficient and suitable for hydrological monitoring of the areas participating in the PES program, especially considering seasonal changes that can significantly interfere with the quality of surface water.

**Author Contributions:** Conceptualization, R.M.L. and J.M.G.; methodology, J.M.G.; validation, R.M.L., J.M.G. and A.N.N.; formal analysis, J.M.G.; investigation, J.M.G.; resources, J.M.G.; data curation, J.M.G.; writing—original draft preparation J.M.G.; writing—review and editing, R.M.L., R.C.G. and A.N.N.; visualization, J.M.G.; supervision, R.M.L.; project administration, R.M.L.; funding acquisition, R.M.L. All authors have read and agreed to the published version of the manuscript.

**Funding:** This work was supported by the São Paulo Research Foundation—FAPESP (regular research project 2022/05062-3 and technical training 2021/03002-0) and the Coordination of Superior Level Staff Improvement—CAPES (Process 88887.691909/2022-00-Postgraduate Development Program-Strategic Postdoctoral). The authors also gratefully acknowledge the support provided by the Centre for Studies in Geography and Spatial Planning (CEGOT), financed by national funds through the Foundation for Science and Technology (FCT), under the reference UIDB/04084/2020.

**Data Availability Statement:** The data supporting this study's findings are openly available at https://github.com/ra-gomes/spatial-temporal-monitoring-wq/blob/2f9164741df1cd689b4d3f5051d4aeedfdf2c5d7/tables%20excel%20water%20quality.xlsx, accessed on 17 September 2024.

**Acknowledgments:** The authors hereby acknowledge the Pontifical Catholic University of Campinas for providing the necessary infrastructure to carry out this study.

**Conflicts of Interest:** The authors declare no conflicts of interest.

## References

1. Araújo, I.S.; Maciel, A.B.C.; Ferreira, J.C.V.; Lima, Z.M.C. Identificação de serviços ecossistêmicos na praia de Ponta Negra, Natal-RN. *Rev. Geociências Do Nordeste* **2018**, *4*, 298–313. [CrossRef]
2. Burkhard, B.; Kandziora, M.; Hou, Y.; Müller, F. Ecosystem service potential, flows and demands-concepts for spatial localisation, indication and quantification. *Landsc. Online* **2014**, *34*, 1–32. [CrossRef]
3. Castro, B.S.; Young, C.E.F.; Pereira, V.S. Iniciativas estaduais de Pagamentos por Serviços Ambientais: Análise legal e seus resultados. *Rev. Iberoam. Econ. Ecológica* **2018**, *28*, 44–71.
4. Mamedes, I.; Guerra, A.; Rodrigues, D.B.B.; Garcia, L.C.; de Faria Godoi, R.; Oliveira, P.T.S. Brazilian payment for environmental services programs emphasize water-related services. In *International Soil and Water Conservation Research*; KeAi Communications Co.: Beijing, China, 2023; Volume 11, pp. 276–289. [CrossRef]
5. Gregori, M.S. Limites e possibilidades do sistema de Pagamento por Serviços Ambientais como instrumento de promoção da sustentabilidade socioambiental no Brasil. Ph.D. Thesis, Universidade Federal de Santa Maria, Santa Maria, Brazil, 2017.
6. Calvet-Mir, L.; Corbera, E.; Martin, A.; Fisher, J.; Gross-Camp, N. Payments for ecosystem services in the tropics: A closer look at effectiveness and equity. *Curr. Opin. Environ. Sustain.* **2015**, *14*, 150–162. [CrossRef]
7. Huang, Y.; Zhang, X.; Sheng, X.; Wang, Y.; Leung, K.M.Y. The impact of payment for ecosystem service schemes on participants' motivation: A global assessment. *Ecosyst. Serv.* **2024**, *65*, 101595. [CrossRef]
8. Martin-Ortega, J.; Ojea, E.; Roux, C. Payments for water ecosystem services in Latin America: A literature review and conceptual model. *Ecosyst. Serv.* **2013**, *6*, 122–132. [CrossRef]
9. Grima, N.; Singh, S.J.; Smetschka, B.; Ringhofer, L. Payment for ecosystem services (PES) in Latin America: Analysing the performance of 40 case studies. *Ecosyst. Serv.* **2016**, *17*, 24–32. [CrossRef]
10. Chiodi, R.E.; Marques, P.E.M. Políticas públicas de pagamento por serviços ambientais para a conservação dos recursos hídricos: Origens, atores, interesses e resultados da ação institucional. *Desenvolv. E Meio Ambiente* **2018**, *45*, 81–104. [CrossRef]
11. Jardim, M.H.; Bursztyn, M.A. Pagamento por serviços ambientais na gestão de recursos hídricos: O caso de Extrema (MG). *Eng. Sanit. Ambient.* **2015**, *20*, 353–360. [CrossRef]
12. Guedes, F.B.; Seehusen, S.E. (Eds.) *O PSA na Mata Atlântica—Situação Atual, Desafios e Recomendações*; Ministério do Meio Ambiente: Brasília, Brazil, 2011.

13. Balvanera, P.; Uriarte, M.; Almeida-Leñero, L.; Altesor, A.; DeClerck, F.; Gardner, T.; Hall, J.; Lara, A.; Laterra, P.; Peña-Claros, M.; et al. Ecosystem services research in Latin America: The state of the art. *Ecosyst. Serv.* **2012**, *2*, 56–70. [CrossRef]
14. Arriagada, R.A.; Sills, E.O.; Ferraro, P.J.; Pattanayak, S.K. Do payments pay off? Evidence from participation in Costa Rica's PES program. *PLoS ONE* **2015**, *10*, e0131544.
15. Bennett, G.; Carroll, N. *Gaining Depth: State of Watershed Investment 2014*; Forest Trends: Washington, DC, USA, 2014. Available online: https://www.foresttrends.org/publications/gaining-depth-2/ (accessed on 13 March 2020).
16. Ureta, J.U.; Ureta, J.C.; Bower, L.M.; Peoples, B.K.; Motallebi, M. The value of improving freshwater ecosystem services: South Carolina residents' willingness to pay for improved water quality. *J. Environ. Manag.* **2024**, *353*, 120260. [CrossRef] [PubMed]
17. Campinas. Lei no. 15.046 de 23 de Julho de 2015. Institui o Programa de Pagamento por Serviços Ambientais, Autoriza o Poder Executivo Municipal a Prestar Apoio aos Proprietários Rurais e Urbanos Determinados pelo Programa, e dá Outras Providências. Campinas, SP. Diário Oficial do Município. Available online: https://portal-api.campinas.sp.gov.br/sites/default/files/publicacoes-dom/dom/2062178756.pdf (accessed on 17 September 2024).
18. Campinas. Decreto Municipal no. 19.441 de 15 de Março de 2017. Regulamenta o Subprograma de Pagamento pela Conservação das Águas e dos Recursos Hídricos-PSA Água, Instituído pela Lei nº 15.046, de 23 de Julho de 2015. Campinas, SP. Diário Oficial do Município. Available online: https://bibliotecajuridica.campinas.sp.gov.br/index/visualizaroriginal/id/130978 (accessed on 17 September 2024).
19. SVDS. Secretaria Municipal do Verde, Meio Ambiente e Desenvolvimento Sustentável. GEOAMBIENTAL: Informações Ambientais Espacializadas. Recuperado em 14 de maio de 2020. Available online: https://geoambiental.campinas.sp.gov.br/pmapper/map_svds.phtml?config=svds (accessed on 17 September 2024).
20. Taffarello, D.; do Carmo, C.M.; Viani, R.A.G.; Marengo, J.A.; Mendiondo, E.M. Hydrological services in the Atlantic Forest, Brazil: An ecosystem-based adaptation using ecohydrological monitoring. *Clim. Serv.* **2017**, *8*, 1–16. [CrossRef]
21. Izquierdo-Tort, S.; Alatorre, A.; Arroyo-Gerala, P.; Shapiro-Garza, E.; Naime, J.; Dupras, J. Exploring local perceptions and drivers of engagement in biodiversity monitoring among participants in payments for ecosystem services schemes in southeastern Mexico. *Conserv. Biol.* **2024**, *38*, e14282. [CrossRef] [PubMed]
22. Young, C.E.F.; de Bakker, L.B. Payments for ecosystem services from watershed protection: A methodological assessment of the Oasis Project in Brazil. *Nat. Conserv.* **2014**, *12*, 71–78. [CrossRef]
23. Brownson, K.; Fowler, L. Evaluating how we evaluate success: Monitoring, evaluation and adaptive management in Payments for Watershed Services programs. *Land Use Policy* **2020**, *94*, 104505. [CrossRef]
24. Longo, R.M.; Garcia, J.M.; Gomes, R.C.; Nunes, A.N. Identifying Key Indicators for Monitoring Water Environmental Services Payment Programs—A Case Study in Brazil. *Sustainability* **2023**, *15*, 9593. [CrossRef]
25. Fiore, F.A.; Bardini, V.S.S.; Novaes, R.C. Monitoramento da qualidade de águas em programas de pagamento por serviços ambientais hídricos: Estudo de caso no município de São José dos Campos/SP. *Eng. Sanit. Ambient.* **2017**, *22*, 1141–1150. [CrossRef]
26. Brasil. *Conselho Nacional do Meio Ambiente–CONAMA. Resolução CONAMA nº357, de 17 de março de 2005. Dispõe Sobre a Classificação dos Corpos de Água e Diretrizes Ambientais para seu Enquadramento, bem Como Estabelece as Condições e Padrões de Lançamento de Efluentes, e dá Outras Providências*; Diário Oficial [da] República Federativa do Brasil: Brasília, Brazil, 2005.
27. Brasil. *Ministério da Saúde. Portaria nº2.914, de 12 de dezembro de 2011. Dispõe Sobre os Procedimentos de Controle e de Vigilância da Qualidade da Água para Consumo Humano e seu Padrão de Potabilidade*; Diário Oficial [da] República Federativa do Brasil: Brasília, Brazil, 2011.
28. Brasil. *Ministério da Saúde. Anexo XX: Do Controle e da Vigilância da Qualidade da Água para Consumo Humano e seu Padrão de Potabilidade. Portaria de Consolidação nº 5, de 28 de Setembro de 2017. Consolidação das Normas sobre as ações e os Serviços de Saúde do Sistema Único de Saúde*; Diário Oficial [da] República Federativa do Brasil: Brasília, Brazil, 2017.
29. CIIAGRO. Centro Integrado de Informações Agrometeorológicas. CIIAGRO online. Recuperado em 20 de outubro de 2020. Available online: http://www.ciiagro.sp.gov.br/ciiagroonline/ (accessed on 20 October 2020).
30. Valladares Neto, J.; Santos, C.B.; Torres, E.M.; Estrela, C. Boxplot: Um recursos gráfico para a análise e interpretação de dados quantitativos. *Rev. Odontológica Do Bras. Cent.* **2017**, *26*, 1–6.
31. Canteri, M.G. SASM-AGRI: Sistema para análise e separação de médias em experimentos agrícolas pelos métodos Scott-Knott, Tukey e Duncan. *Rev. Bras. Agrocomputação* **2001**, *1*, 18–24.
32. Villa, M.E.C.D.; Sampaio, T.V.M. Variabilidade espacial e temporal dos parâmetros físico-químico em ambiente lêntico. *Bol. De Geogr.* **2016**, *34*, 69–80. [CrossRef]
33. Oliveira, C.N.; Campos, V.P.; Medeiros, Y.D.P. Avaliação e identificação de parâmetros importantes para a qualidade de corpos d'água no semiárido baiano. Estudo de caso: Bacia hidrográfica do Rio Salitre. *Química Nova* **2010**, *33*, 1059–1066. [CrossRef]
34. Poersch, A.C.C.; Sebastien, N.Y.; Taborda, J.; Colombari Neto, J.; Caovilla, M.M.; Fonseca, C.A. A demanda bioquímica de oxigênio (DBO) e possíveis fatores de influência nos processos em águas superficiais do Rio São Francisco Verdadeiro–Paraná. *Acta Iguazu* **2019**, *8*, 45–51.

35. Garcia, J.M.; Longo, R.M. Análise de impactos ambientais em Área de Preservação Permanente (APP) como instrumento de gestão em rios urbanos. *Cerrados (UNIMONTES)* **2020**, *18*, 107–128. [CrossRef]
36. Esteves, F.A. *Fundamentos de Limnologia*, 2nd, ed.; Editora Interciência: Rio de Janeiro, Brazil, 1998.
37. Sisinno, C.L.S.; Moreira, J.C. Avaliação da contaminação e poluição ambiental na área de influência do aterro controlado do Morro do Céu, Niterói, Brasil. *Cad. De Saúde Pública* **1996**, *12*, 515–523. [CrossRef] [PubMed]
38. Lamparelli, C.C. Dinâmica da Serapilheira em Manguezais de Bertioga, Região Sudeste do Brasil. Ph.D. Thesis, Universidade de São Paulo, São Paulo, Brazil, 1995.
39. Von Sperling, M. *Introdução à Qualidade das Águas e ao Tratamento de Esgotos*, 3rd, ed.; Editora UFMG: Belo Horizonte, Brazil, 2005.
40. Rocha, H.M.; Cabral, J.B.P.; Braga, C.C. Avaliação espaço-temporal das águas dos afluentes do Reservatório da UHE Barra dos Coqueiros/Goiás. *Rev. Bras. De Recur. Hídricos* **2014**, *19*, 131–142. [CrossRef]
41. Baggio, H.; Freitas, M.O.; Araújo, A.D. Análises dos parâmetros físico-químicos oxigênio dissolvido, condutividade elétrica, potencial hidrogeniônico e temperatura, suas correlações com o uso e ocupação da terra no baixo curso do Rio das Velhas-MG. *Caminhos De Geogr.* **2016**, *17*, 105–117. [CrossRef]
42. Arcova, F.C.S.; Cicco, V. Qualidade da água de microbacia com diferentes usos do solo na região de Cunha, Estado de São Paulo. *Sci. For.* **1999**, *56*, 125–134.
43. Marmontel, C.V.F.; Rodrigues, V.A. Parâmetros indicativos para qualidade da água em nascentes com diferentes coberturas de terra e conservação da vegetação ciliar. *Floresta Ambiente* **2015**, *22*, 171–181. [CrossRef]
44. Donadio, N.M.M.; Galbiatti, J.A.; De Paula, R.C. Qualidade da água de nascentes com diferentes usos do solo na bacia hidrográfica do córrego Rico, São Paulo, Brasil. *Eng. Agrícola* **2005**, *25*, 115–125. [CrossRef]
45. Damasceno, M.C.S.; Ribeiro, H.M.C.; Takiyama, L.R.; De Paula, M.T. Avaliação sazonal da qualidade das águas superficiais do Rio Amazonas na orla da cidade de Macapá, Amapá, Brasil. *Ambiente E Água* **2015**, *10*, 598–613. [CrossRef]
46. Branco, C.C.Z.; Krupek, R.A.; Peres, C.K. Variação sazonal de alguns parâmetros físicos e químicos em três rios pertencentes a uma bacia de drenagem na região centro-sul do Estado do Paraná, Sul do Brasil. *Acta Sci. Biol. Sci.* **2008**, *30*, 431–438. [CrossRef]
47. Soares, R.D.B.; Cruz, R.W.L.; Silva, C.E. A influência da precipitação na variabilidade da qualidade da água do rio Parnaíba. *Braz. J. Dev.* **2019**, *5*, 16643–16672. [CrossRef]
48. Fonseca, H.S.; Salvador, N.N.B. Estudo integrado da qualidade das águas da bacia hidrográfica do Rio Bonito em Descalvado-SP. *Rev. Do Inst. Adolfo Lutz* **2005**, *64*, 179–185. [CrossRef]
49. Silva, D.D.; Pruski, F.F.; Schaefer, C.E.G.R.; Amorim, R.S.S.; Paiva, K.W.N. Efeito da cobertura nas perdas de solo em um Argissolo Vermelho-Amarelo utilizando simulador de chuva. *Eng. Agrícola* **2005**, *25*, 409–419. [CrossRef]
50. Vanzela, L.S.; Hernandez, F.B.T.; Franco, R.A.M. Influência do uso e ocupação do solo nos recursos hídricos do Córrego Três Barras, Marinópolis. *Rev. Bras. De Eng. Agrícola E Ambient.* **2010**, *14*, 55–64. [CrossRef]
51. FUNASA. *Fundação Nacional de Saúde. Manual Prático de Análise de Água*, 4th ed.; FUNASA: Brasília, Brazil, 2013; 150p.
52. Brasil. *Ministério da Saúde. Secretaria de Vigilância em Saúde. Inspeção Sanitária em Abastecimento de Água*; Ministério da Saúde: Brasília, Brazil, 2007.
53. Nunes, L.M. Pesquisa de coliformes totais e termotolerantes no rio Salgadinho no município de Juazeiro do Norte, CE. *Rev. Eletrônica Acervo Científico* **2019**, *7*, e2243. [CrossRef]
54. Santos, E.P.P.; Veiga, W.A.; Gonçalves, M.R.S.; Thomé, M.P.M. Coliformes Totais e Termotolerantes em água de nascentes utilizadas para o consumo humano na zona rural do município de Varre-Sai, RJ. *Sci. Plena* **2015**, *11*, 052401.
55. Scapin, D.; Rossi, E.M.; Oro, D. Qualidade microbiológica da água utilizada para consumo humano na região do extremo oeste de Santa Catarina, Brasil. *Rev. Inst. Adolfo Lutz* **2012**, *71*, 593–596. [CrossRef]
56. Santos, A.A.; Silva, D.P.; Costa, I.A.; Cuba, R.M.F. Avaliação sazonal de índices de estado trófico para fósforo total para mananciais de abastecimento público do Município de Goiânia, Estado do Goiás, Brasil. *Rev. Bras. De Gestão Ambient. E Sustentabilidade* **2019**, *6*, 883–895. [CrossRef]
57. Hossomi, A.M.T. *Avaliação Temporal (2006 a 2016) da Qualidade de Água do Ribeirão Anhumas–Campinas/SP*; (Monografia de Especialização); Escola Superior de Agricultura Luiz de Queiroz ESALQ/USP: Piracicaba, Brazil, 2017.
58. Leite, D.A.N.O.; Ramos, M.A.G.; Godói, D.R.; Mariano, A.P.; Pião, A.C.S.; Angelis, D.F. Avaliação dos parâmetros do índice de Qualidade de Água segundo o modelo estatístico ARIMA. *HOLOS Environ.* **2013**, *13*, 24–39. [CrossRef]
59. Souza, F.A.R. Modelagem Espacial da Qualidade da Água Usando Geotecnologias na Bacia Hidrográfica de Santana-RJ. Dissertação de Mestrado, Universidade Federal do Rio de Janeiro, Rio de Janeiro, Brazil, 2017.
60. Meyer, S.T.; Franceschinelli, E.V. Influência de variáveis limnológicas sobre a comunidade das macrófitas aquáticas em rios e lagoas da Cadeia do Espinhaço, Minas Gerais, Brasil. *Rodriguésia* **2011**, *62*, 743–758. [CrossRef]
61. Carvalho, A.R.; Marinho, A.M.; Silva, A.C.R.S.; Bernardes, M.E.C. Comparação entre os Índices de Qualidade da Água (IQA) dos estuários dos rios Itamambuca e Juqueriquerê–litoral norte de São Paulo–sob mesma sazonalidade. In *Anais do IV Congresso Brasileiro de Gestão Ambiental*; IBEAS: Salvador, Brazil, 2013.

62. Barbieri, E.; Marques, H.L.A.; Bondioli, A.C.V.; Campolim, M.B.; Ferrarini, A.T. Concentrações do nitrogênio amoniacal, nitrito e nitrato em áreas de engorda de ostras no município de Cananeia-SP. *O Mundo Da Saúde* **2014**, *38*, 105–115. [CrossRef]
63. Mota, P.K.; da Costa, A.M.; Prado, R.B.; Fernandes, L.F.S.; Pissarra, T.C.T.; Pacheco, F.A.L. Payment for Environmental Services: A critical review of schemes, concepts, and practice in Brazil. In *Science of the Total Environment*; Elsevier B.V.: Amsterdam, The Netherlands, 2023; Volume 899. [CrossRef]

*water*

MDPI

*Article*

# Hydrochemical Dynamics and Water Quality Assessment of the Ramsar-Listed Ghodaghodi Lake Complex: Unveiling the Water-Environment Nexus

**Ganga Paudel [1], Ramesh Raj Pant [1], Tark Raj Joshi [2], Ahmed M. Saqr [3], Bojan Đurin [4], Vlado Cetl [5,*], Pramod N. Kamble [6] and Kiran Bishwakarma [7]**

[1] Central Department of Environmental Science, Tribhuvan University, Kathmandu 46000, Nepal; paudelganga303@gmail.com (G.P.); rpant@cdes.edu.np (R.R.P.)
[2] Faculty of Science & Technology, Far Western University, Bhimdatta 10400, Nepal; tarkraj2001@gmail.com
[3] Irrigation and Hydraulics Department, Faculty of Engineering, Mansoura University, Mansoura 35516, Egypt; ahmedsaqr@mans.edu.eg
[4] Department of Civil Engineering, University North, 42000 Varaždin, Croatia; bojan.durin@unin.hr
[5] Department of Geodesy and Geoinformatics, University North, 42000 Varaždin, Croatia
[6] Department of Environmental Science, School of Earth Sciences, Central University of Rajasthan, Ajmer 305817, India; pramod.kamble@curaj.ac.in
[7] Institute of Tibetan Plateau Research, Chinese Academy of Sciences, Beijing 100101, China; kiranghatani11@gmail.com
* Correspondence: vlado.cetl@unin.hr

**Abstract:** Human activities and climate change increasingly threaten wetlands worldwide, yet their hydrochemical properties and water quality are often inadequately studied. This research focused on the Ghodaghodi Lake Complex (GLC) and associated lakes in Nepal, a Ramsar-listed site known for its biodiversity and ecological significance. The study was conducted to assess seasonal water quality, investigate the factors influencing hydrochemistry, and assess the lakes' suitability for irrigation. Forty-nine water samples were collected from the GLC in pre-monsoon and post-monsoon periods. Nineteen physicochemical parameters, such as dissolved oxygen (DO), total dissolved solids (TDS), and major ions (calcium '$Ca^{2+}$', magnesium '$Mg^{2+}$', and bicarbonate '$HCO_3^-$'), were analyzed using standard on-site and laboratory methods. Statistical methods, including analysis of variance (ANOVA), T-tests, and hydrochemical diagrams, e.g., Piper, were adopted to explore spatial and seasonal variations in water quality, revealing significant fluctuations in key hydrochemical indicators. Results showed marked seasonal differences, with pre-monsoon TDS levels averaging 143.1 mg/L compared to 78.9 mg/L post-monsoon, underscoring evaporation and dilution effects. The hydrochemical analysis identified $Ca^{2+}$-$HCO_3^-$ as the dominant water type, highlighting the influence of carbonate weathering on GLC's water composition. Gibbs, mixing, and Piper diagram analysis supported these findings, confirming the predominance of $HCO_3^-$, with $Ca^{2+}$ and $Mg^{2+}$ as the main cations. Additionally, sodium adsorption ratio (SAR) values were consistently below 1, confirming excellent irrigation quality. These findings provided critical data for policymakers and stakeholders, supporting sustainable wetland management and aligning with the United Nations' Sustainable Development Goals relevant to environmental conservation, i.e., clean water and life on land.

**Keywords:** wetland hydrochemistry; water quality; irrigation suitability; piper diagram; environmental sustainability; Ghodaghodi Lake

## 1. Introduction

Wetlands are experiencing rapid degradation due to climate change, urbanization, and increased agricultural practices, undermining their ability to deliver vital environmental services globally [1]. Through carbon sequestration, these ecosystems are critically essential for preserving biodiversity, regulating water quality, and stabilizing local climates [2]. Wetlands are organically linked to sustainable development goals (SDGs), which stress the protection of ecosystems and biodiversity [3]. The degradation of wetlands significantly jeopardizes local biodiversity and compromises sustainability goals, therefore increasing the requirement of conservation actions meeting biological and socioeconomic needs [4]. Having lost over 50% in the past century, wetlands maintain over 40% of all species globally, even if they rank among the most vulnerable ecosystems [5]. This decline underlines the significance of preserving wetlands, which are essential defenses against climate change, filters of pollutants, and habitats for many species [6].

Covering nearly 7435 $km^2$, wetlands make up over 5% of Nepal's total land area. [7]. Local populations depend on these ecosystems as they provide resources for farming, fishing, and other industries. The Ghodaghodi Lake Complex (GLC) is a well-known wetland categorized as a Ramsar site for its special biodiversity and ecological values [8]. Comprising many connected lakes, marshes, and seasonally flooded grasslands, this lake complex supports a variety of ecosystems able to sustain a great range of flora and animals [9]. The GLC's major environmental problems are human activities altering the natural surroundings of the lake: deforestation, agricultural expansion, and urbanization [10]. Agricultural runoff and domestic trash have caused nutrient loading that has boosted phosphate levels and lowered dissolved oxygen (DO) in the lake, compromising water quality and negatively affecting aquatic and terrestrial species [11]. Furthermore, the normal hydrochemical processes of the lake have been altered by sedimentation brought on by land expansion and deforestation, disturbing the ecological equilibrium and lowering its potential to offer neighboring people necessary services [12]. The deterioration of GLC stresses the need for thorough management measures since it exposes its biodiversity and affects its capacity to ensure regional water and food security [13].

Hydrochemical analysis has proven its effectiveness in evaluating the condition of aquatic environments through several physicochemical parameters [14,15]. Fundamental indicators of water quality in hydrochemical studies are DO, total dissolved solids (TDS), potential of hydrogen (pH), and major ions: calcium ($Ca^{2+}$) and magnesium ($Mg^{2+}$) [16]. T-tests and Analysis of Variance (ANOVA) enable statistical methods to detect significant seasonal and spatial fluctuations in these parameters, offering information on temporal and locational changes in water quality [17]. Moreover, hydrochemical diagrams, including Piper and Gibbs diagrams, allow the classification of water types and streamline the fundamental processes regulating lake chemistry [18]. Piper diagrams enable scientists to determine the main hydrochemical facies through cation and anion concentration, classifying water samples [19]. Gibbs diagrams, on the other hand, emphasize the primary processes affecting water chemistry, which would help to elucidate both natural and artificial consequences on water quality [20]. Although hydrochemical assessments are extensively employed in river and reservoir studies across Nepal, there is a shortage of research that notably addresses lake systems such as GLC, underscoring the need for targeted studies to establish baseline data for lake preservation [21].

This study attempts to fulfill this research gap in hydrochemical dynamics and water quality evaluation of GLC and associated lakes. The major objectives of this research are to measure these hydrochemical differences and evaluate the appropriateness of GLC's water for agriculture by directly addressing local sustainability issues that align with environmental SDGs. Utilizing statistical analysis and hydrochemical evaluation, this work provides a detailed grasp of the seasonal variations in water quality that affect both ecological health and human use. Ultimately, by providing essential baseline data for sustainable wetland management and conservation techniques, this study promotes evidence-based policy and enhances efforts to protect Nepal's significant wetland ecosystems, including Ramsar sites. This not only benefits the local and regional levels but also contributes to global conservation initiatives, highlighting the universal importance of preserving these vital ecosystems.

## 2. Materials and Methods

### 2.1. Study Area

The GLC is a significant ecological and socio-economic resource, comprising multiple oxbow lakes. It is situated within the geographical coordinates of latitude 28°41′ 21″ N to 28°42′ 21″ N and longitude 80°56′ 23″ E to 80°56′ 27″ E, in Ghodaghodi Municipality, Kailali District, Nepal (Figure 1). This study mainly focuses on six lakes within the complex: Ghodaghodi Lake, sometimes referred to as the main lake; Ojahuwa Lake; Budhiya Nakhrod Lake; Ramphal Lake; Bichka Chaitya Lake; and Sanopokhari Lake, also known as Sunpokhari Lake. The two lakes, Ghodaghodi and Ojahuwa, encompass 75.32 and 3.29 hectares, respectively, in Ward 1. Conversely, the remaining four lakes, Budhiya Nakhrod, Ramphal, Bichka Chaitya, and Sanopokhari, with areas of 5.54, 3.68, 0.5, and 2.38 hectares, respectively, are located in ward-8 of Ghodaghodi Municipality. The Ojahuwa Lake is distinguished for facilitating boating activities, while the four lakes have been employed for aquaculture. The Ghodaghodi Lake and its adjacent lakes, located at an altitude of 205 m above sea level, encompass springs, swamps, marshes, and human-made wetlands, including ponds, irrigated fields, and canals, all situated within a tropical forest of the Siwalik hills' lower slopes [13,22]. The lakes obtain water from multiple sources, such as precipitation, surface runoff, and flooding from adjacent rivers, enhancing their significant ecological diversity and supporting many flora and fauna [22–24]. The local inhabitants rely on the wetland area for their livelihoods, where the lakes' periphery is marked by layers of conglomerates and soft shale, with soil types ranging from clay to alluvial [25].

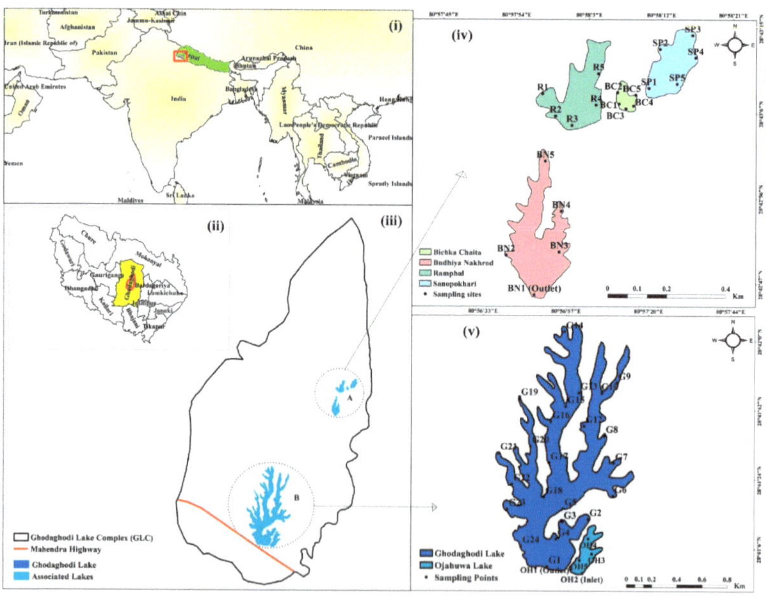


**Figure 1.** Study area region illustrating Ghodaghodi Lake and its adjacent lakes, including sampling sites: (**i**) A global map illustrating the study area, marked by a red polygon; (**ii**) A map of the Kailali District highlighting Ghodaghodi Municipality in yellow and the Ramsar site encompassing the Ghodaghodi Lake complex (GLC) in red; (**iii**) A map of the GLC-Ramsar site, depicting the locations of Ghodaghodi Lake and its associated lakes, classified into Section 'A' and Section 'B' with delineations; (**iv**) Locations of Bichka Chaita, Budhiya Nakhrod, Ramphal, and Sanopokhari Lakes along with their respective sampling sites BC1–BC5, BN1–BN5, R1–R5, and SP1–SP5, and (**v**) Locations of Ghodaghodi and Ojahuwa Lakes with their corresponding sampling sites G1–G24 and OH1–OH5.

Figure 2 illustrates the land use/land cover classes adjacent to the points of lakes where samples were taken. The Ghodaghodi Lake is surrounded by a forest-covered area; the north and south parts of the lake have been exposed to human activities, while in the middle region of the lake, there is a crocodile breeding center. Ojahuwa Lake is also surrounded by forest and settlement areas; this lake has been used for boating and recreational activities. The other four associated lakes, Budhiya Nakhrod, Bichka Chaitya, Sanopokhari, and Ramphal, are surrounded by forest land in the west direction and agricultural land along with a few settlements in the east direction.

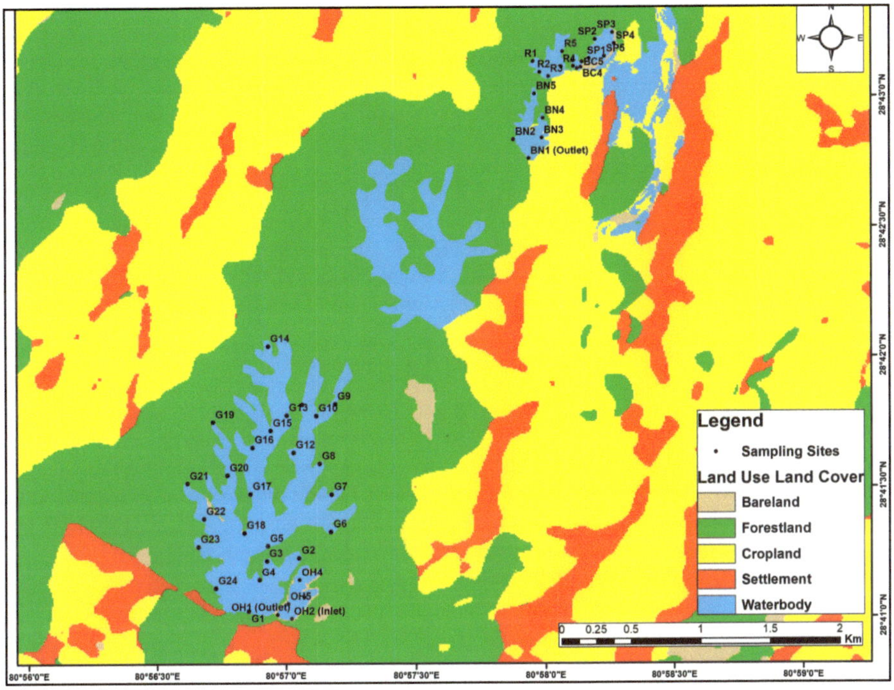


**Figure 2.** Land use/land cover map of the study area region illustrating different categories adjacent to sampling points of the lakes.

## 2.2. Data Collection

Fieldwork and sampling were executed in the pre-monsoon period of March 2022 and the post-monsoon period of November 2022, adhering to standard protocols [26]. Systematic random sampling was utilized to gather 49 water samples from the coastal zones of Ghodaghodi Lake (24 samples), Ojahuwa Lake (5 samples), Budhiya Nakhrod Lake (5 samples), Ramphal Lake (5 samples), Bichka Chaitya Lake (5 samples), and Sanopokhari (5 samples), as depicted in Figure 1. At each of the 49 sites, five subsamples were collected and subsequently combined to form a single composite sample per site for analysis. Water samples from Budhiya Nakhrod Lake and Ramphal Lake were not obtained during the pre-monsoon season due to dry conditions caused by fish farmers' water extraction.

Initially, sediment resuspension in coastal lake zones was seen as a potential factor affecting water quality. The water at the sampling locations demonstrated clean clarity, signifying negligible disturbance from sediment resuspension. To further mitigate any possible influence on sample composition, macrophytes located at specific sites were meticulously extracted using a fine mesh net.

### 2.3. Field and Laboratory Analysis

High-density polyethylene (HDPE) bottles, meticulously rinsed with lake water, were employed to gather water samples. The samples were carefully transferred to the Central Department of Environmental Science (CDES) laboratory at Tribhuvan University in Kirtipur, Kathmandu. They were preserved under freezing conditions to maintain their integrity for future examination.

On-site measurements were conducted for parameters such as water temperature (WT), TDS, DO, pH, chloride ($Cl^-$), magnesium hardness (MgH), calcium hardness (CaH), total hardness (TH), turbidity (Tur.), electrical conductivity (EC), alkalinity (AL), and free carbon dioxide ($CO_2$), adhering to the APHA standard guidelines [26].

The YSI Pro Quatro Multiparameter was utilized to evaluate water temperature, pH, DO, TDS, and EC. Turbidity (Tur) was measured using the Oakton TN-100 Turbidity Meter. The Oakton TN-100 Turbidity Meter is typically more effective than the Secchi disk for assessing water clarity, especially in scientific and professional settings, owing to its higher accuracy and precision. In contrast to the Secchi disk, which depends on subjective visual assessment, the TN-100 delivers objective numerical measurements in Nephelometric Turbidity Units (NTU), hence minimizing human error. Furthermore, TH, CaH, and MgH were assessed utilizing the EDTA Titration Method. $Cl^-$ was quantified using the Argentometric Titration Method, AL was evaluated via the Acid-base Titration Method, bicarbonates ($HCO_3^-$) were derived from AL, and free $CO_2$ was measured through phenolphthalein titration.

Additional parameters such as potassium ($K^+$), sodium ($Na^+$), ammonium ($NH_4^+$), phosphate ($PO_4^{3-}$), nitrate ($NO_3^-$), sulfate ($SO_4^{2-}$), and iron ($Fe^{2+}$) were examined in the laboratory. $K^+$ and $Na^+$ ions were measured with flame photometry. $NH_4^+$ concentrations were evaluated using the phenate method, $PO_4^{3-}$ levels were quantified via the ammonium molybdate solution method, $NO_3^-$ concentrations were ascertained through the phenol disulphonic acid method, $SO_4^{2-}$ levels were measured using the barium chloride method, and $Fe^{2+}$ concentrations were determined by the phenanthroline method. All assessments were completed following the procedures [26]. $Ca^{2+}$ and $Mg^{2+}$ were determined from the assessments of CaH and MgH, respectively.

### 2.4. Data Analysis

The raw data obtained from fieldwork and laboratory analyses underwent thorough assessment. Descriptive statistics, the Piper diagram, Gibbs plot, mixing diagram, parametric and non-parametric tests, and irrigation appropriateness indices were employed to evaluate the comprehensive hydrochemistry and seasonal fluctuations in water quality. This research was performed utilizing MS Excel, IBM SPSS 26, Origin 2019, and R-Studio 2024.04.2 Build 764, guaranteeing a thorough evaluation and interpretation of the data [15,27,28].

#### 2.4.1. Statistical Data Analysis

Descriptive statistics, including maximum, minimum, standard deviation, and mean values for all parameters, were calculated using the statistical package (IBM SPSS version 26). This methodology facilitated the assessment of regional and temporal fluctuations in water quality metrics and comparisons to analogous studies [29]. The Shapiro–Wilk test was adopted to evaluate normality for each parameter in each season and each lake separately. After the results, parametric and non-parametric tests were conducted in R-Studio to compare data over two seasons of the same lake or among various lakes within the same season. A *t*-test (for normally distributed data) or Mann-Whitney U test (for non-normally distributed data) was conducted to compare data between two seasons of the same lake [27]. A one-way ANOVA was conducted to evaluate the same variables across many analyzed lakes for the same season, contingent upon the normality of the data. The Kruskal–Wallis test (applicable for non-normally distributed data) was employed to compare identical variables across various examined lakes for the same season [29]. The parametric test was conducted when variables measured in both seasons or in all lakes were normally distributed; either in one season or one lake, there was non-normally distributed data; the comparison was carried out by a non-parametric test.

#### 2.4.2. Evaluation of Controlling Mechanisms of Hydrochemistry

- Piper diagram

A trilinear diagram utilizing the classification system established by Piper [30] was created with Origin Pro to classify the hydrochemical properties of water. This graphic consists of two triangles, one depicting cation and the other anions, accompanied by a central diamond plot, all within a triangular boundary. This diagram enables the categorization of water kinds into six unique classifications. The classes comprise (1) $Ca^{2+}$-$HCO_3^-$, (2) $Na^+$-$Cl^-$, (3) mixed $Ca^{2+}$-$Na^+$-$HCO_3^-$, (4) mixed $Ca^{2+}$-$Mg^{2+}$-$Cl^-$, (5) $Ca^{2+}$-$Cl^-$, and (6) $Na^+$-$HCO_3^-$ types, offering significant insights into the composition and features of the studied water samples [31,32].

- Gibbs plot and mixing diagram

The governing mechanism of the lake's hydrochemical dynamics was elucidated through meticulous Gibbs plot analysis, a technique that juxtaposes TDS with the ratios of $Na^+/(Na^+ + Ca^{2+})$ and $Cl^-/(Cl^- + HCO_3^-)$ to comprehend the intricate interplay of precipitation, evaporation, and rock weathering [20,33]. Samples with low TDS (<1 mg/L) and elevated $Na^+/(Na^+ + Ca^{2+})$ and $Cl^-/(Cl^- + HCO_3^-)$ ratios (0.5–1) were classified as end-member precipitation, congregating in the lower region of the figure. Centrally grouped samples exhibiting moderate TDS levels (70–300 mg/L) and reduced $Na+/(Na^+ + Ca^{2+})$ and $Cl^-/(Cl^- + HCO^{3-})$ ratios (<0.5) predominantly indicated the impact of rock weathering. Furthermore, samples situated in the upper right quadrant, distinguished by elevated TDS (>300 mg/L) and a high $Na^+/(Na^+ + Ca^{2+})$ and $Cl^-/(Cl^- + HCO_3^-)$ ratio (0.5–1), exhibited the effects of evaporation. This thorough investigation offered significant insights into the fundamental mechanisms influencing lake water chemistry [34]. The mixing diagram was also constructed to demonstrate the impact of carbonate, silicate, and evaporite weathering on hydrochemistry.

#### 2.4.3. Evaluation of Irrigation Suitability

The assessment of irrigation water and lake water quality involved evaluating several factors to ascertain their appropriateness for designated purposes. Critical parameters for assessing irrigation water quality include sodium adsorption ratio (SAR), sodium percentage (Na %), EC, Kelly's ratio (KR), magnesium adsorption ratio (MAR), cation ratio of soil structural stability (CROSS), and permeability index (PI), as these factors significantly influence soil and plant health [33,35,36]. Furthermore, the Wilcox diagram was constructed to assess sodium salinity and the potential risks associated with lake water, providing insights into its appropriateness for diverse applications [37]. All these indexes were computed as follows:

- Electrical Conductivity (EC)

Water with an EC less than 700 µS/cm, between 3000 and 700 µS/cm, and greater than 3000 µS/cm was classified as 'Excellent', 'Good', and 'Fair' in terms of irrigation appropriateness, respectively [38].

- Sodium Percentage (Na %)

The ratio of sodium and potassium to total cationic concentrations (Equation (1)) is computed to assess sodium hazards. The Na % range for irrigation suitability was categorized as follows: 'Unsuitable' > 80, 'Doubtful' 60–80, 'Permissible' 40–60, 'Good' 20–40, 'Excellent' < 20 [39].

$$\mathrm{Na\%} = \frac{(\mathrm{Na^+} + \mathrm{K^+}) * 100}{\left(\mathrm{Ca^{2+}} + \mathrm{Mg^{2+}} + \mathrm{Na^+} + \mathrm{K^+}\right)} \tag{1}$$

- Sodium Adsorption Ratio (SAR)

Soil cation exchange responses were assessed using SAR, indicating that water with an elevated SAR value may adversely impact the soil's physical structure, resulting in its degradation. The irrigation suitability following SAR was categorized as 'Excellent' for values less than 10 meq/L, 'Good' for 10–18 meq/L, 'Fair' for 18–26 meq/L, and 'Poor' for values beyond 26 meq/L [40]. SAR was presented in Equation (2).

$$SAR = \frac{Na^{+}}{\sqrt{\left(Ca^{2+} + Mg^{2+}\right)/2}} \quad (2)$$

- Magnesium Adsorption Ratio (MAR)

The MAR was computed using Equation (3) to assess the alkaline danger of water, with elevated MAR values signifying increased AL and diminished agricultural productivity. The irrigation suitability was classified according to the MAR, with values greater than 50 being 'Unsuitable' and those less than 50 considered 'Suitable' [41].

$$MAR = \frac{Mg^{2+} * 100}{\left(Ca^{2+} + Mg^{2+}\right)} \quad (3)$$

- Kelly's Ratio (KR)

The calculation of KR, which accounted for $Na^{+}$, about $Ca^{2+}$, and $Mg^{2+}$, was performed using Equation (4). The range of KR for irrigation suitability is categorized as 'Unsuitable' when greater than 1 and 'Suitable' when less than 1 [42].

$$KR = \frac{Na^{+}}{\left(Ca^{2+} + Mg^{2+}\right)} \quad (4)$$

- Permeability Index (PI)

The irrigation suitability was categorized according to the PI range as follows: 'Suitable Class I' for values greater than 75, 'Good Class II' for values between 25 and 75, and 'Unsuitable Class III' for values less than 25 [40]. The formula for PI was given by Equation (5) [35].

$$PI = \frac{\left(Na^{+} + \sqrt{HCO_3^{-}}\right) * 100}{Ca^{2+} + Mg^{2+} + Na^{+}} \quad (5)$$

- Cation Ratio of Soil Structural Stability (CROSS)

The suitability of irrigation based on SAR was categorized as follows: 'Unsuitable' for values exceeding 26, 'Fair' for values between 19 and 26, 'Good' for values between 10 and 18, and 'Excellent' for values less than 10. The CROSS formula was presented in Equation (6) [43].

$$CROSS = \frac{(Na^{+} + 0.56K^{+}) * 100}{\sqrt{\left(Ca^{2+} + 0.6Mg^{2+}\right)/2}} \quad (6)$$

- Wilcox Diagram.

The Wilcox classification was utilized to categorize the quality of irrigation water in GLC [39], according to the percentage of $Na^{+}$, in the EC. Various levels of irrigation salinity were the basis for categorizing water quality using the Wilcox diagram [33,35].

## 3. Results and Discussion

### 3.1. General Hydrochemistry and Variation

The hydrochemical evaluation of the GLC complex indicated seasonal and spatial fluctuations influenced by climatic, biological, and anthropogenic factors. Tables 1 and 2 presented a summary of the recorded physicochemical characteristics (i.e., ionic and non-ionic parameters), indicating a significant rise in WT throughout the post-monsoon season across the lakes, with Ghodaghodi and Ojahuwa Lakes exhibiting the highest WTs. The seasonal increase in WT was associated with elevated microbial and planktonic activity, affecting parameters such as TDS, EC, and BOD. Elevated post-monsoon WTs, observed in tropical lake systems, enhanced microbial decomposition and nutrient cycling, hence directly influencing TDS and EC [44].

**Table 1.** Descriptive statistical investigation of ionic parameters during the pre-monsoon and post-monsoon seasons in the lakes: Ojahuwa, Bichka Chaita, Budhiya Nakhrod, Ramphal, and Sanopokhari.

| Lake | Season | Statistics | WT | pH | Free $CO_2$ | EC | TDS | Tur | DO |
|---|---|---|---|---|---|---|---|---|---|
| Ghodaghodi | PRM | M | 28.89 | 6.98 | 26.86 | 251.17 | 143.14 | 10.58 | 5.12 |
| | | SD | 5.05 | 1.35 | 16.68 | 156.34 | 105.91 | 7.93 | 1.05 |
| | POM | M | 19.96 | 6.98 | 14.32 | 156.11 | 78.9 | 6.33 | 5.1 |
| | | SD | 1.41 | 1.35 | 7.60 | 78.05 | 39.84 | 2.61 | 1.49 |
| Ojahuwa | PRM | M | 31.48 | 7.36 | 15.40 | 344.04 | 193.98 | 22.68 | 5.22 |
| | | SD | 0.69 | 0.31 | 6.41 | 9.13 | 5.42 | 2.48 | 0.38 |
| | POM | M | 19.80 | 7.48 | 12.76 | 344.04 | 196.37 | 20.80 | 7.10 |
| | | SD | 3.25 | 0.26 | 10.25 | 10.51 | 119.20 | 6.78 | 2.04 |
| Bichka Chaita | PRM | M | 31.32 | 8.31 | 12.76 | 264.46 | 156.35 | 15.05 | 6.28 |
| | | SD | 1.78 | 0.57 | 4.77 | 8.90 | 6.68 | 9.46 | 1.37 |
| | POM | M | 16.26 | 7.65 | 8.36 | 248.96 | 137.06 | 12.35 | 8.34 |
| | | SD | 0.18 | 0.05 | 1.84 | 11.43 | 15.32 | 0.67 | 0.48 |
| Sanopokhari | PRM | M | 27.66 | 8.20 | 13.64 | 173.00 | 100.71 | 43.58 | 5.60 |
| | | SD | 0.49 | 0.42 | 3.94 | 7.97 | 6.65 | 16.40 | 0.52 |
| | POM | M | 15.90 | 7.74 | 6.16 | 279.40 | 149.23 | 6.30 | 8.60 |
| | | SD | 0.30 | 0.05 | 0.98 | 13.96 | 8.67 | 0.76 | 0.43 |
| Budhiya Nakhrod | POM | M | 16.98 | 7.72 | 17.60 | 323.80 | 167.27 | 10.39 | 8.01 |
| | | SD | 0.13 | 0.22 | 3.48 | 16.63 | 11.32 | 6.19 | 0.57 |
| Ramphal | POM | M | 16.60 | 7.60 | 12.76 | 331.60 | 176.23 | 6.59 | 7.26 |
| | | SD | 0.10 | 0.07 | 5.48 | 107.16 | 49.91 | 0.52 | 0.61 |
| Whole lakes | PRM | M | 29.38 | 7.35 | 21.89 | 250.30 | 145.91 | 16.93 | 5.34 |
| | | SD | 4.18 | 1.21 | 14.76 | 130.62 | 85.99 | 14.15 | 1.03 |
| Whole lakes | POM | M | 18.50 | 7.64 | 12.90 | 224.07 | 117.24 | 8.86 | 6.51 |
| | | SD | 2.22 | 0.25 | 7.10 | 95.35 | 51.06 | 5.58 | 1.91 |

Note: Tur is in NTU, WT is in °C, EC is in μS/cm, while other parameters are in mg/L, PRM = pre-monsoon, POM = post-monsoon, SD = standard deviation, and M = mean.

**Table 2.** Descriptive statistical investigation of non-ionic parameters during the pre-monsoon and post-monsoon seasons in the lakes: Ojahuwa, Bichka Chaita, Budhiya Nakhrod, Ramphal, and Sanopokhari.

| Lake | Season | Statistics | TH | $Ca^{2+}$ | $Mg^{2+}$ | $K^+$ | $Na^+$ | $NH_4^+$ | $Cl^-$ | $Fe^{2+}$ | $NO_3^-$ | $SO_4^{2-}$ | $HCO_3^-$ | $PO_4^{3-}$ |
|---|---|---|---|---|---|---|---|---|---|---|---|---|---|---|
| Ghodaghodi | PRM | M | 59.58 | 17.07 | 4.13 | 4.16 | 8.4 | 0.1 | 9.93 | 0.58 | 5.23 | 0.91 | 85.04 | 0.43 |
| | | SD | 10.63 | 3.32 | 1.73 | 2.06 | 2.14 | 0.07 | 5.60 | 0.37 | 5.17 | 0.49 | 23.01 | 0.23 |
| | POM | M | 99.07 | 26.62 | 7.94 | 1.77 | 5.19 | 0.1 | 12.7 | 0.31 | 4.78 | 0.64 | 99.27 | 0.2 |
| | | SD | 24.11 | 7.93 | 1.93 | 0.93 | 0.89 | 0.07 | 7.17 | 0.11 | 3.26 | 0.27 | 54.37 | 0.07 |
| Ojahuwa | PRM | M | 97.20 | 25.28 | 8.30 | 0.75 | 5.52 | 0.02 | 14.48 | 0.48 | 2.13 | 0.52 | 145.80 | 0.33 |
| | | SD | 5.40 | 2.23 | 0.49 | 0.37 | 2.69 | 0.02 | 8.72 | 0.11 | 0.12 | 0.07 | 37.80 | 0.16 |
| | POM | M | 177.60 | 46.72 | 14.84 | 2.40 | 9.58 | 0.08 | 17.32 | 0.71 | 3.37 | 1.84 | 164.40 | 0.07 |
| | | SD | 37.40 | 10.18 | 3.06 | 0.31 | 7.03 | 0.01 | 15.01 | 0.47 | 0.44 | 0.79 | 131.66 | 0.00 |
| Bichka Chaita | PRM | M | 76.00 | 20.96 | 5.76 | 0.45 | 5.86 | 0.07 | 4.83 | 0.30 | 1.88 | 0.57 | 147.00 | 0.19 |
| | | SD | 5.10 | 1.91 | 1.11 | 0.01 | 0.05 | 0.00 | 1.62 | 0.07 | 0.19 | 0.02 | 22.93 | 0.06 |
| | POM | M | 136.40 | 35.52 | 11.61 | 2.52 | 7.82 | 0.06 | 4.83 | 0.26 | 2.86 | 0.30 | 115.60 | 0.13 |
| | | SD | 12.84 | 4.85 | 0.53 | 0.33 | 0.62 | 0.00 | 1.62 | 0.05 | 0.35 | 0.07 | 15.69 | 0.04 |
| Sanopokhari | PRM | M | 80.00 | 22.24 | 5.95 | 0.60 | 2.89 | 0.20 | 5.40 | 1.93 | 1.94 | 2.38 | 93.00 | 0.23 |
| | | SD | 6.78 | 2.49 | 0.80 | 0.52 | 1.78 | 0.01 | 1.19 | 0.22 | 0.24 | 0.41 | 26.38 | 0.02 |
| | POM | M | 174.00 | 53.12 | 10.05 | 1.82 | 7.66 | 0.08 | 14.20 | 0.31 | 3.26 | 0.33 | 124.00 | 0.19 |
| | | SD | 7.87 | 1.34 | 1.12 | 1.15 | 1.95 | 0.00 | 4.49 | 0.06 | 0.40 | 0.02 | 4.12 | 0.08 |
| Budhiya Nakhrod | POM | M | 118.00 | 32.00 | 9.27 | 2.04 | 9.22 | 0.06 | 16.76 | 0.47 | 3.01 | 0.63 | 34.00 | 0.10 |
| | | SD | 5.10 | 1.26 | 1.14 | 0.15 | 1.62 | 0.01 | 2.73 | 0.06 | 0.16 | 0.04 | 4.18 | 0.01 |
| Ramphal | POM | M | 172.00 | 46.40 | 13.66 | 0.91 | 4.68 | 0.05 | 8.24 | 0.23 | 3.05 | 0.40 | 151.00 | 0.07 |
| | | SD | 55.28 | 12.29 | 6.03 | 0.15 | 1.97 | 0.00 | 2.11 | 0.04 | 0.67 | 0.11 | 57.76 | 0.01 |
| Whole lakes | PRM | M | 69.13 | 19.28 | 5.11 | 2.79 | 7.00 | 0.10 | 9.28 | 0.70 | 3.98 | 1.00 | 101.74 | 0.36 |
| | | SD | 16.19 | 4.19 | 2.03 | 2.39 | 2.79 | 0.07 | 5.98 | 0.57 | 4.33 | 0.69 | 36.26 | 0.21 |
| Whole lakes | POM | M | 127.91 | 34.85 | 9.95 | 1.86 | 6.52 | 0.08 | 12.48 | 0.35 | 3.93 | 0.67 | 106.67 | 0.15 |
| | | SD | 42.26 | 12.32 | 3.48 | 0.85 | 2.94 | 0.03 | 7.68 | 0.21 | 2.43 | 0.52 | 57.64 | 0.08 |

Note: All parameters are in mg/L; PRM = pre-monsoon, POM = post-monsoon, SD = standard deviation, and M = mean.

WT exhibited significant variation in the pre-monsoon period, which was affected by daily changes and sample periods. These variations likely led to elevated TDS and EC since evaporative processes concentrate ions in the water [8]. Due to their reduced water volumes, the evaporation-induced ionic concentration was especially pronounced in smaller lakes, such as Sanopokhari and Bichka Chaita. The WT variations immediately influenced other factors, as elevated WT frequently coincides with aquatic species' heightened metabolic rates, exacerbating organic matter breakdown [45].

The pH levels in the GLC lakes exhibited notable seasonal variations, fluctuating from mildly acidic to alkaline (Table 1). In the pre-monsoon period, Ghodaghodi Lake had increased acidity, presumably due to intensified biological breakdown near thick macrophyte regions, releasing dissolved $CO_2$. Moreover, $CO_2$ is combined with water to generate carbonic acid, reducing pH. This process can be controlled post-monsoon by the influx of rainwater, which neutralizes and buffers the water, stabilizing pH at slightly alkaline levels. The buffering effect of $HCO_3^-$, often sourced from carbonate weathering, aids in stabilizing pH fluctuations throughout the post-monsoon season, averting sudden pH declines that could adversely affect aquatic organisms [45].

The inverse correlation between pH and free $CO_2$ concentrations was particularly evident in pre-monsoon samples. In GLC, elevated $CO_2$ concentrations were observed, especially in areas rich in macrophytes, where breakdown and biological respiration contribute substantial $CO_2$ emissions [46]. Elevated free $CO_2$ concentrations signified heightened organic decomposition activities, thus leading to water acidification and reduced pH values. $CO_2$ generation and pH affected the carbonate-bicarbonate buffering system, impacting lake alkalinity, especially in carbonate-rich areas [21].

EC and TDS exhibited considerable seasonal and geographical variability among the lakes (Table 1). The readings reached their zenith during the pre-monsoon period, particularly at Ghodaghodi Lake, where diminished water levels and elevated WTs resulted in increased ionic concentration. Increased EC and TDS during the pre-monsoon period signified the impacts of evapo-concentration, mineral dissolution, and potential human influences, which became more pronounced when lake water levels declined. These results corresponded with research conducted in tropical and subtropical lakes, wherein diminished water volume from elevated evapotranspiration led to the concentration of dissolved ions in lakes situated within carbonate-rich watersheds [17,44]. In contrast, post-monsoon measurements were diminished in all lakes due to dilution from precipitation, demonstrating that monsoon-induced influx significantly lowered TDS and EC by introducing water with reduced ion concentration [47].

The results obtained for the whole lakes did not yield significant new findings compared to the individual lake analyses (Tables 1 and 2). The overall hydrochemical patterns, including TH, $Ca^{2+}$, $Mg^{2+}$, and $HCO_3^-$, reflected similar seasonal trends influenced by carbonate weathering and rainfall-induced dilution [8]. While these parameters demonstrated the expected seasonal variations, the aggregated data for all lakes did not provide additional insights beyond the localized variations previously described [13]. This suggested that examining each lake individually remained the most informative approach for understanding the hydrochemical dynamics of the Ghodaghodi Lake Complex, aligning with findings from previous studies that have emphasized localized influences over regional patterns [11].

Since data were collected in the coastal zones of the lakes, sediment resuspension was considered a potential factor influencing the observed water quality. The role of internal recharge, involving the upward movement of nutrient-rich groundwater or mixing from sediment pore water, likely contributed to the elevated concentrations of certain ions and nutrients. This internal recharge process can amplify nutrient levels, such as $PO_4^{3-}$ and $NO_3^-$, and increase tur., especially during periods of wind-induced mixing or bioturbation by aquatic organisms. Similar findings have been noted in other wetland studies where internal nutrient cycling significantly impacted hydrochemical variations [12,13]. In many cases, this internal process may have a dominant effect compared to pollutant inflows from the catchment, impacting the hydrochemical balance [11].

### 3.2. Comparison of Pre- and Post-Monsoon Data for Identical Lakes and Inter-Lake Comparison Within the Same Season

A comprehensive comparative investigation of hydrochemical parameters across seasons and lakes revealed significant patterns, especially with important ions such as $Ca^{2+}$ and $HCO_3^-$ (Table 2). Most ions, particularly $Ca^{2+}$ and $HCO_3^-$, attained elevated concentrations during the pre-monsoon period due to evapo-concentration and reduced precipitation. Ghodaghodi, Ojahuwa, and Sanopokhari Lakes demonstrated the highest ion concentrations, a pattern typical with tropical lakes where pre-monsoon circumstances enhanced dissolved ion concentration [48]. In carbonate-rich lakes such as Ghodaghodi, $Ca^{2+}$, and $HCO_3^-$ concentrations were directly affected by the dissolution of adjacent carbonate rock, which was exacerbated under reduced water levels [49].

Conversely, post-monsoon data indicated reduced ion concentrations (Table 3), illustrating the dilutive impact of monsoon precipitation. TH, $Ca^{2+}$, and $Mg^{2+}$ concentrations were markedly diminished after the monsoon, suggesting that precipitation influx considerably decreased ionic levels [50]. Statistical results employing ANOVA and Kruskal–Wallis U tests (Table 4) validated considerable seasonal variations, especially in $Ca^{2+}$, $Mg^{2+}$, and TH, highlighting the influence of monsoonal dilution in reducing ionic concentrations in tropical lake systems [51].

**Table 3.** Comparison of pre-monsoon and post-monsoon data of four lakes: Ghodaghodi, Ojahuwa, Sanopokhari, and Bichka Chaita.

| Ghodaghodi | | | Ojahuwa | | | Sanopokhari | | | Bichka Chaita | | |
|---|---|---|---|---|---|---|---|---|---|---|---|
| *t*-test | md | *p*-Value | *t*-test | md | *p*-Value | *t*-test | md | *p*-Value | *t*-test | md | *p*-Value |
| DO | 0.017 | 0.960 | WT | 11.680 | 0.002 | WT | 11.76 | <0.001 | pH | 0.666 | 0.049 |
| **M-WU test** | **W** | ***p*-Value** | pH | −0.122 | 0.236 | TDS | −48.52 | 0.001 | WT | 15.060 | <0.001 |
| pH | 237.5 | 0.302 | EC | 81.240 | 0.007 | Tur | 37.28 | 0.006 | Free $CO_2$ | 4.400 | 0.034 |
| WT | 568 | <0.001 | Tur | 1.884 | 0.608 | DO | −2.99 | 0.001 | EC | 15.500 | 0.038 |
| Free $CO_2$ | 477.5 | <0.001 | $Mg^{2+}$ | −6.539 | 0.008 | TH | −94.00 | <0.001 | TDS | 19.283 | 0.019 |
| EC | 397 | 0.025 | $K^+$ | −1.650 | 0.001 | $Ca^{2+}$ | −30.88 | <0.001 | Tur | 2.700 | 0.554 |
| TDS | 431.5 | 0.003 | $NH_4^+$ | −0.056 | 0.008 | $K^+$ | −1.22 | 0.065 | DO | −2.058 | 0.015 |
| Tur | 365.5 | 0.112 | $NO_3^-$ | −1.243 | 0.001 | $Na^+$ | −4.77 | 0.002 | TH | −60.400 | 0.001 |
| TH | 0 | <0.001 | $PO_4^{3-}$ | 0.261 | 0.023 | $NH_4^+$ | 0.12 | <0.001 | $Ca^{2+}$ | −14.560 | 0.004 |
| $Ca^{2+}$ | 14 | <0.001 | **M-W U test** | **W** | ***p*-Value** | $Cl^-$ | −8.80 | 0.008 | $Mg^{2+}$ | −5.856 | 0.001 |
| $Mg^{2+}$ | 39.5 | <0.001 | Free $CO_2$ | 19 | 0.203 | $Fe^{2+}$ | 1.61 | <0.001 | $K^+$ | −2.072 | <0.001 |
| $K^+$ | 481.5 | <0.001 | TDS | 25 | 0.012 | $NO_3^-$ | −1.32 | 0.006 | $Cl^-$ | 0.000 | 1.000 |
| $Na^+$ | 524 | <0.001 | DO | 5 | 0.144 | $SO_4^{2-}$ | 2.05 | <0.001 | $Fe^{2+}$ | 0.044 | 0.401 |
| $NH_4^+$ | 263 | <0.001 | TH | 0 | 0.012 | $HCO_3^-$ | −31.00 | 0.060 | $NO_3^-$ | −0.982 | 0.006 |
| $Cl^-$ | 200.5 | <0.001 | $Ca^{2+}$ | 0 | 0.012 | $PO_4^{3-}$ | 0.04 | 0.253 | $SO_4^{2-}$ | 0.269 | 0.001 |
| $Fe^{2+}$ | 441.5 | 0.002 | $Na^+$ | 5.5 | 0.173 | **M-W U test** | **W** | ***p*-Value** | $PO_4^{3-}$ | 0.064 | 0.014 |
| $NO_3^-$ | 286 | 0.975 | $Cl^-$ | 12 | 1.000 | pH | 20 | 0.138 | **M-W U test** | **W** | ***p*-Value** |
| $SO_4^{2-}$ | 422 | 0.006 | $Fe^{2+}$ | 8 | 0.403 | Free $CO_2$ | 25 | 0.009 | $Na^+$ | 0 | 0.011 |
| $HCO_3^-$ | 240.5 | 0.330 | $SO_4^{2-}$ | 0 | 0.012 | EC | 0 | 0.012 | $NH_4^+$ | 25 | 0.011 |
| $PO_4^{3-}$ | 506.5 | <0.001 | $HCO_3^-$ | 16 | 0.531 | $Mg^{2+}$ | 0 | 0.010 | $HCO_3^-$ | 23 | 0.037 |

Note: md = mean difference, *p*-value < 0.05 = there is a significant difference in variable measured during two different seasons in the same lake, W = the sum of the ranks of the first sample, and M-W = Mann-Whitney.

The data indicated notable regional variations in tur. and $NO_3^-$ concentrations, particularly in lakes such as Sanopokhari and Bichka Chaita, which exhibited the highest tur. levels during the pre-monsoon period (Tables 1 and 2). Elevated tur. in Sanopokhari was likely attributable to agricultural runoff, which introduced particulate matter and $NO_3^-$ that increased suspended particle concentrations [52]. The elevated tur. was associated with nutrient concentrations, as both $NO_3^-$ and $NH_4^+$ levels were greater in Sanopokhari, indicating nutrient enrichment from adjacent agricultural land. Nutrient concentrations in GLC lakes suggested that agricultural practices significantly impact them, particularly in the pre-monsoon period when runoff could not be mitigated by substantial rainfall [53].

Increased $NO_3^-$ concentrations in lakes such as Ojahuwa further substantiated the influence of nutrient runoff from adjacent agricultural areas, especially following the monsoon season when rainfall transported fertilizers into the lakes (Table 2). The variations in nutrients underscored the seasonal susceptibility of lakes next to agricultural land to nutrient enrichment, which, if not well managed, may result in eutrophic conditions, hence compromising water quality and ecosystem health over time [53].

**Table 4.** Comparison of pre-monsoon and post-monsoon data among the lakes: Ghodaghodi, Ojahuwa, Bichka Chaita, Sanopokhari, Budhiya Nakhrod, and Ramphal.

| Pre-Monsoon | | | | Post-Monsoon | | | |
|---|---|---|---|---|---|---|---|
| **ANOVA Test** | **df** | **f-Value** | ***p*-Value** | **ANOVA Test** | **df** | **f-Value** | ***p*-Value** |
| DO | 3 | 2.032 | 0.127 | WT | 5 | 14.34 | <0.001 |
| TH | | 27 | <0.001 | Tur | | 17.2 | <0.001 |
| $Ca^{2+}$ | | 13.28 | <0.001 | $K^+$ | | 2.808 | <0.001 |
| **K-W test** | **df** | **chi-squared** | ***p*-value** | **K-W test** | **df** | **chi-squared** | ***p*-value** |
| pH | 3 | 11.66 | 0.009 | pH | 5 | 7.73 | 0.172 |
| WT | | 3.23 | 0.358 | Free $CO_2$ | | 23.85 | <0.001 |
| Free $CO_2$ | | 11.19 | 0.011 | EC | | 30.27 | <0.001 |
| EC | | 8.39 | 0.039 | TDS | | 29.14 | <0.001 |
| TDS | | 8.16 | 0.043 | DO | | 32.95 | <0.001 |
| Tur | | 17.03 | 0.001 | TH | | 34.00 | <0.001 |
| $Mg^{2+}$ | | 18.32 | <0.001 | $Ca^{2+}$ | | 30.75 | <0.001 |
| $K^+$ | | 15.55 | 0.001 | $Mg^{2+}$ | | 29.89 | <0.001 |
| $Na^+$ | | 18.29 | <0.001 | $Na^+$ | | 25.38 | <0.001 |
| $NH_4{}^+$ | | 19.65 | <0.001 | $NH_4{}^+$ | | 33.62 | <0.001 |
| $Cl^-$ | | 11.30 | 0.010 | $Cl^-$ | | 25.48 | <0.001 |
| $Fe^{2+}$ | | 16.09 | 0.001 | $Fe^{2+}$ | | 22.03 | 0.001 |
| $NO_3{}^-$ | | 10.64 | 0.014 | $NO_3{}^-$ | | 3.66 | 0.599 |
| $SO_4{}^{2-}$ | | 21.54 | <0.001 | $SO_4{}^{2-}$ | | 29.07 | <0.001 |
| $HCO_3{}^-$ | | 17.34 | 0.001 | $HCO_3{}^-$ | | 28.14 | <0.001 |
| $PO_4{}^{3-}$ | | 12.34 | 0.006 | $PO_4{}^{3-}$ | | 33.83 | <0.001 |

Note: df = degree of freedom, *p*-value < 0.05 = a significant difference in variable measured in lakes during the same season, K-W = Kruskal–Wallis, and ANOVA = analysis of variance.

### *3.3. Comparative Assessment of Hydrochemical Facies*

The hydrochemical facies of GLC lakes, depicted in Piper diagrams, revealed a predominance of Ca–$HCO_3$ and Mg–$HCO_3$ types, characteristic of carbonate weathering settings (Figures 3 and 4). During the pre-monsoon season, Ghodaghodi Lake exhibited a dominating Ca–$HCO_3$ type, indicating the influence of $HCO_3{}^-$ dissolution from the underlying lithology. Increased $Cl^-$ concentrations in Ojahuwa and Sanopokhari, especially during the pre-monsoon period, indicated anthropogenic impacts, including agricultural and domestic runoff from adjacent regions. Increased $Cl^-$ concentrations, aligned with patterns noted in other lakes, were affected by anthropogenic activities, as $Cl^-$ was often derived from fertilizers and household garbage [54].

Cation analysis revealed that $Ca^{2+}$ was the predominant ion in all lakes, succeeded by $Mg^{2+}$, underscoring the impact of carbonate weathering as the principal ionic contributor (Table 2). The Piper diagrams illustrated seasonal variations, indicating that although Ca–$HCO_3$ predominated, lakes such as Sanopokhari and Bichka Chaita displayed transitions to mixed Ca–Mg–Cl facies in the post-monsoon period (Figures 3 and 4). This transition indicated that human activities, such as runoff from adjacent agricultural areas, influenced water chemistry. Mixed facies patterns frequently occurred in lakes exposed to external nutrient inputs, as runoff could inject supplementary ions that modified the original water type [46].

Table 5 summarizes the classifications of water types derived from the Piper diagrams, indicating that Ghodaghodi Lake regularly exhibited a Ca–$HCO_3$ type in both seasons. Conversely, Sanopokhari and Bichka Chaita's transition to mixed Ca–Mg–Cl facies after the monsoon indicated that nutrient and ion contributions from agricultural regions significantly influenced water composition. Seasonal facies variations had consequences for lake management since they signified regions where external activities might be modifying natural hydrochemical processes, potentially impacting water suitability for various applications [21].

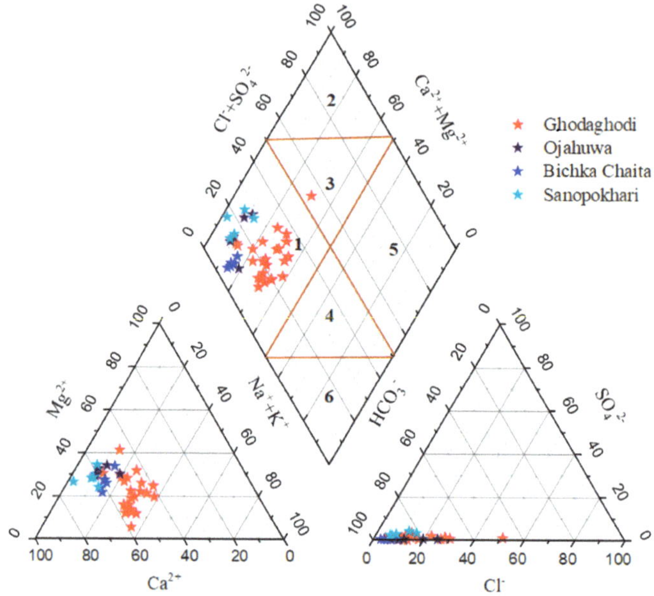


**Figure 3.** Piper diagram for the classification of lake water types in Ghodaghodi and its associated lakes (Ojahuwa, Bichka Chaita, and Sanopokhari) during the pre-monsoon season, featuring three plots: anionic, cationic, and diamond plots.

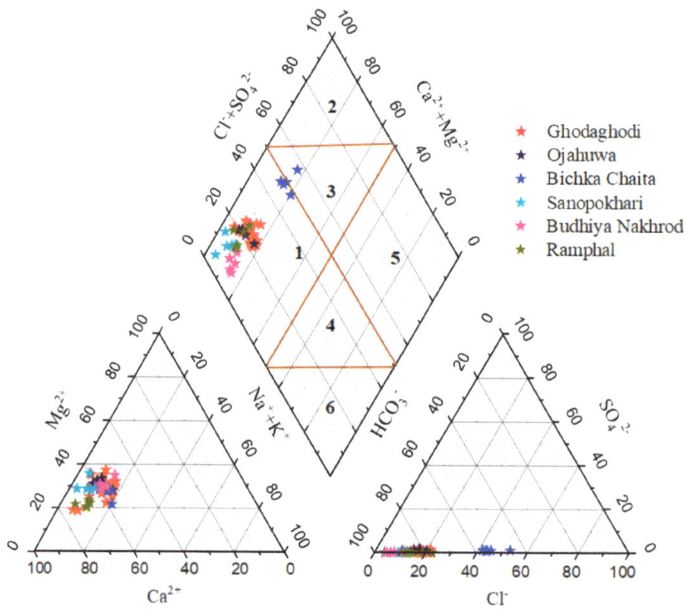


**Figure 4.** Piper diagram for the classification of lake water types in Ghodaghodi and its related lakes (Ojahuwa, Bichka Chaita, Budhiya Nakhrod, Ramphal, and Sanopokhari) during the post-monsoon season, featuring three plots: anionic, cationic, and diamond plots.

**Table 5.** Classification of lake water types (Ghodaghodi, Ojahuwa, Bichka Chaita, Budhiya Nakhrod, Sanopokhari, and Ramphal) utilizing the diamond plot derived from the Piper trilinear diagram.

| Box | Types [32] | Pre-Monsoon (%) | Post-Monsoon (%) |
|---|---|---|---|
| 1 | $Ca^{2+}$-$HCO_3^-$ | 95.83 (All Lakes) | 95.83 (All Lakes) |
| 2 | $Ca^{2+}$-$Cl^-$ | Nill | Nill |
| 3 | Mixed type $Ca^{2+}$-$Mg^{2+}$-$Cl^-$ | 4.17 (Ghodaghodi) | 4.17 (Bichka Chaita) |
| 4 | Mixed type $Ca^{2+}$-$Na^+$-$HCO_3^-$ | Nill | Nill |
| 5 | $Na^+$-$Cl^-$ | Nill | Nill |
| 6 | $Na^+$-$HCO_3^-$ | Nill | Nill |

### *3.4. Controlling Mechanisms of Hydrochemistry*

The primary mechanisms affecting hydrochemical fluctuations in the GLC were rock weathering, evaporation, and anthropogenic contributions [22]. Gibbs plots indicated that pre-monsoon samples predominantly aggregated inside the rock-weathering dominance zone, implying that lithological sources are the principal determinants of ion concentrations in these lakes (Figures 5 and 6). The elevated levels of $Ca^{2+}$ and $HCO_3^-$ in lakes, such as Ghodaghodi and Ojahuwa, indicated that $HCO_3^-$ dissolution is a primary contributor, with dissolved ions mirroring the mineralogical composition of the adjacent geology [35].

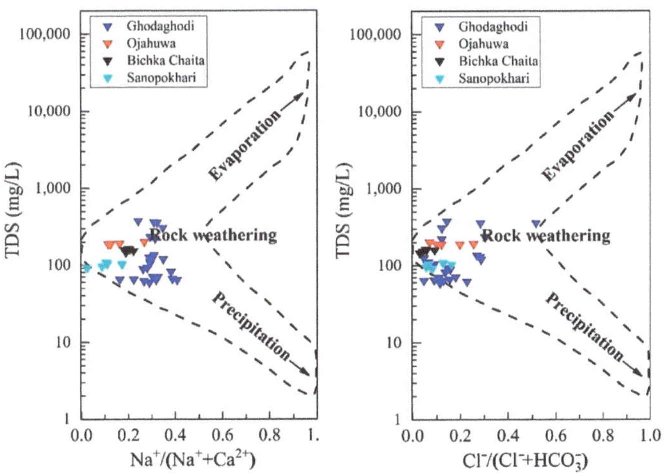


**Figure 5.** Gibbs diagrams illustrating the fluctuation of the weight ratio of $Na^+/(Na^+ + Ca^{2+})$ and $Cl^-/(Cl^- + HCO^{3-})$ concerning TDS (pre-monsoon) throughout all examined lakes (Ghodaghodi, Ojahuwa, Bichka Chaita, and Sanopokhari).

Evaporation influenced hydrochemistry, especially in the pre-monsoon period, since elevated evaporation rates concentrated dissolved salts, resulting in increased TDS and EC. This phenomenon was especially evident in smaller lakes such as Sanopokhari, where reduced water volumes result in more significant evapo-concentration. Increased pre-monsoon $Na^+$ and $Cl^-$ concentrations further illustrated the impact of evaporation, which intensified salt concentration in the absence of dilution [55]. Monsoon-induced rainfall dilution reduced TDS and EC levels after the monsoon, demonstrating that seasonal precipitation substantially lowered ionic concentration in lakes [47].

Anthropogenic influences, particularly agricultural runoff, significantly contributed to elevated nutrient levels (e.g., $NH_4^+$, $NO_3^-$) in lakes such as Sanopokhari and Budhiya Nakhrod, especially following the monsoon season. Increased nutrient concentrations indicated the influence of nitrogen fertilizers and animal waste from adjacent agricultural land, which heightened the risk of eutrophication if nutrient management practices were not used. Nutrient enrichment, especially in agricultural regions, posed dangers to lake ecosystems, as excessive nitrogen and phosphorus could result in algal blooms, reduced oxygen levels, and deteriorated water quality [53].

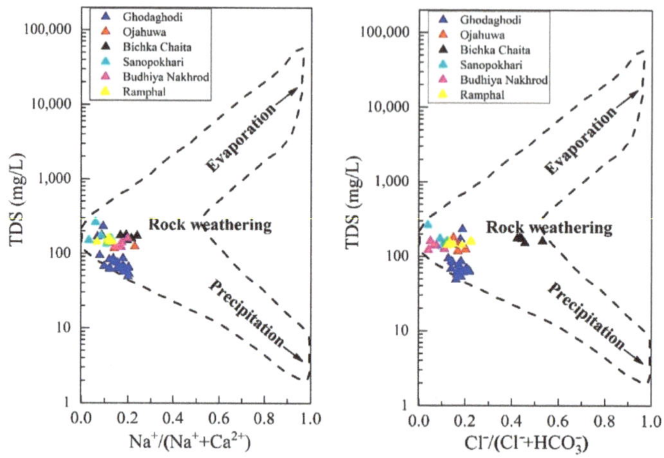


**Figure 6.** Gibbs diagrams illustrating the fluctuation of the weight ratio of $Na^+/(Na^+ + Ca^{2+})$ and $Cl^-/(Cl^- + HCO^{3-})$ concerning TDS (post-monsoon) throughout all examined lakes (Ghodaghodi, Ojahuwa, Bichka Chaita, Budhiya Nakhrod, Ramphal, and Sanopokhari).

The mixing diagrams validated the preeminent influence of carbonate weathering, with silicate weathering playing a subordinate role in determining hydrochemistry (Figures 7 and 8). Pre-monsoon samples, distinguished by elevated levels of $Ca^{2+}$ and $HCO_3^-$, indicated the impact of $HCO_3^-$ weathering, exacerbated during periods of reduced water levels, leading to the concentration of weathering byproducts through mineral dissolution processes. Lakes such as Sanopokhari, displaying silicate influence, demonstrated variety in lithological influence, probably attributable to the heterogeneous mineral composition of their surrounding catchments [21].

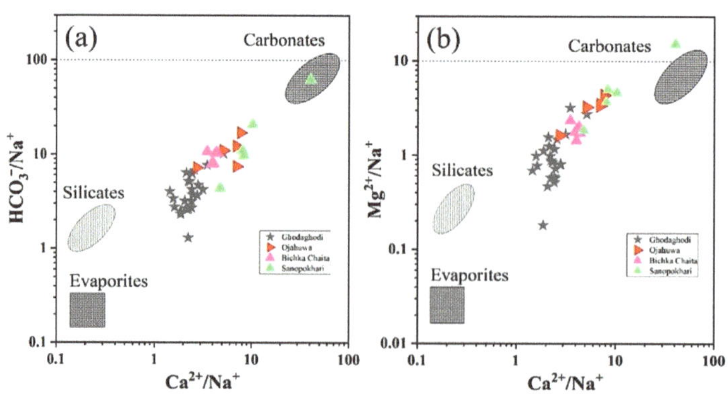


**Figure 7.** Mixing diagrams illustrating the roles of carbonate, silicate, and evaporates in the hydrochemistry of Ghodaghodi and associated lakes (Ojahuwa, Bichka Chaita, and Sanopokhari) during the pre-monsoon season. (**a**) represents $HCO_3^-/Na^+$ vs $Ca^{2+}/Na^+$ and (**b**) represents $Mg^{2+}/Na^+$ vs $Ca^{2+}/Na^+$ of mixing diagram.

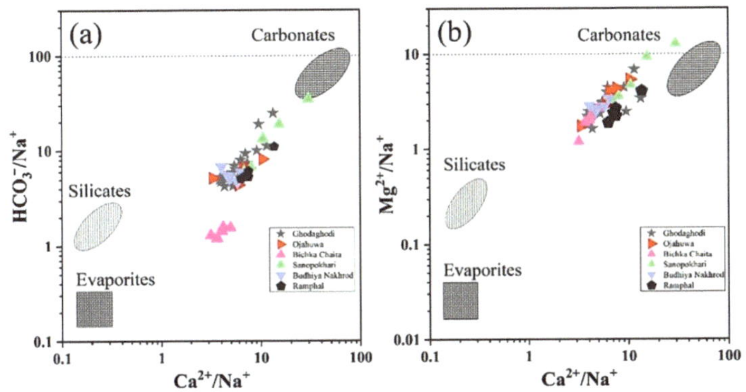


**Figure 8.** Mixing diagrams illustrating the roles of carbonate, silicate, and evaporates in the hydrochemistry of Ghodaghodi and its associated lakes (Ojahuwa, Bichka Chaita, Budhiya Nakhrod, Ramphal, and Sanopokhari) during the post-monsoon season. (**a**) represents $HCO_3^-/Na^+$ vs $Ca^{2+}/Na^+$ and (**b**) represents $Mg^{2+}/Na^+$ vs $Ca^{2+}/Na^+$ of mixing diagram.

### 3.5. Water Suitability for Irrigation

The suitability of water for irrigation was assessed using SAR, MAR, Na%, and other indices, concentrating on salinity, permeability, and sodium risks. SAR values for all lakes remained beneath the crucial threshold (<10), categorizing the water as 'excellent' for irrigation during both pre- and post-monsoon seasons, according to the guidelines [40] (Table 6). Low SAR values signified a negligible sodium risk, indicating that water from these lakes was appropriate for diverse soil types without endangering soil integrity due to salt-induced dispersion.

**Table 6.** Results of various indices for irrigation water quality (SAR, Na %, MAR, KR, PI, and CROSS) calculated for six studied lakes for both seasons.

| Lake | Season | SAR: M ± SD (WQ) | Na%: M ± SD (WQ) | MAR: M ± SD (WQ) | KR: M ± SD (WQ) | PI: M ± SD (WQ) | Cross: M ± SD (WQ) |
|---|---|---|---|---|---|---|---|
| Ghodaghodi | PRM | 0.68 ± 0.16 (E) | 28.47 ± 6.53 (G) | 28.46 ± 9.4 (S) | 0.32 ± 0.08 (S) | 101.48 ± 18.05 (S) | 0.44 ± 0.11 (E) |
| | POM | 0.33 ± 0.07 (E) | 12.66 ± 3.89 (E) | 33.65 ± 5.21 (S) | 0.12 ± 0.04 (S) | 68.12 ± 8.15 (G) | 0.21 ± 0.05 (E) |
| Ojahuwa | PRM | 0.35 ± 0.17 (E) | 11.68 ± 4.21 (E) | 35.47 ± 2.68 (S) | 0.12 ± 0.06 (S) | 81.96 ± 12.26 (S) | 0.21 ± 0.10 (E) |
| | POM | 0.43 ± 0.25 (E) | 11.23 ± 3.62 (E) | 34.69 ± 1.90 (S) | 0.11 ± 0.05 (S) | 47.66 ± 4.13 (G) | 0.27 ± 0.14 (E) |
| Bichka Chaita | PRM | 0.42 ± 0.05 (E) | 15.05 ± 0.76 (E) | 31.38 ± 5.34 (S) | 0.17 ± 0.01 (S) | 102.35 ± 3.64 (S) | 0.24 ± 0.01 (E) |
| | POM | 0.42 ± 0.04 (E) | 13.08 ± 1.55 (E) | 35.52 ± 3.3 (S) | 0.13 ± 0.02 (S) | 56.79 ± 8.13 (G) | 0.26 ± 0.03 (E) |
| Sanopokhari | PRM | 0.2 ± 0.12 (E) | 8.05 ± 3.99 (E) | 30.95 ± 4.00 (S) | 0.08 ± 0.05 (S) | 79.49 ± 12.07 (S) | 0.12 ± 0.07 (E) |
| | POM | 0.36 ± 0.09 (E) | 9.86 ± 2.25 (E) | 23.95 ± 1.56 (S) | 0.10 ± 0.02 (S) | 46.6 ± 2.35 (G) | 0.21 ± 0.05 (E) |
| Budhiya Nakhrod | PRM | - | - | - | - | - | - |
| | POM | 0.53 ± 0.10 (E) | 16.24 ± 2.61 (E) | 32.54 ± 3.12 (S) | 0.17 ± 0.04 (S) | 41.84 ± 4.05 (G) | 0.32 ± 0.05 (E) |
| Ramphal | PRM | - | - | - | - | - | - |
| | POM | 0.22 ± 0.10 (E) | 6.48 ± 2.90 (E) | 32.22 ± 3.08 (S) | 0.06 ± 0.03 (S) | 50.11 ± 7.06 (G) | 0.14 ± 0.06 (E) |

Note: WQ = water quality, E = excellent, G = good, S = suitable, PRM = pre-monsoon, POM = post-monsoon, SD = standard deviation, and M = mean.

Na% levels similarly denoted appropriateness, with values below 20% categorized as 'good' to 'excellent' according to Wilcox standards [39]. Low MAR values in the lakes indicated less alkalinity danger for sensitive crops, supporting irrigation appropriateness. Wilcox diagrams categorized the majority of samples within the C1-S1 and C2-S1 zones, indicating low sodium and moderate salinity risks appropriate for agricultural use [35] (Figures 9 and 10).

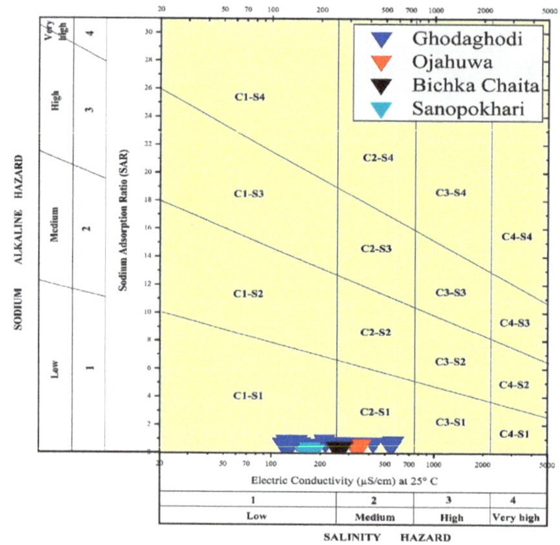


**Figure 9.** Wilcox diagram depicting the irrigation water quality based on SAR and EC for Ghodaghodi Lake and three related lakes (Ojahuwa, Bichka Chaita, and Sanopokhari) during the pre-monsoon period.

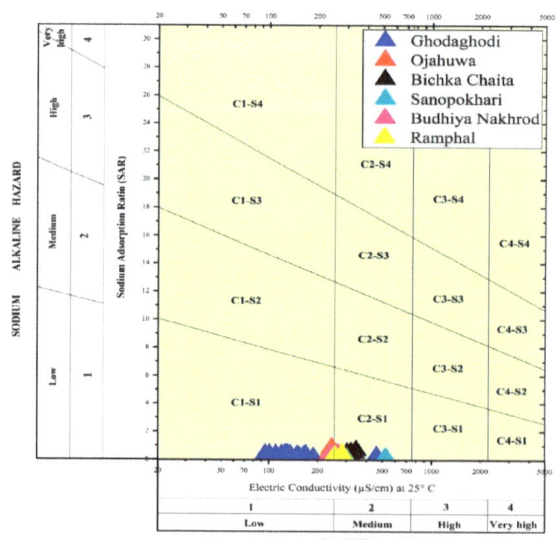


**Figure 10.** Wilcox diagram depicting the irrigation water quality based on SAR and EC for Ghodaghodi Lake and five related lakes (Ojahuwa, Bichka Chaita, Sanopokhari, Budhiya Nakhrod, and Ramphal) during the post-monsoon period.

The PI analysis categorized pre-monsoon water as Class I ('Suitable') and post-monsoon as Class II ('Good'), indicating greater permeability in pre-monsoon conditions due to concentrated ions. These classifications underscored the necessity of ongoing monitoring and nutrient management, especially in lakes such as Sanopokhari, to maintain water quality, aligning with both ecological integrity and agricultural productivity [21].

## 4. Comparison of the Study Findings with the Literature

The hydrochemical dynamics of the GLC offered significant insights concerning the current literature and national water quality standards. The principal sources of pollutant influx at each lake were limited, primarily originating from agricultural runoff in proximity to agriculture, with substantial macrophyte proliferation and organic matter further exacerbating contamination at all locations [8]. Aquaculture operations in specific lakes also contributed to fertilizers, affecting water quality [11].

The pH, EC, TDS, $Na^+$, $Cl^-$, and $Fe^{2+}$ indicators were typically aligned with the National Water Quality Guidelines for Irrigation Water (NWQGIW), signifying the lakes' appropriateness for agricultural utilization [13]. Moreover, critical parameters such as WT, pH, free $CO_2$, DO, TH, $Mg^{2+}$, and $NO_3^-$ predominantly conformed to the National Water Quality Guidelines for Aquaculture (NWQGA), leading to fostering a conducive habitat for aquatic organisms [8]. Nonetheless, tur. levels in the majority of lakes, especially during the pre-monsoon phase, surpassed the thresholds established by the National Water Quality Guidelines for Recreation (NWQGR), indicating that these waters might not consistently be suitable for recreational activities [11].

The study results corresponded with previous studies in multiple aspects. Comparable pH values were noted in Ghodaghodi Lake relative to prior research by Bhatta et al. [11] and Pant et al. [10]. In contrast, the study data indicated a higher post-monsoon pH, diverging from earlier findings that reported more alkaline conditions during the pre-monsoon season [8,11]. EC and TDS values were significantly elevated during the pre-monsoon season, aligning with the evaporative concentration effects reported in other tropical lakes [11]. TH levels exhibited seasonal variation, with the study results showing elevated post-monsoon values, a trend that diverged from some historical accounts while corroborating others, indicative of site-specific hydrological and anthropogenic factors [8]. Table 7 illustrated these comparisons, indicating consistencies in critical parameters, such as $Cl^-$ levels, with the current study [10]. The existing study emphasized the need for ongoing monitoring and efficient management to mitigate the tur. and nutrient discharge, which can represent essential elements for preserving ecological balance and sustainability.

**Table 7.** Comparative Analysis of Hydrochemical Findings of the Ghodaghodi Lake Complex with Previous Studies.

| Study Area | Major Findings | Comparison to Our Study | Recommendations | Reference |
|---|---|---|---|---|
| Ghodaghodi Lake | Water was slightly alkaline, low in ionic strength, and primarily influenced by rock weathering and anthropogenic activities. | Similar TH, $Cl^-$ values. | Nitrate concentrations required careful monitoring due to rising human activities. | [10] |
| Ghodaghodi Lake | Water was suitable for irrigation and aquaculture. | Similar pH values were observed in Ghodaghodi Lake. | Continuous monitoring was required to ensure the sustainability of using the lake's water for irrigation and aquaculture. | [11] |
| Ghodaghodi Lake and Rara Lake | Both lakes were dominated by carbonate weathering with higher $Na^+$ and $Cl^-$ post-monsoon, suggesting marine transport. | Matches in seasonal variation in Tur of Ghodaghodi Lake water. | There is a need for continuous assessment of hydrochemical dynamics to ensure the sustainability of these Ramsar wetlands. | [49] |

## 5. Impact of the Study Finding on Environmentally Sustainable Development Goals (SDGs)

The results of this study on the water quality dynamics and hydrochemical properties of the GLC had significant consequences for environmental SDGs 6 (Clean Water and Sanitation) and 15 (Life on Land), as illustrated in Figure 11. The study indicated seasonal variations in water quality over this Ramsar-listed site, showing how factors including agricultural runoff, evapo-concentration, and natural water cycles can affect key hydrochemical parameters, including TDS, EC, and ion concentrations, notably $Ca^{2+}$ and $Mg^{2+}$. These differences directly supported SDG 6, which seeks to ensure the availability and sustainable management of water and sanitation, influencing local irrigation, water consumption, and community health [56]. The study's finding that the lake water's SAR stayed low and, hence, appropriate for irrigation was crucial for the livelihoods of the surrounding inhabitants, who depended on GLC's resources for agricultural productivity and food security.

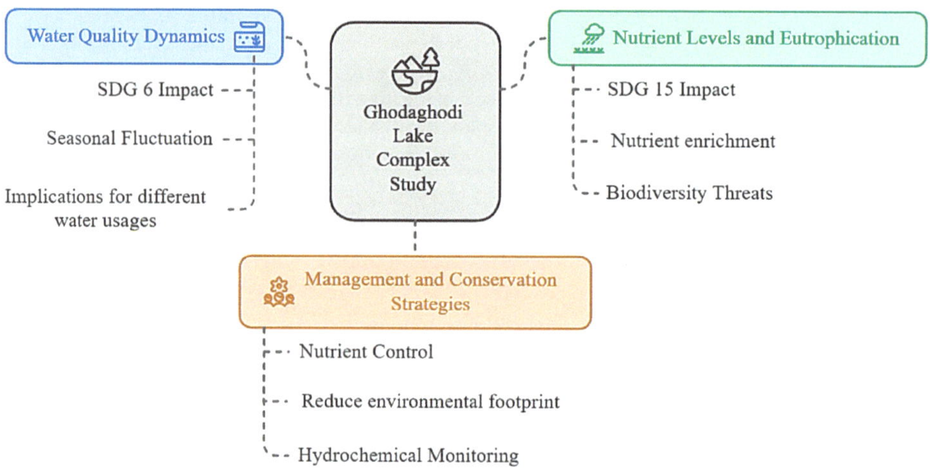


**Figure 11.** Hydrochemical dynamics, sustainable development goals (SDGs) impact, and conservation strategies for Ghodaghodi Lake Complex (GLC).

Beyond water quality, the high nutrient levels in GLC, especially $NO_3^-$ and $PO_4^{3-}$ from agricultural runoff, raised concerns over the likelihood of eutrophication, a process that may considerably diminish biodiversity by stimulating algal blooms and depleting oxygen in aquatic habitats [57]. Substantial runoff provided many nutrients and suspended materials; hence, this nutrient enrichment was especially evident in areas of the lake close to agricultural areas. These environmental pressures compromised the diverse flora and fauna of the lake, thereby connecting the significance of the research with SDG 15, which strived at the preservation, rehabilitation, and sustainable use of land and freshwater ecosystems [3]. Particularly, SDG Target 15.1 emphasized the preservation of inland freshwater ecosystems since it was necessary to retain biodiversity, ecosystem health, and the critical functions that wetlands and the GLC offer intact using the protection of these resources.

The study's results underlined the need for targeted management activities, including control of nitrogen input from surrounding agricultural land and implementation of conservation measures to reduce the environmental impact on the lake system. As this work revealed, continuous hydrochemical monitoring can produce valuable data to guide these management plans and ensure sustainable use of lake resources. By advocating science-based policies and practices, this study helped to encourage the development of a thorough conservation plan for GLC, thereby harmonizing ecological sustainability with the socioeconomic needs of the local population. Employing an integrated approach that guaranteed GLC would remain a vital ecological and community resource suited for SDGs 6 and 15, and such initiatives helped to achieve more SDGs.

## 6. Limitations and Future Research Directions

This study of the hydrochemical dynamics and water quality of the GLC has several limitations. Data collection was confined to coastal areas due to accessibility challenges, including marshy and unstable soil conditions, and safety concerns arising from a crocodile breeding facility. Thus, the representativeness of our findings may be constrained, as central lake regions may exhibit distinct hydrochemical properties. The study's temporal focus was limited to pre-monsoon and post-monsoon seasons, potentially overlooking hydrochemical alterations during other periods, such as the dry season or extreme weather events. The paucity of trace metal and emerging pollutant analysis, coupled with insufficient biological monitoring, constrains our comprehension of the overall ecological health and potential pollution hazards impacting the lake complex.

Future studies should rectify these constraints by broadening the spatial scope to encompass central lake regions and employing modern technology, such as drones and remote sensing, to surmount physical obstacles. Gathering data throughout all seasons would yield a more thorough comprehension of hydrochemical variability. Examining trace metals and developing contaminants in conjunction with biological evaluations of aquatic organisms would provide an enhanced understanding of the lake's ecological integrity. Furthermore, simulating hydrochemical variations under diverse climatic and land-use scenarios may inform future management methods. Involving local populations in participatory research would facilitate translating scientific results into effective, sustainable conservation initiatives.

While sediment resuspension appeared minimal, we acknowledge that internal recharge and nutrient release from sediments could still influence hydrochemical parameters. Future studies should investigate these processes more comprehensively, incorporating direct measurements of sediment-water interactions and internal nutrient cycling to understand their impact on water quality better.

## 7. Conclusions

The hydrochemical evaluation of the GLC in this work demonstrated notable seasonal fluctuations in water quality driven by both human activity and natural processes. TDS averaged 143.1 mg/L in the pre-monsoon period, compared with a post-monsoon drop to 78.9 mg/L, stressing the significant influence of evaporation and consequent dilution. Driven mainly through carbonate weathering processes, the GLC's hydrochemistry was defined by a $Ca^{2+}$-$HCO_3^-$ water type. Leading anions were $HCO_3^-$; principal cations were $Mg^{2+}$ (4.13 mg/L to 7.94 mg/L) and $Ca^{2+}$ (range from 17.07 mg/L in pre-monsoon to 26.62 mg/L in Ghodaghodi Lake). Piper diagram studies validated these profiles, therefore highlighting the inherent lithological contributions to the water composition of the lake. DO levels also showed seasonal fluctuations, averaging 5.12 mg/L pre-monsoon and somewhat raised post-monsoon, presumably under the influence of decreased organic breakdown with rainfall. Statistical studies revealed significant seasonal variations across pH, TDS, and DO levels, emphasizing the lake's dynamic response to climatic events. SAR across all sites was consistently low, below 1, and confirmed great irrigation appropriateness with zero salinity hazards. Although the irrigation quality was good, fertilizer was loaded from nearby farms. Moreover, $NO_3^-$ levels up to 5.23 mg/L pre-monsoon posed a possible eutrophication danger, compromising water quality and biodiversity. Future studies should track nutrient influx throughout time, evaluate seasonal agricultural runoff effects, and combine biological evaluations to offer a clear ecological picture. These actions are essential for guiding SDG development plans, ensuring GLC stays a strong environmental, ecological, and community resource. This study was limited by access constraints to central lake areas and the exclusion of certain contaminants. This suggests that future research should incorporate broader spatial coverage and a more comprehensive analysis of pollutants.

**Author Contributions:** The authors confirm their contribution to the paper: G.P. and R.R.P.: Data collection, experiments, analysis and interpretation of results, draft manuscript preparation. P.N.K., T.R.J., A.M.S., B.Đ., V.C. and K.B.: Writing—review and co-wrote the paper. All authors have read and agreed to the published version of the manuscript.

**Funding:** The research was partially funded by University North, Croatia, Scientific Project UNIN—TEH-24-1-16. The role of geodesy and geomatics in the development of smart spaces (2024).

**Data Availability Statement:** The datasets generated and analyzed during the current study are available from the authors upon reasonable request.

**Acknowledgments:** The first author expresses deep gratitude to the Ministry of Industry, Tourism, Forest and Environment, Sudurpaschim Province, Nepal, for their invaluable support. Special thanks to the Department of Forests and Soil Conservation, Ministry of Forests and Environment, Government of Nepal, for their field support. Additionally, heartfelt appreciation goes to the Central Department of Environmental Science (CDES) for the opportunity to conduct this research.

**Conflicts of Interest:** The authors declare no conflicts of interest.

## References

1. Mitsch, W.J.; Gosselink, J.G. *Wetlands*, 5th ed.; John Wiley Sons, Inc.: Hoboken, NJ, USA, 2015.
2. George, S.J.; Harper, R.J.; Hobbs, R.J.; Tibbett, M.A. Sustainable agricultural landscape for Australia: A review of interlacing carbon sequestration, biodiversity and salinity management in agroforestry systems. *Agric. Ecosyst. Environ.* **2012**, *163*, 28–36. [CrossRef]
3. Saqr, A.M.; Nasr, M.; Fujii, M.; Yoshimura, C.; Ibrahim, M.G. Monitoring of Agricultural Expansion Using Hybrid Classification Method in Southwestern Fringes of Wadi El-Natrun, Egypt: An Appraisal for Sustainable Development. In *Environment and Sustainable Development*; Ujikawa, K., Ishiwatari, M., Hullebusch, E.V., Eds.; ACESD 2022; Environmental Science and Engineering; Springer: Singapore, 2023; pp. 349–362. [CrossRef]
4. Zheng, H.; Cao, S. Threats to China's Biodiversity by Contradictions Policy. *Ambio* **2015**, *44*, 23–33. [CrossRef] [PubMed]
5. Xu, T.; Weng, B.; Yan, D.; Wang, K.; Li, X.; Bi, W.; Li, M.; Cheng, X.; Liu, Y. Wetlands of international importance: Status, threats, and future protection. *Int. J. Environ. Res. Public Health* **2019**, *16*, 1818. [CrossRef]
6. Cantonati, M.; Poikane, S.; Pringle, C.M.; Stevens, L.E.; Turak, E.; Heino, J.; Richardson, J.S.; Bolpagni, R.; Borrini, A.; Cid, N.; et al. Characteristics, main impacts, and stewardship of natural and artificial freshwater environments: Consequences for biodiversity conservation. *Water* **2020**, *12*, 260. [CrossRef]
7. IUCN. *A Review of the Status and Threats to Wetlands in Nepal*; IUCN: Gland, Switzerland, 2004; p. 80.
8. Pant, R.R.; Pal, K.B.; Bishwakarma, K.; Thapa, L.B.; Dangol, A.; Dawadi, B.; Poudel, P.; Bhattarai, B.; Joshi, T.R.; Bhatt, Y.R. Application of Multivariate Approaches to the Hydro-chemical Assessment of the Ghodaghodi Lake, Sudurpaschim Province, Nepal. *Nepal J. Sci. Technol.* **2021**, *19*, 46–54. [CrossRef]
9. Shrestha, B.; Shrestha, S.; Shrestha, A.; Khadka, U.R. Ramsar sites in Nepal: Conservation, present scenario, biodiversity value and threats. *J. Wetl. Ecol.* **2020**, *2020*, 15. [CrossRef]
10. Pant, R.R.; Pal, K.B.; Pathak, G.; Bishwakarma, K.; Thapa, L.B.; Dhital, Y.P.; Bhatta, Y.R.; Joshi, G.R.; Rijal, K. Water quality and health risk assessment of trace elements contamination in Ghodaghodi Lake, Sudurpaschim Province, Nepal. *Nepal J. Environ. Sci.* **2021**, *9*, 29–40. [CrossRef]
11. Bhatta, R.; Tuladhar, S.; Gurung, S.; Raut, N. Water quality of Ghodaghodi Lake: A Ramsar site in western Nepal. In Proceedings of the International Conference on Natural Resources, Agriculture and Society in Changing Climate, Kathmandu, Nepal, 17–19 August 2019; pp. 73–84.
12. Krauze, K.; Wagner, I. An ecohydrological approach for the protection and enhancement of ecosystem services. In *Use of Landscape Sciences for the Assessment of Environmental Security*; Springer: Berlin/Heidelberg, Germany, 2007; pp. 177–207. [CrossRef]
13. Lamsal, P.; Atreya, K.; Ghosh, M.K.; Pant, K.P. Effects of population, land cover change, and climatic variability on wetland resource degradation in a Ramsar listed Ghodaghodi Lake Complex, Nepal. *Environ. Monit. Assess.* **2019**, *191*, 415. [CrossRef]
14. Alao, J.O.; Ayejoto, D.A.; Fahad, A.; Mohammed, M.A.A.; Saqr, A.M.; Joy, A.O. Environmental Burden of Waste Generation and Management in Nigeria. In *Technical Landfills and Waste Management*; Souabi, S., Anouzla, A., Eds.; Springer Water; Springer: Cham, Switzerland, 2024; pp. 27–56. [CrossRef]
15. Noman, M.; Ullah, I.; Khan, M.A.; Qazi, A.; Farooq, W.; Saqr, A.; Elsheikh, A. Analysis of overcurrent protective relaying as minimum adopted fault protection for small-scale hydropower plants. *Int. J. Environ. Sci. Technol.* **2024**, *21*, 4457–4470. [CrossRef]
16. Lawal, A.; Tijani, M.N.; Snow, D.; D'Alessio, M. Quality and hydrochemical assessment of groundwater in geological transition zones: A case study from N.E. Nigeria. *Environ. Sci. Pollut. Res.* **2023**, *30*, 10643–10663. [CrossRef]
17. Boehrer, B.; Schultze, M. Stratification of lakes. *Rev. Geophys.* **2008**, *46*. [CrossRef]

18. Boualem, B.; Egbueri, J.C. Graphical, statistical and index-based techniques integrated for identifying the hydrochemical fingerprints and groundwater quality of In Salah, Algerian Sahara. *Environ. Geochem. Health* **2024**, *46*, 158. [CrossRef] [PubMed]
19. Touhari, F.; Meddi, M.; Mehaiguene, M.; Razack, M. Hydrogeochemical assessment of the Upper Cheliff groundwater (North West Algeria). *Environ. Earth Sci.* **2015**, *73*, 3043–3061. [CrossRef]
20. Gibbs, R.J. Mechanisms Controlling World Water Chemistry. *Science* **1970**, *170*, 1088–1090. [CrossRef] [PubMed]
21. Pant, R.R.; Bishwakarma, K.; Pal, K.B.; Thapa, L.B.; Shrestha, R.G.; Karuppannan, S.; Garu, L.; Bista, S.; Singh, V.B. Comparative analysis of hydrochemical variables of two Ramsar-listed lakes in Pokhara Valley, Nepal. *Int. J. Energy Water Resour.* **2023**, *7*, 1–13. [CrossRef]
22. Lamsal, P.; Pant, K.P.; Kumar, L.; Atreya, K. Diversity, Uses, and Threats in the Ghodaghodi Lake Complex, a Ramsar Site in Western Lowland Nepal. *ISRN Biodivers.* **2014**, *2014*, 680102. [CrossRef]
23. Bhandari, R.; Miya, M.S.; Timilsina, S. Status and Diversity of Terrestrial plants in Ghodaghodi Lake Complex, Nepal. *Int. J. Appl. Sci. Biotechnol.* **2021**, *9*, 38–44. [CrossRef]
24. Siwakoti, M.; Karki, J.B. Conservation status of Ramsar sites of Nepal Tarai: An overview. *Bot. Orient. J. Plant Sci.* **2010**, *6*, 76–84. [CrossRef]
25. Lamichhane, S.; Bhattarai, D.; Karki, J.B.; Gautam, A.P.; Pandeya, P.; Tirpathi, S.; Mahat, N. Population status, habitat occupancy and conservation threats to *Mugger crocodile* (*Crocodylus palustris*) in Ghodaghodi lake complex, Nepal. *Glob. Ecol. Conserv.* **2021**, *33*, e01977. [CrossRef]
26. APHA. *Standard Methods for the Examination of Water and Wastewater*; American Public Health Association: Washington, WA, USA, 2005.
27. Zidan, A.; Abdalla, M.; Khalaf, S.; Saqr, A. Kinetic Energy and Momentum Coefficients for Egyptian Irrigation Canals. *Bull. Fac. Eng. Mansoura Univ.* **2020**, *41*, 1–16. [CrossRef]
28. Saqr, A.M.; Nasr, M.; Fujii, M.; Yoshimura, C.; Ibrahim, M.G. Delineating suitable zones for solar-based groundwater exploitation using multi-criteria analysis: A techno-economic assessment for meeting sustainable development goals (SDGs). *Groundw. Sustain. Dev.* **2024**, *25*, 101087. [CrossRef]
29. Muijs, D. *Doing Quantitative Research in Education with IBM SPSS Statistics*; Sage Publications: Thousand Oaks, CA, USA, 2022; pp. 1–100.
30. Piper, A.M. A graphic procedure in the geochemical interpretation of water-analyses. *Eos. Trans. Am. Geophys. Union.* **1944**, *25*, 914–928.
31. Chalaune, T.B.; Dangol, A.; Sharma, J.; Sharma, C.M. First results on physico-chemical status and bathymetry of lakes in Ramaroshan Wetland, Far-West Nepal. *Nepal J. Environ. Sci.* **2020**, *8*, 17–27. [CrossRef]
32. Khadka, U.R.; Ramanathan, A.L. Major ion composition and seasonal variation in the Lesser Himalayan lake: Case of Begnas Lake of the Pokhara Valley, Nepal. *Arab. J. Geosci.* **2013**, *6*, 4191–4206. [CrossRef]
33. Prasanna, M.V.; Chidambaram, S.; Gireesh, T.V.; Ali, T.V.J. A study on hydrochemical characteristics of surface and sub-surface water in and around Perumal Lake, Cuddalore district, Tamil Nadu, South India. *Environ. Earth Sci.* **2011**, *63*, 31–47. [CrossRef]
34. Pant, R.R.; Zhang, F.; Rehman, F.U.; Wang, G.; Ye, M.; Zeng, C.; Tang, H. Spatiotemporal variations of hydrogeochemistry and its controlling factors in the Gandaki River Basin, Central Himalaya Nepal. *Sci. Total Environ.* **2018**, *622–623*, 770–782. [CrossRef]
35. Acharya, A.; Sharma, M.L.; Bishwakarma, K.; Dahal, P.; Chaudhari, S.K.; Adhikari, B.; Neupane, S.; Pokhrel, B.N.; Pant, R.R. Chemical Characteristics of the Karmanasha River Water and Its Appropriateness for Irrigational Usage. *J. Nepal Chem. Soc.* **2020**, *41*, 94–102. [CrossRef]
36. Zhang, D.; Lu, D.; Yang, B.; Zhang, J.; Ning, Z.; Yu, K. Influence of natural and anthropogenic factors on spatial-temporal hydrochemistry and the susceptibility to nutrient enrichment in a subtropical estuary. *Mar. Pollut. Bull.* **2019**, *146*, 945–954. [CrossRef]
37. Imran, U.; Ullah, A.; Mahar, R.B.; Shaikh, K.; Khokhar, W.A.; Weidhaas, J. An integrated approach for evaluating freshwater ecosystems under the influence of high salinity: A case study of Manchar Lake in Pakistan. *Environ. Monit. Assess.* **2023**, *195*, 1340. [CrossRef]
38. Zaman, M.; Shahid, S.A.; Heng, L. Irrigation Water Quality. In *Guideline for Salinity Assessment, Mitigation and Adaptation Using Nuclear and Related Techniques*; Springer: Berlin/Heidelberg, Germany, 2018. [CrossRef]
39. Wilcox, L.V. *Classification and Use of Irrigation Waters*; US Department of Agriculture: Washington, WA, USA, 1955.
40. Richards, L.A. Diagnosis and Improvement of Saline and Alkali Soils. *Agron. J.* **1954**, *46*, 290. [CrossRef]
41. Kaddah, M.T. Irrigation with Saline Water. *Soil Sci. Soc. Am. J.* **1975**, *39*, 1973. [CrossRef]
42. Kelley, W.P. Use of saline irrigation water. *Soil Sci.* **1963**, *95*, 385–391. [CrossRef]
43. Rengasamy, P.; Marchuk, A. Cation ratio of soil structural stability (CROSS). *Soil Res.* **2011**, *49*, 280–285. [CrossRef]
44. Oyem, H.H.; Oyem, I.M.; Ezeweali, D. Temperature, pH, Electrical Conductivity, Total Dissolved Solids and Chemical Oxygen Demand of Groundwater in Boji-BojiAgbor/Owa Area and Immediate Suburbs. *Res. J. Environ. Sci.* **2014**, *8*, 444–450. [CrossRef]
45. Liu, H.; Yuan, H.; Wang, S.; Zheng, L.; Liao, M. Spatiotemporal dynamics of water body changes and their influencing factors in the seasonal lakes of the poyang lake region. *Water* **2021**, *13*, 1539. [CrossRef]
46. Bhatta, R.; Gurung, S.; Joshi, R.; Tuladhar, S.; Regmi, D.; Kafle, B.K.; Dahal, B.M.; Raut, N.; Kafle, K.R.; Kayastha, R.; et al. Spatio-Temporal Hydrochemistry of Two Selected Ramsar Sites (Rara and Ghodaghodi) of West Nepal. *SSRN Electron. J.* **2022**, *8*, e11243. [CrossRef]

47. Gurung, S.; Gurung, A.; Sharma, C.M.; Jüttner, I.; Tripathee, L.; Bajracharya, R.M.; Raut, N.; Pradhananga, P.; Sitaula, B.K.; Zhang, Y.; et al. Hydrochemistry of Lake Rara: A high mountain lake in western Nepal. *Lakes Reserv. Sci. Policy Manag. Sustain. Use* **2018**, *23*, 87–97. [CrossRef]
48. Brett, M.T.; Bunn, S.E.; Chandra, S.; Galloway, A.W.E.; Guo, F.; Kainz, M.J.; Kankaala, P.; Lau, D.C.P.; Moulton, T.P.; Power, M.E.; et al. How important are terrestrial organic carbon inputs for secondary production in freshwater ecosystems? *Freshw. Biol.* **2017**, *62*, 833–853. [CrossRef]
49. Yan, J.; Chen, J.; Zhang, W. Study on the groundwater quality and its influencing factor in Songyuan City, Northeast China, using integrated hydrogeochemical method. *Sci. Total Environ.* **2021**, *773*, 144958. [CrossRef]
50. Akhtar, S.; Equeenuddin, S.M. Spatio-temporal variation of major ion chemistry and nutrient stoichiometry in a tropical monsoonal estuary: Insight into biogeochemical processes. *Environ. Earth Sci.* **2023**, *82*, 48. [CrossRef]
51. Cheng, G.; Li, X.; Zhao, W.; Xu, Z.; Feng, Q.; Xiao, S.; Xiao, H. Integrated study of the water-ecosystem-economy in the Heihe River Basin. *Natl. Sci. Rev.* **2014**, *1*, 413–428. [CrossRef]
52. Hoess, R.; Geist, J. Effect of fish pond drainage on turbidity, suspended solids, fine sediment deposition and nutrient concentration in receiving pearl mussel streams. *Environ. Pollut.* **2021**, *274*, 116520. [CrossRef] [PubMed]
53. Lawniczak, A.E.; Zbierska, J.; Nowak, B.; Achtenberg, K.; Grześkowiak, A.; Kanas, K. Impact of agriculture and land use on nitrate contamination in groundwater and running waters in central-west Poland. *Environ. Monit. Assess.* **2016**, *188*, 1727. [CrossRef] [PubMed]
54. Tripathee, L.; Kang, S.; Sharma, C.M.; Rupakheti, D.; Paudyal, R.; Huang, J.; Sillanpää, M. Preliminary Health Risk Assessment of Potentially Toxic Metals in Surface Water of the Himalayan Rivers, Nepal. *Bull. Environ. Contam. Toxicol.* **2016**, *97*, 855–862. [CrossRef]
55. Das, A.K.; Shrivastava, N.P. Ecology of Sarni reservoir (M.P.) in the context of fisheries. *Pollut. Res.* **2003**, *22*, 533–539.
56. Saqr, A.M.; Nasr, M.; Fujii, M.; Yoshimura, C.; Ibrahim, M.G. Optimal Solution for Increasing Groundwater Pumping by Integrating MODFLOW-USG and Particle Swarm Optimization Algorithm: A Case Study of Wadi El-Natrun, Egypt. In *Proceedings of the 2022 12th International Conference on Environment Science and Engineering (ICESE 2022)*; Environmental Science and Engineering; Springer: Singapore, 2023; pp. 59–73. [CrossRef]
57. Glibert, P.M. Eutrophication, harmful algae and biodiversity—Challenging paradigms in a world of complex nutrient changes. *Mar. Pollut. Bull.* **2017**, *124*, 591–606. [CrossRef]

water

MDPI

Article

# Health Risk Assessment of Trace Elements in Surface Water from Dayat Roumi Lake, Morocco

**Ihsane Ougrad [1,*], Zahra Elassassi [2], Abdessamad Mrabet [3], Ibrahim Mssillou [4,*], Adrian Lim [5], Abdelaaty Abdelaziz Shahat [6], Sanae Rezouki [1] and Tarik Moubchir [7]**

1 Laboratory of Plant, Animal and Agro-Industry Productions, Department of Biology, Faculty of Sciences, Ibn Tofail University, Kenitra 14000, Morocco; rezoukisanae@gmail.com
2 Organic Chemistry, Catalysis and Environment, Department of Chemistry, Faculty of Sciences, Ibn Tofail University, Kenitra 14000, Morocco; zahra.elassassi@gmail.com
3 Laboratory of Biology and Health, Unit of Neurosciences and Nutrition, Department of Biology, Faculty of Sciences, Ibn Tofail University, B.P. 133, Kenitra 14000, Morocco; abdessamad_mrabet@yahoo.fr
4 Laboratory of Natural Substances, Pharmacology, Environment, Modeling, Health & Quality of Life (SNAMOPEQ), Faculty of Sciences Dhar El Mahraz, Sidi Mohamed Ben Abdellah University, Fez 30000, Morocco
5 Department of Medicine, Cedars-Sinai Medical Center, Los Angeles, CA 90048, USA; adrian.lim@cshs.org
6 Pharmacognosy Department, College of Pharmacy, King Saud University, Riyadh 11451, Saudi Arabia; ashahat@ksu.edu.sa
7 Polyvalent Team in Research and Development, Polydisciplinary Faculty, Sultan Moulay Slimane University, Beni Mellal 23000, Morocco; tarik.moubchir@usms.ac.ma
* Correspondence: oihsane@hotmail.com (I.O.); mssillouibrahim@gmail.com (I.M.)

**Abstract:** To assess the human impact on the water of Dayat Roumi Lake and to develop effective management strategies to protect and restore this vital ecosystem in the region, seasonal sampling was carried out at six stations distributed around the lake. During these sampling campaigns, 24 parameters were measured, including 20 trace elements. Results showed that measured levels of trace elements increased in the following order: Cd < Be < Tl < Co < Sb < Mo < Cu < Zn < Ni < V < Rb < Mn < As < Cr < Pb < Li < Ba < Se < Pd < Sr in the lake water and that these recorded values were lower than those recommended by the Moroccan standard and the World Health Organization, except for Pb and Se. Correlation analysis revealed two principal water-contamination sources: natural geological origins and anthropogenic inputs. In addition, the Water Quality Index WQI showed that the lake's water quality is poor, and its use can be dangerous for human and animal health. Health risk assessment associated with prolonged exposure to trace elements in lake water revealed that the Hazard quotient HQ and Hazard index HI of certain elements, such as Tl, Sb, V, As, Cr, Pb, Li, and Se, are higher than 1 in adult and children, indicating a significant risk for people living near the lake. Children are particularly vulnerable, with higher levels of HQ and HI, and selenium poses a substantial risk to their health through ingestion and skin absorption. In both adults and children, the total risk of cancer due to metals is classified as follows: CI (Cr) > CI (Ni) > CI (As) > CI (Pb) > CI (Cd). The Cr presents the highest carcinogenic risk—by ingestion or dermal route—in both groups. The total risk for these five metals exceeds $1 \times 10^{-4}$, indicating a danger for residents who drink or swim in the lake.

**Keywords:** water quality index; trace elements; health risk indexes; Dayat Roumi Lake; contamination

## 1. Introduction

Trace elements are chemical elements naturally present in the environment in low concentrations (air, water, soil, and the tissues of living organisms) [1–3]. They are distributed throughout the aquatic ecosystem and may exist as complexes, particles, or solutions. Their mobility and transfer in the different environmental compartments depend on the chemical properties of the metal, especially their volatility, their solubility in water, and the metal partition coefficient between water and particles. These factors determine their fate, transport, and interactions with other ecosystem compartments [4,5]. Metals naturally present in the soil pose few problems in trace amounts. However, their accumulation in particular areas (such as wetlands) pose a risk to the environment and its organisms [1].

Trace elements are considered as serious contaminants in aquatic ecosystems due to the fact that they are persistent and have a tendency to bioaccumulate in fish tissue and other aquatic organisms [6]. Some metals, such as Cu, Zn, Ni, Co, and Mn, are oligo-elements essential to the development and growth of living organisms and are involved in numerous biochemical processes at low concentrations [7]. However, they can become toxic depending on the nature, content, mobility, and bioavailability of the element and the medium in question [8]. Certain chemical elements like lead have no known physiological role in aquatic species and are thought to have only harmful effects [9]. At low concentrations, many elements, including Hg, Cd, Pb, As, and Cu, inhibit photosynthesis and the growth of phytoplankton, leading to delayed embryonic development, malformations, and poorer adult growth in fish, mollusks, and crustaceans [10]. Aquatic ecosystems and their living beings are subject to numerous stress factors that trigger a range of adaptive responses (genetic, metabolic, physiological, and behavioral) and enable the organism to overcome the threat and maintain its integrity. Hepatic steatosis associated with a reduction in liver glycogen reserves has been demonstrated in bony fish subjected to heavy metal intoxication. This observation was interpreted as a reaction to the stress caused by the intoxication [11]. Similar results were obtained by exposing eels to cadmium concentrations over a long period [12]. In addition, studies focusing on the effects of heavy metals on bivalves have revealed structural alterations to the gills [13].

Similar effects in freshwater organisms can occur in humans following the use of polluted lake water and products [14]. Arsenic has severe effects on the skin (hyperpigmentation), respiratory system, cardiovascular system (arrhythmia, pericarditis), and cancers in humans [15]. Hexavalent chromium, ingested or inhaled, can also cause nasal and pulmonary lesions, inflammation of the digestive system, and cancer [16]. Exposure to certain elements, such as lead, can affect various organs, including the nervous system, cardiovascular system, liver, kidneys, and blood. These effects vary according to the duration and dose of exposure and the person affected [17].

Faced with this problem and in the absence of studies on the distribution of trace elements in Dayat Roumi Lake water samples, we carried out this research to assess the quality of the lake water and to supplement and add new elements to previous work. Our objectives are to study trace element distribution in lake water, establish relationships between the concentrations of the various trace elements measured to determine their natural and/or anthropogenic sources in lake water, and finally, to determine the human health hazards attributed to the consumption of these chemical substances through ingestion and dermal exposure.

## 2. Materials and Methods

### 2.1. Study Area

Located in the Rabat-Salé-Kénitra region in northwest Morocco (Figure 1), Dayat Roumi Lake occupies a surface area of 96 hectares and attains a depth of 11 m in the southeastern region of the lake. Rainfall, tributaries, and groundwater sources all contribute to its supply. The area climate is temperate thermo-Mediterranean, with semi-arid conditions, mild winters, and hot, dry summers. Annual rainfall varies between 500 and 600 mm [18], and ambient air temperature fluctuates between 7 and 38 °C. Dayat Roumi Lake is classified as a Site of Biological and Ecological Interest (SBEI) [19]. It harbours a wide variety of aquatic plants, crustaceans, heteropterans, amphibians, and reptiles. Its fish reserves and avifauna have declined in recent years compared with other Moroccan aquatic ecosystems. This dysfunction results from natural pressures such as erosion, alternate drought periods, and human pressures [20]. The lake is frequented for recreational activities (fishing, kayaking, ski boats, and walks along the banks) and pastoral activities, and it is undergoing urban expansion. Previous studies have determined that leaching from rocks and agricultural land in the catchment area, runoff water, domestic wastewater, household waste, and excrements from humans and animals are the main causes of declining water quality [21,22].

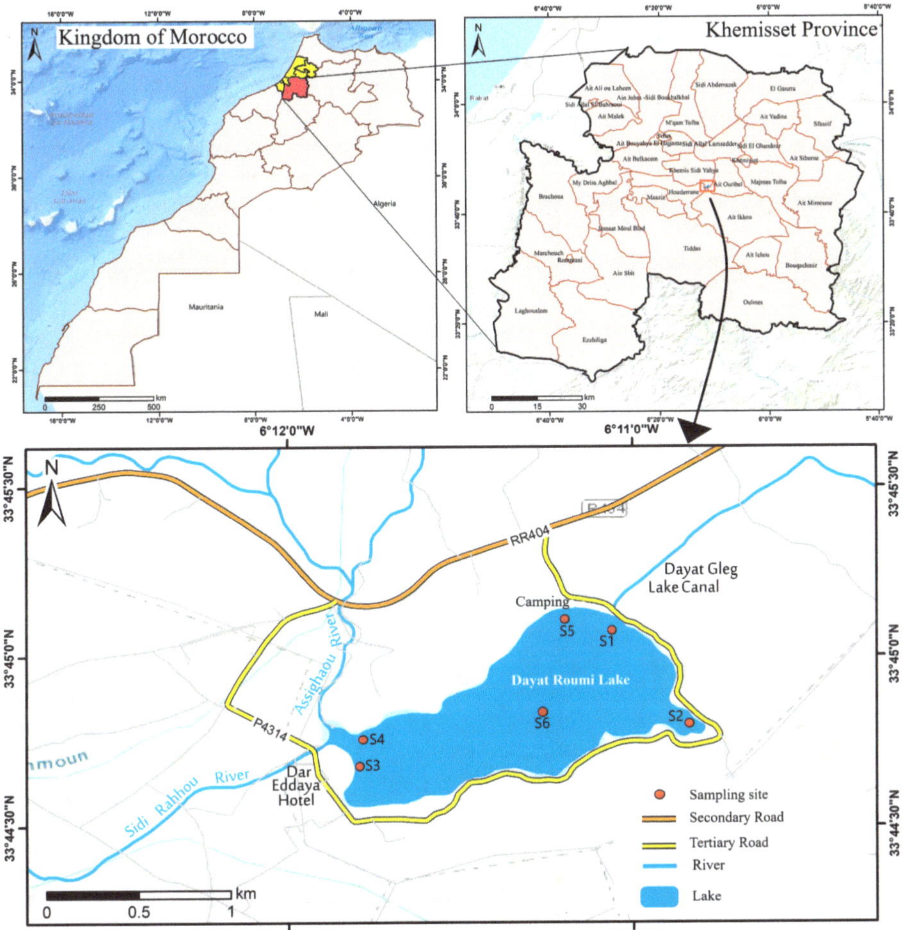


**Figure 1.** Location map of Dayat Roumi Lake and sampling stations.

## *2.2. Sampling*

Surface water samples were collected from six stations. They were chosen to take into account both the receiving points for tributary water and runoff and the practices of residents and visitors likely to have a negative impact on the study area: S1 (33°45′06.4″ N 06°11′06.3″ W): a bathing area that receives a stream from a marsh (Dayat Glag) located 1–2 km from the lake; S2 (33°44′49.5″ N 06°10′52.6″ W): a station located next to a meadow and drained by a stream that runs through the southeastern part of the lake area; S3 (33°44′38. 0″ N 06°11′48.6″ W): a station located next to the Dar Eddaya hotel, which covers an area of 4800 $m^2$ and has a total capacity of 100 beds; S4 (33°44′45.2″ N 06°11′46.6″ W): an area crossed by the Sidi Rahhou River, a tributary of the Assighaou river; S5 (33°45′07. 5″ N 06°11′13.6″ W): a camping area with recreational water activities; and S6 (33°44′48.9″ N 06°11′18.6″ W): a lake center. Sampling was done seasonally, in summer (Su) and autumn (A) (2017) and in winter (W) and spring (Sp) (2018). Samples were collected at a depth of 0.3 to 0.5 m. Water samples were stored in 50 mL high-density polyethylene (HDPE) bottles, pre-cleaned with ultra-pure $HNO_3$, washed with deionized water in the laboratory, then stored at 4 °C and transported immediately to the laboratory. Temperature (T), electrical conductivity (EC), pH, and dissolved oxygen (DO) were measured in situ at each sampling point using a HACH model HQ40d multiparameter instrument, (Hach Company, Loveland, CO, USA).

### 2.3. Dosage Methods

Twenty chemical elements were measured, including Arsenic (As), Copper (Cu), Cadmium (Cd), Lead (Pb), Zinc (Z), Chromium (Cr), Nickel (Ni), Cobalt (Co), Antimony (Sb), Selenium (Se), Lithium (Li), Manganese (Mn), Molybdenum (Mo), Strontium (Sr), Vanadium (V), Rubidium (Rb), Palladium (Pd), Beryllium (Be), Barium (Ba), and Thallium (Tl). From each water sample, 10 mL was digested in acid (0.2 mL ultra-pure $HNO_3$ 65%). The ultra-pure nitric acid was purchased from Merck (Merck Group, Darmstadt, Germany). The measurements were carried out using a Thermo Scientific XSERIES 2 ICP-MS (Thermo Fisher Scientific, Bremen, Germany) at the Centre National de l'Energie des Sciences et des Techniques Nucléaires (CNESTEN). The spectrometer was optimised to provide minimum CeO+/Ce+ and Ba2+/Ba+ ratios and optimum analyte intensity. External calibration was performed using SPEX CertiPrep monoelement solutions. Rh 0.01 mg/L and Ir 0.01 mg/L were used as internal standards.

### 2.4. Water Quality Index

The global quality of the lake's water was assessed using the water quality index (WQI) calculated according to the following Formula (Equation (1)):

$$WQI = \Sigma\,[Pi \times (Ci/Si) \times 100] \tag{1}$$

$$Pi = pi/\Sigma pi \tag{2}$$

where Pi is the relative weight calculated by Equation (2), pi is the weight of each parameter between 1 and 5 depending on the importance of the chemical element in determining water quality [23], Σ pi is the sum of the weights of all parameters, Ci is the average concentration of each trace element measured in the water samples, and Si is the WHO recommended trace element value for drinking water [24]. Only 14 elements were included in the WQI analysis without quality guidelines for drinking water for Sr, Be, Rb, Li, Pd, and V. Based on the calculated WQI values, five classes were established: Class I: excellent (WQI < 50), Class II: good (WQI = 50–100), Class III: poor (WQI = 100–200), Class IV: very poor (WQI = 200–300) and Class V: water unfit for consumption (WQI > 300) [25].

### 2.5. Health Risk Assessment

Trace elements are essential to health. They are involved in many vital functions. However, high concentrations can have harmful effects on health. An assessment of the risks associated with contamination by these elements, present in the lake's water by two main routes (ingestion and skin absorption), was carried out on adults and children living in the lake. To evaluate the toxic risks to human health from trace elements resulting from the use of lake water, we used non-carcinogenic and carcinogenic health risk indices.

#### 2.5.1. Non-Carcinogenic Risk Index

a. Hazard quotient HQ

The hazard quotient is the ratio between the average daily dose of the chemical substance absorbed by ingestion or through the skin (ADD) and the daily dose of the substance likely to have no harmful effects on the consumer's life (RfD). These two parameters are expressed in μg/kg/day. An HQ value less than 1 indicates that the risk level of the substance is acceptable, while a value greater than 1 means that the non-carcinogenic risk level is unacceptable [26]. The hazard quotient is calculated using the following Formula (Equation (3)) [27]. The meanings and values of the above parameters are given in Tables 1 and 2.

$$HQ = ADD/RfD \tag{3}$$

$$HQ_{ing} = \frac{ADD_{ing}}{RfD_o} \tag{4}$$

$$HQ_{der} = \frac{ADD_{der}}{RfD_{der}} \tag{5}$$

$$ADD_{ing} = \frac{CWxIRxEFxED}{BWxAT} \tag{6}$$

$$ADD_{der} = \frac{CWxSAxKpxETxEFxEDxCF}{BWxAT} \tag{7}$$

**Table 1.** Hazard quotient measurement parameters.

| Parameters | Description | Unit | Value | |
|---|---|---|---|---|
| | | | Adult | Children |
| ADD | Average daily dose of the chemical substance taken by ingestion (ADDing) or by skin contact (ADDder) | µg/kg-jour | Site specific | |
| RfD | Reference dose, specific to the chemical element | µg/kg-jour | Site specific | |
| Cw | Chemical average concentration in water | µg/L | Site specific | |
| IR | Ingestion rate | L/day | 2 | 0.64 |
| EF | Exposure Frequency | days/year | 350 | 350 |
| ED | Exposure Duration | Year | 30 | 6 |
| BW | Body Weight | Kg | 70 | 15 |
| AT | Averaging Time<br>For non-carcinogens<br>For carcinogens | Days | ED × 350<br>365 × 70 | ED × 350<br>365 × 70 |
| SA | Exposed skin area | $cm^2$ | 18 | 6.6 |
| ET | Exposure time during bathing and shower | h/day | 0.58 | 1 |
| Kp | Skin Permeability Coefficient in water | cm/hr | Chemical specific | |
| CF | Conversion factor | $L/cm^3$ | $10^{-3}$ | $10^{-3}$ |

**Table 2.** Kp and RfD values for the elements studied.

| Parameters | Kp (cm/h) | $RfD_o$ (µg/kg/day) | | $RfD_{der}$ (µg/kg/day) | |
|---|---|---|---|---|---|
| | | Value | Source | Value | Source |
| Cd | $1.0 \times 10^{-3}$ | 0.5 | [28] | 0.025 | [28] |
| Be | $1.0 \times 10^{-3}$ | 2 | [29] | 1.4 [a] | - |
| Tl | $1.0 \times 10^{-3}$ | 0.01 | [30] | 0.01 | [30] |
| Co | $4.0 \times 10^{-4}$ | 0.3 | [28] | 0.06 | [28] |
| Sb | $1.0 \times 10^{-3}$ | 0.4 | [31,32] | 0.008 | [31,32] |
| Mo | $1.0 \times 10^{-3}$ | 5 | [31,32] | 1.9 | [31,32] |
| Cu | $1.0 \times 10^{-3}$ | 40 | [28] | 8 | [28] |
| Zn | $6.0 \times 10^{-4}$ | 300 | [28] | 60 | [28] |
| Ni | $2.0 \times 10^{-4}$ | 20 | [28] | 0.8 | [28] |
| V | $1.0 \times 10^{-3}$ | 1 | [31,32] | 0.01 | [31,32] |
| Rb | $1.0 \times 10^{-3}$ | n.d | - | n.d | - |
| Mn | $1.0 \times 10^{-3}$ | 24 | [28] | 0.96 | [28] |
| As | $1.0 \times 10^{-3}$ | 0.3 | [28] | 0.123 | [28] |
| Cr | $2.0 \times 10^{-3}$ | 3 | [28] | 0.075 | [28] |
| Pb | $1.0 \times 10^{-3}$ | 1.4 | [28] | 0.42 | [28] |
| Li | $1.0 \times 10^{-3}$ | 2 | [33] | 2 | [33] |

**Table 2.** *Cont.*

| Parameters | Kp (cm/h) | $RfD_o$ (µg/kg/day) | | $RfD_{der}$ (µg/kg/day) | |
|---|---|---|---|---|---|
| | | Value | Source | Value | Source |
| Ba | $1.0 \times 10^{-3}$ | 200 | [28] | 14 | [28] |
| Se | $1.0 \times 10^{-3}$ | 5 | [34] | 2.2 | [34] |
| Pd | $1.0 \times 10^{-3}$ | n.d | - | n.d | - |
| Sr | $1.0 \times 10^{-3}$ | 600 | [34] | 120 | [34] |

Notes: [a] RfD dermal calculated. n.d: not detected.

b. Hazard index HI

Assessing the health risks of a mixture of chemical elements in water requires the application of a hazard index, which is the sum of the non-carcinogenic health risks HQ associated with several substances and/or several contamination routes. An HI value greater than 1 indicates the presence of an unacceptable non-carcinogenic health risk [26]. HI is measured using the following formula (Equations (8)–(10)):

$$HI_{ing} = \Sigma\, HQ_{ing} \tag{8}$$

$$HI_{der} = \Sigma\, HQ_{der} \tag{9}$$

$$HI = HI_{ing} + HI_{der} \tag{10}$$

2.5.2. Carcinogenic Risk Indexes

The carcinogenic health hazards associated with exposure to chemical elements are determined by the cancer risk CR. This indicates the probability that an individual will develop a type of cancer following exposure to carcinogenic substances during their lifetime. It is calculated by multiplying the average daily dose of the chemical substance ADD by the cancer slope factor CSF (Equation (11)). The CSF, expressed in $(mg/kg\text{-}day)^{-1}$, is specific for each element and each route of exposure. The CSF values are shown in Table 3. For this study, the carcinogenic risk is calculated for five elements Cd, Cr, Pb, As, and Ni known to be carcinogenic, and for the two exposure routes (Equations (12) and (13)). The cancer index CI is calculated by summing the carcinogenic risks CR of the five elements and/or the two routes of contamination (Equation (14)). A CR and CI value greater than $10^{-4}$ indicates a high risk of cancer in humans [26].

$$CR = ADD \times CSF \tag{11}$$

$$CR_{ing} = ADD_{ing} \times CSF_O \tag{12}$$

$$CR_{der} = ADD_{der} \times CSF_{ABS} \tag{13}$$

$$CI = \Sigma\, CR \tag{14}$$

**Table 3.** CSF values by ingestion and dermal exposure.

| Parameters | $CSF_o$ | | $CSF_{ABS}$ | |
|---|---|---|---|---|
| | Value $(mg/kg\text{-}jour)^{-1}$ | Source | Value $(mg/kg\text{-}jour)^{-1}$ | Source |
| Cd | $6.30 \times 10$ | [35] | $7.63 \times 10^{1}$ | a |
| Ni | $1.70 \times 10$ | [36] | $4.25 \times 10^{1}$ | [36] |
| As | $1.50 \times 10$ | [35] | $1.50 \times 10$ | [37] |
| Cr | $5.00 \times 10^{-1}$ | [35] | $2.00 \times 10^{1}$ | [36] |
| Pb | $8.50 \times 10^{-3}$ | [35] | $8.50 \times 10^{-2}$ | a |

Notes: a: calculated using the following formula: $CSF_{ABS} = CSF_o/ABS_{gi}$. $ABS_{gi}$: gastrointestinal absorption factor [38]: $ABS_{gi}(Cd) = 0.08$; $ABS_{gi}(Pb) = 0.1$.

### 2.6. Data Statistical Analysis

The data set was submitted to analytical and statistical techniques. All results were processed using Microsoft Office Excel 12.0 and SPSS statistics 23.0. One-way analysis of variance (ANOVA) was used to validate the normality of the data using the Kolmogorov–Smirnov (KS) and Shapiro–Wilk (SW) tests, and the homogeneity of the intragroup variance was verified using Levene's test. Bivariate correlation was calculated using Pearson's correlation coefficient (r).

## 3. Results and Discussion

### 3.1. Spatial and Temporal Distribution of the Parameters Studied

The results of the analyses of the 24 water samples are represented on the graphs in Figure A1 in Appendix A, which shows the spatial and temporal distribution of each variable studied. Table 4 summarizes the average concentrations of trace elements in the waters of Dayat Roumi Lake and in other national and international lakes.

**Table 4.** Average concentration of trace elements (µg/L) in the waters of Lake Dayat Roumi and comparison with values for national and international lakes.

| Trace Elements | Dayat Roumi Lake | Fouarate Lake [39] | Baikal Lake [40] | Ghodaghodi Lake [25] | Phewa Lake [41] | Gosaikunda Lake [41] | East Dongting Lake [42] | Manchar Lake [43] |
|---|---|---|---|---|---|---|---|---|
| Cd | 0.004 | 90.5 | 0.008 | 0.01 | 0.003 | 0.02 | 0.05 | 5.3 |
| Be | 0.05 | - | <0.0014 | - | - | - | - | - |
| Tl | 0.09 | - | $\leq$0.0005 | 0.01 | - | - | - | - |
| Co | 0.14 | - | 0.003 | 0.11 | 1.44 | 0.11 | - | 38.93 |
| Sb | 0.23 | - | 0.031 | - | - | - | - | - |
| Mo | 0.58 | - | 1.34 | - | - | - | - | - |
| Cu | 0.86 | 10.5 | 0.21 | 1.39 | 0.3 | 1.76 | 0.07 | 18.9 |
| Zn | 0.97 | 70.5 | 3.2 | 5.79 | 4.55 | 13.79 | 8.86 | 730.4 |
| Ni | 0.99 | 46 | 0.15 | 0.63 | 0.28 | 0.76 | n.d | 34.96 |
| V | 1.2 | - | 0.37 | 0.22 | 0.13 | 0.26 | - | - |
| Rb | 1.45 | - | 0.55 | 1.9 | - | - | - | - |
| Mn | 1.46 | 145.5 | 0.13 | 0.58 | 393.88 | 4.03 | - | 72.56 |
| As | 1.48 | - | 0.4 | 1.55 | 5.7 | 0.12 | 3.23 | 80.8 |
| Cr | 3.75 | 10.5 | 0.07 | 0.7 | 0.71 | 0.21 | n.d | 7.64 |
| Pb | 10.6 | 4710 | <0.02 | 0.19 | 0.16 | 0.65 | 0.04 | 82.42 |
| Li | 13.48 | - | 2.02 | 0.53 | - | - | - | - |
| Ba | 70.49 | - | 9.9 | 12.78 | - | - | - | |
| Se | 127.56 | - | 0.04 | - | - | - | - | 52.76 |
| Pd | 605.9 | - | - | - | - | - | - | - |
| Sr | 672.93 | - | 101 | 39.09 | - | - | - | - |
| Water quality | - | Polluted water | Clean water | Clean water | Poor quality | Excellent quality | Heavy metal pollution | High pollution |

Temperature values ranged from 14.9 to 27.8 °C, and the mean was 20.5 °C. S2 recorded the lowest value in winter, while S6 recorded the highest value in summer. These temperature fluctuations follow the annual cycle of air temperature in the region. The lake water is slightly alkaline, with a pH ranging from 8.05 to 9.2 and an overall average of 8.65. S4 reported the lowest value during the summer, while S3 recorded the highest value during autumn. A statistically significant difference was found ($p$-value = 0.01) between the stations. DO mean value was 7.84 mg/L. The lowest value was 2.32 mg/L in S2 in summer and the highest was 11.8 mg/L in S5 in winter. The DO levels measured were significantly low ($p$-value = 0.000) during the warm season and in zones with overgrown vegetation (S2, S3, S4, and S5). EC values varied from 2.94 mS/cm to 3.08 mS/cm and revealed a significant difference between seasons ($p$-value = 0.002). The mean EC value was 2.99 mS/cm. Minimal EC was measured in station S8 in summer and maximal in station

S1 in spring. This high level of conductivity reflects the presence of important quantity of mineral salts in the water, resulting from the leaching of the region's soils by runoff water.

The trace element values recorded in this study are below those recommended by the Moroccan standard [44] and the WHO [24], except Pb and Se.

Cd concentrations fluctuated between 0 and 0.02 µg/L and had an overall mean of 0.004 µg/L. All samples except those collected in spring showed the lowest concentration. The highest level was found in S3 in spring. There was a significant difference in Cd values between stations ($p$-value = 0.023) and seasons ($p$-value = 0.03). Spatially, the average Cd concentration increased from east to southwest. This increase remains lower than that measured in Fouarate Lake, which is 90.5 µg/L [39]. The Be interval in the water samples was 0–0.21 µg/L, and its overall mean concentration was 0.05 µg/L. All stations recorded the lowest value during the cold season and the highest values in spring. Maximum Be concentration was observed in S3 in spring. The spatial distribution shows an increase in Be concentration from the Lake center towards the southwestern part and a slight increase in stations S1 and S2. The overall mean Be concentration is higher than that found in Baikal Water Lake [40]. Tl values ranged from 0 to 0.54 µg/L; the lowest concentration was recorded at all stations in autumn and winter. However, the highest level was measured in S2 in summer. An increase in Tl concentration was observed from the west to the southeast. The mean value recorded during the study period of 0.09 µg/L was higher than that measured in other water bodies (0.01 µg/L in Ghodaghodi Lake) [25]. Co concentrations ranged from 0 to 0.7 µg/L. The minimum value was recorded in all samples taken during the cold period, while its maximum value was measured in S2 in summer. High concentrations were recorded in summer. A considerable seasonal difference was observed between the Co averages ($p$ = 0.000). The spatial distribution of Co showed an increasing concentration from the southwest to the southeast. The overall mean concentration (0.14 µg/L) was higher than that found in an isolated lake, Gosaikunda Lake (0.11 µg/L), and lower than that measured in an urban lake, Phewa Lake located in Nepal [41]. Sb concentrations varied between 0.18 and 0.31 µg/L and showed an overall mean concentration of 0.23 µg/L. Minimal level was measured in S2 in summer, and maximal level was observed in S4 in autumn. The highest concentrations were recorded in autumn. Spatially, the Sb averages showed a gradual increase from southeast to southwest. The Mo range in all water samples was 0.31–0.88 µg/L. Its overall mean value was 0.58 µg/L. S1 showed the highest concentration in summer, while S5 noted the lowest concentration in winter. The spatial distribution of Mo showed an increase in its mean concentration in the north-northeast and southwest parts (S4). The high Mo concentrations in the samples taken during the warm season can be explained by the increase in temperature and the drop in water level. These concentrations are lower than those recorded in Baikal Lake [40]. Cu values ranged from 0.56 to 1.34 µg/L. The lowest level was measured in S1 in summer and highest in S4 in autumn. The highest concentrations were recorded in the southeast and southwest parts of the lake, while the lowest concentrations were recorded in the central and north-northeast parts. The overall mean concentration of 0.86 µg/L was lower than that found in Honghu Lake in China (3.09 µg/L) [45] and Rawal Lake in Pakistan (10 µg/L) [46], while it was higher than that observed in East Dongting Lake in China 0.07 µg/L, where copper has both a natural and anthropogenic origin [42]. Zn concentrations ranged from 0 to 5.02 µg/L. The minimum value was recorded in stations S1, S2, S3, S4, S5, and S6 in summer and winter and in stations S3, S4, S5, and S6 in autumn. The maximum value was measured in S2 in autumn. This concentration difference is significant seasonally ($p$-value = 0.001). The spatial distribution showed a gradual increase in annual Zn concentration from the central to the southeastern and southwestern parts of the lake. The overall mean Zn concentration in the water samples was 0.97 µg/L. This value is much lower than that measured in Fouarate Lake, which is 70.5 µg/L [39]. The range of Ni was 0–4.56 µg/L. Its overall mean concentration was 0.99 µg/L. All stations recorded minimum values in autumn and winter, while the maximum value was measured in S2 in summer. High concentrations were significantly recorded during the summer period. Spatially, Ni concentration showed a gradual increase from the southwest to the southeast of the lake. The average Ni concentration is very low compared to that recorded in Fouarate Lake (46 μg/L) [39]. V values ranged from 0.05 to 3.36 μg/L with an overall mean of 1.20 μg/L. The lowest and highest concentrations were measured in S6 in winter and summer, respectively. The concentration of V in the lake water increases from the periphery towards the center of the lake except S4, which showed high values of V. The V concentrations recorded at some stations were higher than those measured in Ghodaghodi Lake [25], whose trace elements have both natural and anthropogenic origins. Rb concentrations ranged from 1.24 to 1.71 μg/L. The minimum value was observed at S2, S5, and S6 in spring, while the maximum value was recorded in S6 in autumn. Rb concentrations revealed a significant difference between seasons ($p$-value = 0.02). Pant et al. [25] measured a higher mean Rb concentration (1.90 μg/L) in Ghodaghodi Lake, Nepal than that observed in our study (1.45 μg/L). They concluded that this increase is natural since Rb is present in the earth's crust (alkali and alkaline-earth metals). The range of Mn in the water samples collected was 0–25.39 μg/L. Its overall average concentration was 1.46 μg/L. All samples recorded the lowest quantity during the cold seasons (autumn and winter). The maximum value was measured in S3 in spring. A significant seasonal difference was noted between the Mn averages ($p$-value = 0.000). The spatial distribution of Mn concentration increased gradually from the eastern to the southwestern part (S3). This increase can be attributed to an anthropogenic source (steel objects in contact with water, wastewater discharge). As concentrations fluctuated between 0 and 4.22 μg/L, and showed an overall average of 0.05 μg/L. All stations recorded minimum values in autumn and spring. Maximum values were recorded in summer. This increase can be explained by the lake's low water levels at this time of year. Spatially, concentrations are high in the northern part, average in the southwestern and central parts, and low in the eastern and southeastern parts. As concentration increases may be associated with the motorized boating activities practiced in this part of the lake. Arsenic accumulation in fish tissues (liver, brain, gills, muscles, kidneys, and gonads) can induce enzymatic, genetic, and immune disorders [47]. Cr concentrations fluctuated between 0 and 13.33 μg/L and showed a mean of 3.75 μg/L. The maximum value was measured in S3 in spring. A significant increase in Cr concentration was noted ($p$ = 0.001) after the rainy period. The spatial distribution of Cr showed a rise in concentration in S2, S3, and S5. This concentration may be associated with the weathering of rocks and the leaching of agricultural soils (containing pesticides). The basalts and Triassic clays constituting the lake's watershed are rich in Cr [48]. Pb interval is 0 to 118.90 μg/L, and its global mean is 10.6 μg/L. The highest values were recorded in summer. The maximum concentration was observed in S3 in spring. The high Pb concentrations recorded in spring and after the rainy season suggest leaching from agricultural soils and roads surrounding the lake. The highest concentrations were found in the southwestern part of the lake, while the lowest concentrations were found in the eastern part. High concentrations have been recorded in other bodies of water, including Manchar Lake in Pakistan (82.42 μg/L) [43]. The Pb concentrations measured in the lake were lower than those recorded in Fouarate Lake in Morocco [39]. The high Pb concentrations could also result from the boating activities practiced in the lake. They are also due to fertilizer and pesticide use in agriculture and wastewater discharges. High concentrations of Pb in water can cause an accumulation of this element in fish tissue. Studies have shown high concentrations of Pb in the tissues of fish such as *Tilapia nilotica* from High Dam Lake in Egypt [49] and *Labeo rohita* and *Ctenopharyngodon idella* from Bhopal Lake [50]. Such a phenomenon can lead to fish poisoning, illness, and even death. Consumption of poisoned fish will, therefore, hurt human health. Li values fluctuated between 10.4 and 17.27 μg/L, averaging 13.48 μg/L. The lowest concentration was observed in S5 in summer, and the highest concentration was measured in S3 in spring. Significant variability between seasonal values was observed ($p$-value = 0.001). A gradual increase in Li concentration was noted from the southwestern to the southeastern part of the lake, which receives water from a stream. Our Li values are higher than the concentration observed in Baikal Lake [40] and lower than the concentration found in Rawal Lake [51]. Ba values ranged from 53.24 to 99.23 μg/L and showed an overall mean of 70.48 μg/L. S3 and S6 registered the lowest concentration in autumn, and S3 the highest in spring. High concentrations were significantly measured in spring in S1, S3, S4, S5, and S6. The spatial distribution of Ba showed a concentration gradient for this element that increases from the east to the southeast. Very high Ba concentrations, 155 ppm, and 294 ppm, found in the rocks of the lake's watershed [48] could explain the Ba values measured in the water. Se interval in the water samples was 1.64–490.4 μg/L. Its overall mean was 127.56 μg/L. The minimum value was measured in S1 in autumn and the maximum in S3 in spring. Spatially, Se values were high in the northeastern and southwestern parts and the central part of the lake. These values were much higher than the results found in Lake Baikal [40,52]. The high Se concentrations recorded in spring and after the rainy season suggest leaching from rocks in the Lake Basin and agricultural soils enriched with Se-based fertilizers. Seliniferous rock waters contain high concentrations of Se ranging from 50 to 300 μg/L [53]. Jurassic limestone formations are enriched in selenium [54]. Selenium is a trace element, but at high concentrations, it becomes toxic and can cause damage to the liver, heart, kidneys, ovaries, and gills of freshwater fish [55]. Pd concentrations fluctuated between 501.2 and 829 μg/L, with an average value of 605.90 μg/L. Its minimum value was recorded in S6 in summer, while the maximum value was measured in S2 in autumn. In autumn, high Pd concentrations were observed to a significant degree in all stations ($p$ = 0.000). The spatial distribution of Pd showed an increasing concentration gradient from the west to the southeast. In the present study, the Pd values measured were higher than those obtained in other bodies of water [56]. This high concentration may have both natural and human sources. Sr concentrations ranged from 532.1 to 922.8 μg/L. The overall mean value was 672.93 μg/L. S6 measured the lowest concentration in summer and S2 the highest in autumn. High values are significantly recorded in all stations except S4. The spatial distribution of this metal shows an increasing concentration gradient from the west to the southeast. The highest Sr concentrations detected in water samples could be attributed to the area's geology.

Globally, the study results showed that two elements, Pd and Sr, were the major contributors, while Cd was the minor contributor in the lake water. The mean concentration of trace elements decreased in this order in Dayat Roumi Lake: Cd (0.004 μg/L) < Be (0.05 μg/L) < Tl (0.09 μg/L) < Co (0.14 μg/L) < Sb (0.23 μg/L) < Mo (0.58 μg/L) < Cu (0.86 μg/L) < Zn (0.97 μg/L) < Ni (0.99 μg/L) < V (1, 20 μg/L) < Rb (1.45 μg/L) < Mn (1.46 μg/L) < As (1.58 μg/L) < Cr (3.75 μg/L) < Pb (10.60 μg/L) < Li (13.48 μg/L) < Ba (70.49 μg/L) < Se (127.56 μg/L) < Pd (605.90 μg/L) < Sr (672.93 μg/L). Given the spatial variability of trace element average concentrations, three groups can be distinguished with different behaviors: (1) elements such as Tl, Co, Zn, Ni, Li, Ba, Pd, and Sr, which tend to be distributed along an increasing gradient from the western to the southeastern part; (2) a group composed mainly of Sb, Mn, and Pb, which show a gradual increase in concentration from the eastern to the western sector; and (3) a group composed of Cd, Be, Mo, Cu, V, Rb, As, Cr, and Se, whose distribution is heterogeneous and does not follow an increasing order in the lake water. The average concentration measured was used to divide these elements into three chemical groups: Dominant trace elements with a concentration greater than or equal to 10 μg/L. Pb, Li, Ba, Se, Pd, and Sr: Pb, Li, Ba, Se, Pd, and Sr, moderate trace elements whose average concentration is between 10 and 0.1 μg/L: Co, Sb, Mo, Cu, Zn, Ni, V, Rb, Mn, As, Cr, and weak trace elements whose average concentration is less than or equal to 0.1 μg/L: Cd, Be, Tl [57].

### 3.2. Correlation Between Variables

The geochemical characterization of the lake water was completed by analyzing the relationships between trace elements and physicochemical parameters using Pearson correlation coefficients. The coefficients obtained are shown in Table 5.

**Table 5.** Physicochemical parameter correlation matrix.

| | Li | Rb | Zn | Cd | V | Cr | Mn | Co | Ni | Cu | Mo | Pd | Sr | Ba | Be | As | Sb | Tl | Pb | Se | pH | EC | DO | T |
|---|---|---|---|---|---|---|---|---|---|---|---|---|---|---|---|---|---|---|---|---|---|---|---|---|
| Li | 1 | | | | | | | | | | | | | | | | | | | | | | | |
| Rb | −0.597 ** | 1 | | | | | | | | | | | | | | | | | | | | | | |
| Zn | 0.766 ** | −0.411 * | 1 | | | | | | | | | | | | | | | | | | | | | |
| Cd | 0.700 ** | −0.323 | 0.681 ** | 1 | | | | | | | | | | | | | | | | | | | | |
| V | −0.076 | −0.246 | −0.164 | −0.016 | 1 | | | | | | | | | | | | | | | | | | | |
| Cr | 0.860 ** | −0.734 ** | 0.720 ** | 0.744 ** | 0.136 | 1 | | | | | | | | | | | | | | | | | | |
| Mn | 0.349 | −0.025 | 0.343 | 0.845 ** | 0.067 | 0.420 * | 1 | | | | | | | | | | | | | | | | | |
| Co | 0.063 | −0.379 | −0.105 | 0.011 | 0.927 ** | 0.17 | 0.087 | 1 | | | | | | | | | | | | | | | | |
| Ni | 0.122 | −0.429 * | −0.041 | 0.088 | 0.933 ** | 0.259 | 0.137 | 0.988 ** | 1 | | | | | | | | | | | | | | | |
| Cu | 0.398 | 0.144 | 0.276 | 0.429 * | −0.490 * | 0.041 | 0.297 | −0.375 | −0.378 | 1 | | | | | | | | | | | | | | |
| Mo | 0.386 | −0.492 * | 0.267 | 0.397 | 0.781 ** | 0.649 ** | 0.241 | 0.694 ** | 0.758 ** | −0.311 | 1 | | | | | | | | | | | | | |
| Pd | 0.19 | 0.317 | 0.436 * | 0.036 | −0.727 ** | −0.191 | −0.077 | −0.611 ** | −0.633 ** | 0.588 ** | −0.659 ** | 1 | | | | | | | | | | | | |
| Sr | 0.163 | 0.266 | 0.361 | 0.004 | −0.856 ** | −0.183 | −0.122 | −0.743 ** | −0.772 ** | 0.554 ** | −0.762 ** | 0.955 ** | 1 | | | | | | | | | | | |
| Ba | 0.636 ** | −0.463 * | 0.797 ** | 0.688 ** | −0.388 | 0.664 ** | 0.468 * | −0.265 | −0.203 | 0.203 | 0.036 | 0.339 | 0.382 | 1 | | | | | | | | | | |
| Be | 0.733 ** | −0.609 ** | 0.632 ** | 0.813 ** | 0.28 | 0.912 ** | 0.633 ** | 0.314 | 0.383 | 0.022 | 0.699 ** | −0.296 | −0.321 | 0.580 ** | 1 | | | | | | | | | |
| As | 0.044 | −0.378 | 0.019 | 0.132 | 0.929 ** | 0.349 | 0.154 | 0.847 ** | 0.891 ** | −0.529 ** | 0.886 ** | −0.746 ** | −0.878 ** | −0.18 | 0.464 * | 1 | | | | | | | | |
| Sb | −0.299 | 0.678 ** | −0.228 | −0.174 | 0.051 | −0.457 * | −0.087 | −0.112 | −0.138 | 0.346 | −0.071 | 0.182 | 0.045 | −0.622 ** | −0.34 | −0.05 | 1 | | | | | | | |
| Tl | 0.296 | −0.550 ** | 0.054 | 0.124 | 0.758 ** | 0.323 | 0.128 | 0.912 ** | 0.896 ** | −0.203 | 0.602 ** | −0.465 * | −0.564 ** | −0.086 | 0.452 * | 0.699 ** | −0.226 | 1 | | | | | | |
| Pb | 0.650 ** | −0.306 | 0.590 ** | 0.956 ** | −0.017 | 0.733 ** | 0.913 ** | 0.018 | 0.092 | 0.309 | 0.363 | −0.047 | −0.058 | 0.682 ** | 0.830 ** | 0.141 | −0.267 | 0.139 | 1 | | | | | |
| Se | 0.874 ** | −0.707 ** | 0.743 ** | 0.750 ** | 0.023 | 0.992 ** | 0.408 * | 0.058 | 0.148 | 0.095 | 0.569 ** | −0.113 | −0.088 | 0.708 ** | 0.883 ** | 0.241 | −0.459 * | 0.222 | 0.735 ** | 1 | | | | |
| pH | 0.183 | −0.077 | 0.209 | 0.103 | −0.311 | 0.218 | −0.007 | −0.378 | −0.37 | −0.06 | −0.084 | 0.079 | 0.219 | 0.254 | 0.087 | −0.263 | −0.218 | −0.346 | 0.153 | 0.27 | 1 | | | |
| EC | 0.632 ** | −0.166 | 0.536 ** | 0.538 ** | −0.193 | 0.441 * | 0.29 | −0.104 | −0.066 | 0.476 * | 0.125 | 0.359 | 0.296 | 0.35 | 0.432 * | −0.146 | 0.106 | 0.059 | 0.431 * | 0.462 * | 0.32 | 1 | | |
| DO | −0.01 | 0.024 | 0.057 | 0.013 | −0.722 ** | −0.004 | −0.052 | −0.713 ** | −0.722 ** | 0.043 | −0.540 ** | 0.276 | 0.508 * | 0.417 * | −0.167 | −0.659 ** | −0.401 | −0.643 ** | 0.06 | 0.087 | 0.609 ** | 0.074 | 1 | |
| T | 0.416 * | −0.607 ** | 0.35 | 0.386 | 0.717 ** | 0.704 ** | 0.19 | 0.648 ** | 0.709 ** | −0.398 | 0.926 ** | −0.632 ** | −0.696 ** | 0.144 | 0.714 ** | 0.855 ** | −0.23 | 0.613 ** | 0.358 | 0.629 ** | 0.035 | −0.107 | −0.399 * | 1 |

Note(s): ** Correlation is significant at the 0.01 level (2-tailed). * Correlation is significant at the 0.05 level (2-tailed).

The associations V-Co-Ni-As, Cd-Pb-Be, Tl-Co-Ni-V, Ni-V-Mo-As, Cr-Be-Se, Cd-Mn-Pb, Li-Cr-Se, Li-Zn, Zn-Ba, and Pd-Sr have a strong positive correlation ($r > 0.75$) indicating that these trace elements have similar origins, in particular, rock weathering phenomena. The associations Li-Ba-Be, Ba-Pb-Se, Zn-Cd-Cr, Cd-Ba-Li, Cr-Pb-Ba, Cu-Pd, Cu-Sr, Rb-Sb, Mn-Be, As-Tl, Zn-Be, Mo-Cr, Mo-Be, and Co-Mo showed a moderate positive correlation ($0.75 < r \leq 0.5$), suggesting that these elements come from natural sources to which may be added a similar anthropogenic input (leaching from roads and agricultural soils). Negative correlations between the trace elements showed their different origins. The elements As and Sr were negatively correlated, indicating that As came from ecotourism activities in the lake area, while Sr came from the weathering of rocks in the catchment area. The results of the correlation matrix showed no significant correlation between trace elements and pH values. EC showed a moderate positive correlation with Li, Zn, and Cd. DO levels showed a moderate positive correlation with Sr and a negative correlation with the elements V, Ni, Co, As, Tl, and Mo. T values showed a strong positive correlation with Mo and As, a moderate correlation with V, Be, Ni, Cr, Se, and Tl, and a negative correlation with Sr, Pd, and Rb. The correlations observed between trace elements and other abiotic factors in the lake depend on the behavior and sources of these elements.

### 3.3. Water Quality Index Calculation

Calculation results for the WQI of Dayat Roumi Lake are presented in Table 6. In this study, the WQI was 167.94, which classifies the lake waters as poor quality. The WQI value was very high because of the high concentrations of Pb and Se included in the calculation of this index. It should be noted that the WQI value of Dayat Roumi Lake was very high compared with that of Ghodaghodi Lake (WQI = 10.08) [25] and Phewa Lake (WQI = 159.2), located in Nepal [41]. Thus, the WQI shows that trace elements are a health hazard for the lake's population, and consequently, its waters cannot be consumed.

**Table 6.** Dayat Roumi Lake water quality index (WQI).

| Elements (µg/L) | pi | Pi = pi/∑pi | Ci | Si | Ci/Si | Pi × Ci/Si |
|---|---|---|---|---|---|---|
| Cd | 5 | 0.119 | 0.004 | 3.000 | 0.001 | 0.000 |
| Tl | 1 | 0.024 | 0.095 | 2.000 | 0.047 | 0.001 |
| Co | 1 | 0.024 | 0.136 | 40.000 | 0.003 | 0.000 |
| Sb | 4 | 0.095 | 0.226 | 20.000 | 0.011 | 0.001 |
| Mo | 1 | 0.024 | 0.580 | 70.000 | 0.008 | 0.000 |
| Cu | 2 | 0.048 | 0.855 | 2000.000 | 0.000 | 0.000 |
| Zn | 1 | 0.024 | 0.973 | 3000.000 | 0.000 | 0.000 |
| Ni | 1 | 0.024 | 0.993 | 70.000 | 0.014 | 0.000 |
| Mn | 4 | 0.095 | 1.464 | 400.000 | 0.004 | 0.000 |
| As | 5 | 0.119 | 1.480 | 10.000 | 0.148 | 0.018 |
| Cr | 5 | 0.119 | 3.750 | 50.000 | 0.075 | 0.009 |
| Pb | 5 | 0.119 | 10.604 | 10.000 | 1.060 | 0.126 |
| Ba | 2 | 0.048 | 70.488 | 700.000 | 0.101 | 0.005 |
| Se | 5 | 0.119 | 127.558 | 10.000 | 12.756 | 1.519 |
| Total | 42 | 1.000 | | | | 1.679<br>WQI = 167.94 |

### 3.4. Health Risk Indexes

The human health risk assessment for metals was used to calculate the HQ hazard quotient and the HI hazard index, the results of which are summarized in Table 7. The HQ and HI risk indices for Pd and Rb were not calculated due to the lack of RfD for these two elements. Table 8 presents the results of the carcinogenic risk index CR for As, Cd, Pb, and Cr and the total carcinogenic risk CI for the two exposure routes (ingestion and skin absorption).

**Table 7.** ADD, HQ, and HI values for adults and children and for both routes of exposure (ingestion and dermal).

| Parameters | $ADD_{ing}$ | | $ADD_{der}$ | | $HQ_{ing}$ | | $HQ_{der}$ | | HI | |
|---|---|---|---|---|---|---|---|---|---|---|
| | Adult | Child | Adult | Child | Adult | Child | Adult | Child | Adult | Child |
| Cd | $1.22 \times 10^{-4}$ | $1.82 \times 10^{-4}$ | $6.38 \times 10^{-4}$ | $1.88 \times 10^{-3}$ | $2.44 \times 10^{-4}$ | $3.65 \times 10^{-4}$ | $2.55 \times 10^{-2}$ | $7.52 \times 10^{-2}$ | $2.57 \times 10^{-2}$ | $7.56 \times 10^{-2}$ |
| Be | $1.24 \times 10^{-3}$ | $1.85 \times 10^{-3}$ | $6.46 \times 10^{-3}$ | $1.91 \times 10^{-2}$ | $6.19 \times 10^{-4}$ | $9.24 \times 10^{-4}$ | $4.61 \times 10^{-3}$ | $1.36 \times 10^{-2}$ | $5.23 \times 10^{-3}$ | $1.45 \times 10^{-2}$ |
| Tl | $2.59 \times 10^{-3}$ | $3.87 \times 10^{-3}$ | $1.35 \times 10^{-2}$ | $3.99 \times 10^{-2}$ | $2.59 \times 10^{-1}$ | $3.87 \times 10^{-1}$ | $1.35 \times 10$ | $3.99 \times 10$ | $1.61 \times 10$ | $4.38 \times 10$ |
| Co | $3.73 \times 10^{-3}$ | $5.57 \times 10^{-3}$ | $1.95 \times 10^{-2}$ | $5.75 \times 10^{-2}$ | $1.24 \times 10^{-2}$ | $1.86 \times 10^{-2}$ | $3.25 \times 10^{-1}$ | $9.58 \times 10^{-1}$ | $3.37 \times 10^{-1}$ | $9.76 \times 10^{-1}$ |
| Sb | $6.20 \times 10^{-3}$ | $9.26 \times 10^{-3}$ | $3.24 \times 10^{-2}$ | $9.55 \times 10^{-2}$ | $1.55 \times 10^{-2}$ | $2.31 \times 10^{-2}$ | $4.05 \times 10$ | $1.19 \times 10^{1}$ | $4.06 \times 10$ | $1.20 \times 10^{1}$ |
| Mo | $1.59 \times 10^{-2}$ | $2.37 \times 10^{-2}$ | $8.30 \times 10^{-2}$ | $2.45 \times 10^{-1}$ | $3.18 \times 10^{-3}$ | $4.75 \times 10^{-3}$ | $4.37 \times 10^{-2}$ | $1.29 \times 10^{-1}$ | $4.68 \times 10^{-2}$ | $1.34 \times 10^{-1}$ |
| Cu | $2.34 \times 10^{-2}$ | $3.50 \times 10^{-2}$ | $1.22 \times 10^{-1}$ | $3.61 \times 10^{-1}$ | $5.86 \times 10^{-4}$ | $8.75 \times 10^{-4}$ | $1.53 \times 10^{-2}$ | $4.51 \times 10^{-2}$ | $1.59 \times 10^{-2}$ | $4.60 \times 10^{-2}$ |
| Zn | $2.67 \times 10^{-2}$ | $3.98 \times 10^{-2}$ | $1.39 \times 10^{-1}$ | $4.10 \times 10^{-1}$ | $8.88 \times 10^{-5}$ | $1.33 \times 10^{-4}$ | $2.32 \times 10^{-3}$ | $6.84 \times 10^{-3}$ | $2.41 \times 10^{-3}$ | $6.97 \times 10^{-3}$ |
| Ni | $2.72 \times 10^{-2}$ | $4.06 \times 10^{-2}$ | $1.42 \times 10^{-1}$ | $4.19 \times 10^{-1}$ | $1.36 \times 10^{-3}$ | $2.03 \times 10^{-3}$ | $1.77 \times 10^{-1}$ | $5.24 \times 10^{-1}$ | $1.79 \times 10^{-1}$ | $5.26 \times 10^{-1}$ |
| V | $3.30 \times 10^{-2}$ | $4.93 \times 10^{-2}$ | $1.72 \times 10^{-1}$ | $5.08 \times 10^{-1}$ | $3.30 \times 10^{-2}$ | $4.93 \times 10^{-2}$ | $1.72 \times 10^{1}$ | $5.08 \times 10^{1}$ | $1.73 \times 10^{1}$ | $5.09 \times 10^{1}$ |
| Rb | $3.98 \times 10^{-2}$ | $5.94 \times 10^{-2}$ | $2.08 \times 10^{-1}$ | $6.13 \times 10^{-1}$ | - | - | - | - | - | - |
| Mn | $4.01 \times 10^{-2}$ | $5.99 \times 10^{-2}$ | $2.09 \times 10^{-1}$ | $6.18 \times 10^{-1}$ | $1.67 \times 10^{-3}$ | $2.50 \times 10^{-3}$ | $2.18 \times 10^{-1}$ | $6.43 \times 10^{-1}$ | $2.20 \times 10^{-1}$ | $6.46 \times 10^{-1}$ |
| As | $4.05 \times 10^{-2}$ | $6.05 \times 10^{-2}$ | $2.12 \times 10^{-1}$ | $6.24 \times 10^{-1}$ | $1.35 \times 10^{-1}$ | $2.02 \times 10^{-1}$ | $1.72 \times 10$ | $5.08 \times 10$ | $1.86 \times 10$ | $5.28 \times 10$ |
| Cr | $1.03 \times 10^{-1}$ | $1.53 \times 10^{-1}$ | $5.36 \times 10^{-1}$ | $1.58 \times 10$ | $3.43 \times 10^{-2}$ | $5.11 \times 10^{-2}$ | $7.15 \times 10$ | $2.11 \times 10^{1}$ | $7.19 \times 10$ | $2.11 \times 10^{1}$ |
| Pb | $2.91 \times 10^{-1}$ | $4.34 \times 10^{-1}$ | $1.52 \times 10$ | $4.47 \times 10$ | $2.08 \times 10^{-1}$ | $3.10 \times 10^{-1}$ | $3.61 \times 10$ | $1.07 \times 10^{1}$ | $3.82 \times 10$ | $1.10 \times 10^{1}$ |
| Li | $3.69 \times 10^{-1}$ | $5.52 \times 10^{-1}$ | $1.93 \times 10$ | $5.69 \times 10$ | $1.85 \times 10^{-1}$ | $2.76 \times 10^{-1}$ | $9.64 \times 10^{-1}$ | $2.84 \times 10$ | $1.15 \times 10$ | $3.12 \times 10$ |
| Ba | $1.93 \times 10$ | $2.88 \times 10$ | $1.01 \times 10^{1}$ | $2.97 \times 10^{1}$ | $9.66 \times 10^{-3}$ | $1.44 \times 10^{-2}$ | $7.20 \times 10^{-1}$ | $2.12 \times 10$ | $7.30 \times 10^{-1}$ | $2.14 \times 10$ |
| Se | $3.49 \times 10$ | $5.22 \times 10$ | $1.82 \times 10^{1}$ | $5.38 \times 10^{1}$ | $6.99 \times 10^{-1}$ | $1.04 \times 10$ | $8.29 \times 10$ | $2.45 \times 10^{1}$ | $8.99 \times 10$ | $2.55 \times 10^{1}$ |
| Pd | $1.66 \times 10^{1}$ | $2.48 \times 10^{1}$ | $8.67 \times 10^{1}$ | $2.56 \times 10^{2}$ | - | - | - | - | - | - |
| Sr | $1.84 \times 10^{1}$ | $2.75 \times 10^{1}$ | $9.62 \times 10^{1}$ | $2.84 \times 10^{2}$ | $3.07 \times 10^{-2}$ | $4.59 \times 10^{-2}$ | $8.02 \times 10^{-1}$ | $2.37 \times 10$ | $8.33 \times 10^{-1}$ | $2.41 \times 10$ |
| Total | $4.15 \times 10^{1}$ | $6.20 \times 10^{1}$ | $2.17 \times 10^{2}$ | $6.39 \times 10^{2}$ | $1.63 \times 10$ | $2.43 \times 10$ | $4.67 \times 10^{1}$ | $1.38 \times 10^{2}$ | $4.83 \times 10^{1}$ | $1.40 \times 10^{2}$ |
| | | | | | | | | | $\Sigma HI = 1.89 \times 10^{2}$ | |

**Table 8.** CR and CI values for adults and children and both exposure routes.

| Parameters | Adult | | Child | | Adult | | | Child | | |
|---|---|---|---|---|---|---|---|---|---|---|
| | $ADD_{ing}$ | $ADD_{der}$ | $ADD_{ing}$ | $ADD_{der}$ | $CR_{ing}$ | $CR_{der}$ | CI | $CR_{ing}$ | $CR_{der}$ | CI |
| Cd | $5.23 \times 10^{-8}$ | $2.73 \times 10^{-7}$ | $1.56 \times 10^{-8}$ | $1.61 \times 10^{-7}$ | $3.30 \times 10^{-7}$ | $2.08 \times 10^{-5}$ | $2.12 \times 10^{-5}$ | $9.54 \times 10^{-8}$ | $1.23 \times 10^{-5}$ | $1.24 \times 10^{-5}$ |
| Ni | $1.17 \times 10^{-5}$ | $6.08 \times 10^{-5}$ | $3.48 \times 10^{-6}$ | $3.59 \times 10^{-5}$ | $1.98 \times 10^{-5}$ | $2.59 \times 10^{-3}$ | $2.61 \times 10^{-3}$ | $5.92 \times 10^{-6}$ | $1.53 \times 10^{-3}$ | $1.53 \times 10^{-3}$ |
| As | $1.74 \times 10^{-5}$ | $9.07 \times 10^{-5}$ | $5.19 \times 10^{-6}$ | $5.35 \times 10^{-5}$ | $2.61 \times 10^{-5}$ | $1.36 \times 10^{-4}$ | $1.62 \times 10^{-4}$ | $7.78 \times 10^{-6}$ | $8.03 \times 10^{-5}$ | $8.81 \times 10^{-5}$ |
| Cr | $4.40 \times 10^{-5}$ | $2.30 \times 10^{-4}$ | $1.32 \times 10^{-5}$ | $1.36 \times 10^{-4}$ | $2.20 \times 10^{-5}$ | $4.60 \times 10^{-3}$ | $4.62 \times 10^{-3}$ | $6.58 \times 10^{-6}$ | $2.71 \times 10^{-3}$ | $2.72 \times 10^{-3}$ |
| Pb | $1.25 \times 10^{-4}$ | $6.50 \times 10^{-4}$ | $3.72 \times 10^{-5}$ | $3.83 \times 10^{-4}$ | $1.06 \times 10^{-6}$ | $5.52 \times 10^{-5}$ | $5.63 \times 10^{-5}$ | $3.16 \times 10^{-7}$ | $3.26 \times 10^{-5}$ | $3.29 \times 10^{-5}$ |
| Total | - | - | - | - | $6.93 \times 10^{-5}$ | $7.40 \times 10^{-3}$ | $7.46 \times 10^{-3}$ | $2.07 \times 10^{-5}$ | $4.36 \times 10^{-3}$ | $4.38 \times 10^{-3}$ |

### 3.4.1. Non-Carcinogenic Health Risks

$HQ_{ing}$ values ranged from $8.88 \times 10^{-5}$ to $1.04 \times 10$. The highest value was recorded for Se in children. $HQ_{der}$ values fluctuated between $2.32 \times 10^{3}$ and $5.08 \times 10^{1}$, with the highest value recorded for V in children. The $HQ_{ing}$ values for all the elements were below 1, except the HQ of Se in children. Selenium, an essential trace element, plays a crucial role in the human body. A selenium deficiency leads to immune, thyroid, and reproductive disorders, while an excess of selenium in the body, or selenosis, can have harmful effects. The most common are gastrointestinal disorders (abdominal pain, nausea, vomiting, and diarrhea), neurological disorders, dermatoses (alopecia, brittle nails, and skin lesions), garlic breath, cardiac, respiratory, and endocrine disorders, and other serious complications [58]. Children are more vulnerable to the risks associated with excess selenium because of their small body size and stage of development. Total HQ values for ingestion and dermal exposure are higher than 1, and as a result, lake residents are exposed to non-carcinogenic health risks. The $HQ_{ing}$ and $HQ_{der}$ values for children are higher than those recorded for adults. This observation has been reported in other studies [25]. The $HQ_{der}$ values for elements Tl, Sb, V, As, Cr, Pb, and Se are higher than 1 in adult and children. These values are higher than those recorded in other aquatic ecosystems [25,28,41,59].

The $HQ_{der}$ values for Li, Ba, and Sr are above 1 in children and less than 1 in adults. $HQ_{der}$ values are higher than $HQ_{ing}$ values in adults and children. The opposite observation was noted in Ghodaghodi Lake [25]. HI values ranged from $2.41 \times 10^{-3}$ to $5.09 \times 10^{1}$. The highest value is recorded for V in children. Total HI values are higher than 1 for both adults and children. The estimated HI value for each element in children is higher than that calculated for adults. Children are more vulnerable to the risks associated with exposure to metals because of their small size and stage of development. The HI values for elements Tl,

Sb, V, As, Cr, Pb, Li, and Se are above 1 in adults and children. As a result, these elements present a toxic risk for people living near the lake.

3.4.2. Carcinogenic Health Risks

In children, $CR_{der}$ and $CR_{ing}$ levels ranged from $1.23 \times 10^{-5}$ to $2.71 \times 10^{-3}$ and $9.54 \times 10^{-8}$ to $7.78 \times 10^{-6}$, respectively. In adults, $CR_{der}$ and $CR_{ing}$ values fluctuated between $2.08 \times 10^{-5}$ and $4.60 \times 10^{-3}$ and between $3.30 \times 10^{-7}$ and $2.61 \times 10^{-5}$, respectively. The CI values for the metal elements (Cd, Ni, As, Cr, and Pb) ranged from $1.24 \times 10^{-5}$ to $2.72 \times 10^{-3}$ for children and from $2.12 \times 10^{-5}$ to $4.62 \times 10^{-3}$ for adults. The order of CI of metals for adults and children is as follows: CI (Cr) > CI (Ni) > CI (As) > CI (Pb) > CI (Cd). The highest carcinogenic risk for children and adults through ingestion or skin contact was observed for Cr. A similar observation was made in the waters of Phewa Lake in Nepal CR = $5.49 \times 10^{-3}$ [41] and in the drinking water of the city of Khorramabad, Iran (CR = $6.54 \times 10^{-3}$) [60]. The CI associated with the five metals studied is greater than $1 \times 10^{-4}$, indicating that residents, both adults and children, are exposed to a carcinogenic risk if they drink or swim in the lake water. The total estimated risk of cancer in adults is higher than that measured in children, whether through ingestion or skin exposure. The same observation was made in a recent study of groundwater in an industrial area [61]. These results show that adults are more vulnerable to the risk of cancer induced by the metals studied than children due to the longer duration of exposure to metals and the gradual accumulation of toxic substances in their tissues.

## 4. Conclusions

This study was carried out to investigate the spatial and temporal distribution of 20 trace elements measured in the lake, their relationships, and their relationships with physicochemical parameters (pH, T, EC, and DO). Measured trace element levels increased in the following order: Cd < Be < Tl < Co < Sb < Mo < Cu < Zn < Ni < V < Rb < Mn < As < Cr < Pb < Li < Ba < Se < Pd < Sr in lake water. The concentrations of these elements are within the permissible limits of the criteria for surface water recommended by the Moroccan standard and drinking water by the WHO, except Pb and Se, which showed high values in spring. High values of these elements cause changes in the tissues of aquatic organisms, especially fish, and therefore hurt human health. Based on the average concentrations of trace elements in the lake, three groups can be distinguished: (1) Dominant trace elements with a concentration of 10 μg/L or more: Pb, Li, Ba, Se, Pd, and Sr; (2) Moderate trace elements with an average concentration of between 10 and 0.1 μg/L: Co, Sb, Mo, Cu, Zn, Ni, V, Rb, Mn, As, and Cr; (3) Weak trace elements with an average concentration of less than or equal to 0.1 μg/L: Cd, Be, and Tl. In addition, the spatial distribution divided the trace elements into three groups: (I) elements such as Tl, Co, Zn, Ni, Li, Ba, Pd, and Sr, which tend to be distributed along an increasing gradient from the western to the southeastern part; (II) a group composed mainly of Sb, Mn, and Pb, which show a gradual increase in concentration from the eastern to the western sector; and (III) a group composed of Cd, Be, Mo, Cu, V, Rb, As, Se, and Cr, whose distribution is heterogeneous and does not follow a direction in the lake water. The correlation analysis between trace elements revealed two principal element sources in the water, namely natural geological pathways and anthropogenic pathways. In addition, the WQI suggests that the lake's water is of poor quality and unsuitable for consumption.

Skin HQ for elements Tl, Sb, V, As, Cr, Pb, and Se and HI for elements Tl, Sb, V, As, Cr, Pb, Li, and Se are higher than 1 in adults and children. Therefore, these elements present a non-negligible risk for lake residents. In addition, HQ and HI levels are higher in children than in adults. Selenium poses a health risk, whether through ingestion or skin absorption, and children are the most affected because of their small size and stage of development. Moreover, cutaneous exposure to trace elements presents a greater health risk than the digestive route for children and adults when swimming.

The order of the total cancer risk due to prolonged metal exposure for adults and children is as follows: CI (Cr) > CI (Ni) > CI (As) > CI (Pb) > CI (Cd). The highest carcinogenic risk was observed for Cr in children and adults by ingestion or dermal route. The total cancer risk for the five elements is above $1 \times 10^{-4}$, indicating that residents are exposed to a carcinogenic risk if they drink or bathe in the lake water. In addition, the estimated total risk of cancer in adults is higher than that measured in children because of the longer duration of exposure to metals and the gradual accumulation of toxic substances in their tissues. Based on the results of this study, it is concluded that regular monitoring of water quality assessment is necessary to draw up an effective management plan for the lake's water resources. To this end, we propose to spread the sampling over a long period and study the distribution of trace elements in the sediments and groundwater of the region in order to better identify and pinpoint the source(s) of each element.

**Author Contributions:** Conceptualization, I.O.; Methodology, I.O., Z.E. and A.M.; Software, S.R.; Validation, T.M.; Formal analysis, Z.E. and A.A.S.; Investigation, Z.E.; Data curation, I.O. and A.M.; Writing—original draft, I.O.; Writing—review & editing, I.O., I.M., A.L., A.A.S. and S.R.; Visualization, A.A.S. and T.M.; Supervision, T.M.; Funding acquisition, A.A.S. All authors have read and agreed to the published version of the manuscript.

**Funding:** The authors are thankful to the researchers supporting project number (RSPD2024R1057), King Saud University, Riyadh, Saudi Arabia.

**Data Availability Statement:** The original contributions presented in this study are included in the article. Further inquiries can be directed to the corresponding author(s).

**Conflicts of Interest:** The authors declare no conflict of interest.

## Appendix A

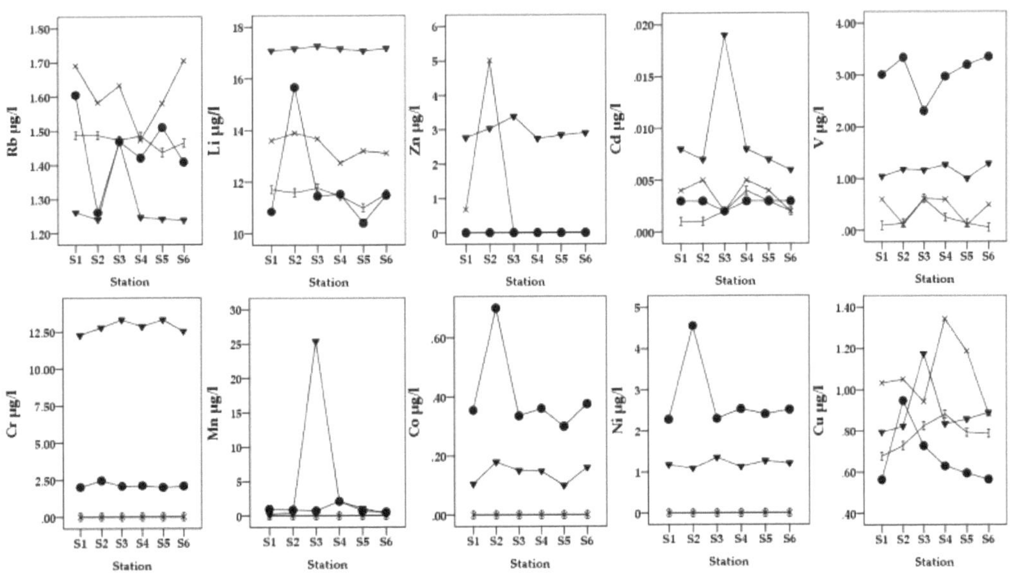


**Figure A1.** *Cont.*

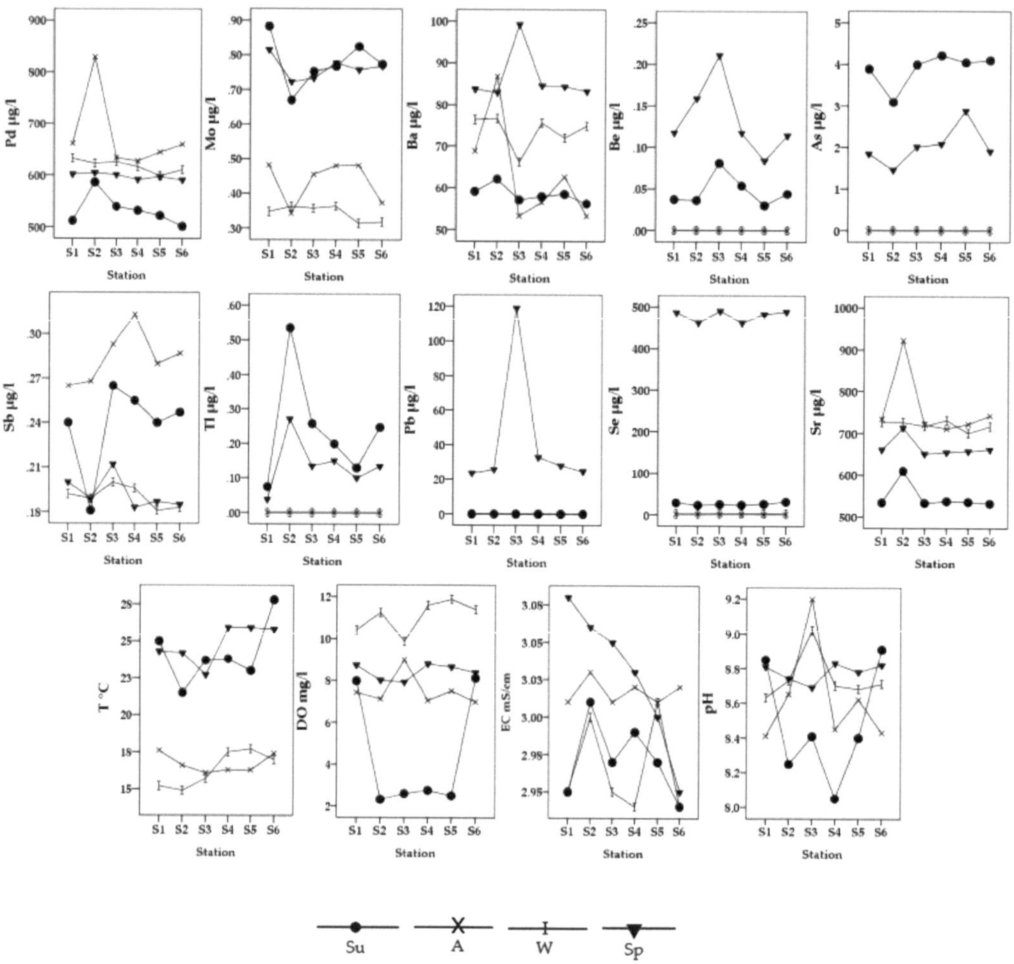


**Figure A1.** Spatial and temporal distribution of 24 physicochemical parameters.

## References

1. Moubchir, T.; Eloutassi, N.; Bendaoud, A.; Belkhiri, A.; Maai, M.; Moubchir, M.; Zahir, I. Heavy metals analysis and quality evaluation in drinking groundwater around an abandoned mine Area of Ouichane (Nador's Province, Morocco). *J. Ecol. Eng.* **2023**, *24*, 118–127. [CrossRef]
2. Rezouki, S.; El-Haji, S.; Moubchir, T.; Faouzi, J.; Bendaoud, A.; Beniaich, G.; Allali, A.; Eloutassi, N. Evaluation of the impact of anthropic activities on the physico-chemical and microbiological parameters of the water of the wadi inaouene and its tributaries. *Ecol. Eng. Environ. Technol.* **2023**, *5*, 282–294.
3. Moubchir, T.; Faouzi, J.; Loukili, E.H.; Bendaoud, A.; Belkhiri, A.; Rezouki, S.; Eloutassi, N.; Zahir, I. Investigation of the Hydrochemistry Quality of the Ouichane Groundwater (Morocco) Using Multivariate Statistical Methods and Diagram Analysis. *Moroc. J. Chem.* **2024**, *12*, 734–746.
4. Casellas, P.C. Caractérisation de L'exposition des Écosystèmes Aquatiques à des Produits Phytosanitaires: Spéciation, Biodisponibilité et Toxicité. Exemple du Cuivre dans les eaux de Ruissellement de Parcelles Viticoles (Roujan, Hérault, France). Ph.D. Thesis, Université Montpellier I, Montpellier, France, University of Geneva, Geneva, Switzerland, 2001. Available online: https://theses.hal.science/tel-00011143v2 (accessed on 21 October 2024).
5. Rezouki, S.; Moubchir, T.; El Hanafi, L.; Flouchi, R.; Zahir, I.; Alzain, M.N.; El Guerrouj, B.; Noman, O.; Shahat, A.A.; Allali, A. Assessment of Ecological Hazards in the Inaouen Wadi and Its Tributaries Using the Presence of Potentially Toxic Elements in Its Sediments. *Water* **2024**, *16*, 2936. [CrossRef]

6. Elkady, A.A.; Sweet, S.T.; Wade, T.L.; Klein, A.G. Distribution and assessment of heavy metals in the aquatic environment of Lake Manzala, Egypt. *Ecol. Indic.* **2015**, *58*, 445–457. [CrossRef]
7. Dhal, S.; Pal, H. Trace Elements in Mitigating Environmental Stress: An Overview. In *Biology and Biotechnology of Environmental Stress Tolerance in Plants*; Apple Academic Press: Palm Bay, FL, USA, 2023.
8. Kolarova, N.; Napiórkowski, P. Trace elements in aquatic environment. Origin, distribution, assessment and toxicity effect for the aquatic biota. *Ecohydrol. Hydrobiol.* **2021**, *21*, 655–668. [CrossRef]
9. Szynkowska, M.I.; Pawlaczyk, A.; Maćkiewicz, E. Bioaccumulation and Biomagnification of Trace Elements in the Environment. In *Recent Advances in Trace Elements*; Chojnacka, K., Saeid, A., Eds.; Wiley: Hoboken, NJ, USA, 2018; pp. 251–276. [CrossRef]
10. Devez, A. Caractérisation des Risques Induits par les Activités Agricoles sur les Écosystèmes Aquatiques. Ph.D. Thesis, Hydro Sciences Montpellier, Montpellier, France, 2004. Available online: https://pastel.hal.science/tel-00006497 (accessed on 21 October 2024).
11. Dixon, D.G.; Hilton, J.W. Effects of Available Dietary Carbohydrate and Water Temperature on the Chronic Toxicity of Waterborne Copper to Rainbow Trout (*Salmo gairdneri*). *Can. J. Fish. Aquat. Sci.* **1985**, *42*, 1007–1013. [CrossRef]
12. Gony-Lemaire, S. Etude Experimentale de L'action du Cadmium sur les Juveniles (Civelles et Anguillettes) de L'anguille Europeenne Anguilla Anguilla (l., 1758): Anayse Structurale et Fonctionnelle de Certaines Organes-Cibles. Ph.D. Thesis, Paris 6, Paris, France, 1990. Available online: https://theses.fr/1990PA066525 (accessed on 21 October 2024).
13. Joshy, A.; Sharma, S.R.K.; Mini, K.G.; Gangadharan, S.; Pranav, P. Histopathological evaluation of bivalves from the southwest coast of India as an indicator of environmental quality. *Aquat. Toxicol.* **2022**, *243*, 106076. [CrossRef] [PubMed]
14. Qadri, R.; Faiq, M.A. Freshwater Pollution: Effects on Aquatic Life and Human Health. In *Fresh Water Pollution Dynamics and Remediation*; Qadri, H., Bhat, R.A., Mehmood, M.A., Dar, G.H., Eds.; Springer: Singapore, 2020; pp. 15–26. [CrossRef]
15. Hughes, M.F.; Beck, B.D.; Chen, Y.; Lewis, A.S.; Thomas, D.J. Arsenic Exposure and Toxicology: A Historical Perspective. *Toxicol. Sci.* **2011**, *123*, 305–332. [CrossRef]
16. Saha, R.; Nandi, R.; Saha, B. Sources and toxicity of hexavalent chromium. *J. Coord. Chem.* **2011**, *64*, 1782–1806. [CrossRef]
17. Charkiewicz, A.E.; Backstrand, J.R. Lead Toxicity and Pollution in Poland. *Int. J. Environ. Res. Public Health* **2020**, *17*, 4385. [CrossRef] [PubMed]
18. MHUAE. *Etude du Plan D'aménagement Spécifique de Dayet Roumi. Pahse II: Rapport D'analyse Diagnostic, Variantes D'aménagement*; Ministère de l'Habitat, de l'Urbanisme et de l'Aménagement de l'Espace: Paris, France, 2009; p. 40.
19. MATEE (Ministère de l'Aménagement du Territoire de l'Eau et de l'Environnement). *Rapport Thématique sur les Zones Protégées du Maroc*; Secrétariat d'Etat chargé de l'Environnement: Rabat, Morocco, 2003; Available online: https://www.cbd.int/doc/world/ma/ma-nr-pa-fr.pdf (accessed on 21 October 2024).
20. Flower, R.J.; Stevenson, A.C.; Dearing, J.A.; Foster, I.D.L.; Airey, A.; Rippey, B.; Wilson, J.P.F.; Appleby, P.G. Catchment disturbance inferred from paleolimnological studies of three contrasted sub-humid environments in Morocco. *J. Paleolimnol.* **1989**, *1*, 293–322. [CrossRef]
21. Ihsane, O.; Zahra, E.; Sanae, R.; Mohammed, E.; Bourhia, M.; Ali, G.A.M.; Ouahmane, L.; Salamatullah, A.M.; Aboul-Soud, M.A.M.; Giesy, J.P.; et al. Physicochemical Characterization and Assessment of Magnitude of Pollution to Contribute to Water Sustainability. *Sustainability* **2022**, *14*, 6689. [CrossRef]
22. Ougrad, I.; Elassassi, Z.; Allali, A.; Elbouch, M.; Abid, A.; Fadli, M. Anthropogenic Activities Impact on the Physicochemical and Bacteriological Quality of Water in Dayat Roumi Lake, Morocco. *Ecol. Eng. Environ. Technol.* **2022**, *23*, 162–172. [CrossRef]
23. Sahu, P.; Sikdar, P.K. Hydrochemical framework of the aquifer in and around East Kolkata Wetlands, West Bengal, India. *Environ. Geol.* **2008**, *55*, 823–835. [CrossRef]
24. WHO (World Health Organization). *Directives sur la Qualité de L'eau de Boisson*, 4th ed.; intégrant le premier additif; Organisation mondiale de la Santé: Genève, Switzerland, 2017. Available online: https://iris.who.int/handle/10665/258887 (accessed on 21 October 2024).
25. Pant, R.R.; Pal, K.B.; Pathak, G.; Bishwakarma, K.; Thapa, L.B.; Dhital, Y.P.; Bhatta, Y.R.; Joshi, G.R.; Rijal, K. Water quality and health risk assessment of trace elements contamination in Ghodaghodi Lake, Sudurpaschim Province, Nepal. *Nepal J. Environ. Sci.* **2021**, *9*, 29–40. [CrossRef]
26. US Environmental Protection Agency. *Risk Assessment Guidance for Superfund Volume I: Human Health Evaluation Manual (Part E, Supplemental Guidance for Dermal Risk Assessment) Final*; EPA/540/R/99/005; US Environmental Protection Agency: Washington, DC, USA, 2004. Available online: https://www.epa.gov/sites/default/files/2015-09/documents/part_e_final_revision_10-03-07.pdf (accessed on 21 October 2024).
27. US Environmental Protection Agency. *Risk Assessment Guidance for Superfund Volume I Human Health Evaluation Manual (Part A) Interim Final*; EPA/540/1-89/002; US Environmental Protection Agency: Washington, DC, USA, 1989. Available online: https://www.epa.gov/sites/default/files/2015-09/documents/rags_a.pdf (accessed on 21 October 2024).
28. Tripathee, L.; Kang, S.; Sharma, C.M.; Rupakheti, D.; Paudyal, R.; Huang, J.; Sillanpää, M. Preliminary Health Risk Assessment of Potentially Toxic Metals in Surface Water of the Himalayan Rivers, Nepal. *Bull. Environ. Contam. Toxicol.* **2016**, *97*, 855–862. [CrossRef] [PubMed]
29. U.S. Environmental Protection Agency et Integrated Risk Information System (IRIS). *Beryllium and Compounds*; CASRN 7440-41-7; U.S. Environmental Protection Agency et Integrated Risk Information System (IRIS): Washington, DC, USA, 1998. Available online: https://iris.epa.gov/static/pdfs/0012tr.pdf (accessed on 21 October 2024).

30. Xiao, J.; Wang, L.; Deng, L.; Jin, Z. Characteristics, sources, water quality and health risk assessment of trace elements in river water and well water in the Chinese Loess Plateau. *Sci. Total Environ.* **2019**, *650*, 2004–2012. [CrossRef]
31. Wang, J.; Liu, G.; Liu, H.; Lam, P.K.S. Multivariate statistical evaluation of dissolved trace elements and a water quality assessment in the middle reaches of Huaihe River, Anhui, China. *Sci. Total Environ.* **2017**, *583*, 421–431. [CrossRef]
32. Wu, B.; Zhao, D.Y.; Jia, H.Y.; Zhang, Y.; Zhang, X.X.; Cheng, S.P. Preliminary Risk Assessment of Trace Metal Pollution in Surface Water from Yangtze River in Nanjing Section, China. *Bull. Environ. Contam. Toxicol.* **2009**, *82*, 405–409. [CrossRef]
33. Chen, L.; Ma, T.; Wang, Y.; Zheng, J. Health risks associated with multiple metal(loid)s in groundwater: A case study at Hetao Plain, northern China. *Environ. Pollut.* **2020**, *263*, 114562. [CrossRef] [PubMed]
34. US Environmental Protection Agency. *Risk-Based Screening Table—Resident Tapwater Table*; US Environmental Protection Agency: Washington, DC, USA, 2015. Available online: https://archive.epa.gov/region9/superfund/web/html/index-23.html (accessed on 21 October 2024).
35. US Environmental Protection Agency. *Guidelines for Carcinogen Risk Assessment*; EPA/630/P-03/001F; US Environmental Protection Agency: Washington, DC, USA, 2005. Available online: https://www.epa.gov/sites/default/files/2013-09/documents/cancer_guidelines_final_3-25-05.pdf (accessed on 21 October 2024).
36. Adimalla, N. Heavy metals contamination in urban surface soils of Medak province, India, and its risk assessment and spatial distribution. *Environ. Geochem. Health* **2020**, *42*, 59–75. [CrossRef] [PubMed]
37. Kamunda, C.; Mathuthu, M.; Madhuku, M. Health Risk Assessment of Heavy Metals in Soils from Witwatersrand Gold Mining Basin, South Africa. *Int. J. Environ. Res. Public Health* **2016**, *13*, 663. [CrossRef] [PubMed]
38. Akoto, O.; Gyimah, E.; Zhan, Z.; Xu, H.; Nimako, C. Evaluation of health risks associated with trace metal exposure in water from the Barekese reservoir in Kumasi, Ghana. *Hum. Ecol. Risk Assess. Int. J.* **2020**, *26*, 1134–1148. [CrossRef]
39. Latifa, E.H.; Rachida, F.; Dalale, M.; Kaoutar, H.; Hafida, A.; Mohamed, F. Content of seven metallic elements in water, sediment and in liver and gonads of Lepomis gibbosus Linnaeus, 1758 (Fish Centrachid, percifome, actinopterygiis) living in fouarate lake (Morocco). *Int. J. Biol. Res.* **2018**, *3*, 6.
40. Vetrov, V.A.; Kuznetsova, A.I.; Sklyarova, O.A. Baseline levels of chemical elements in the water of Lake Baikal. *Geogr. Nat. Resour.* **2013**, *34*, 228–238. [CrossRef]
41. Rupakheti, D.; Tripathee, L.; Kang, S.; Sharma, C.M.; Paudyal, R.; Sillanpää, M. Assessment of water quality and health risks for toxic trace elements in urban Phewa and remote Gosainkunda lakes, Nepal. *Hum. Ecol. Risk Assess. Int. J.* **2017**, *23*, 959–973. [CrossRef]
42. Makokha, V.A.; Qi, Y.; Shen, Y.; Wang, J. Concentrations, Distribution, and Ecological Risk Assessment of Heavy Metals in the East Dongting and Honghu Lake, China. *Expo. Health* **2016**, *8*, 31–41. [CrossRef]
43. Kazi, T.G.; Arain, M.B.; Jamali, M.K.; Jalbani, N.; Afridi, H.I.; Sarfraz, R.A.; Baig, J.A.; Shah, A.Q. Assessment of water quality of polluted lake using multivariate statistical techniques: A case study. *Ecotoxicol. Environ. Saf.* **2009**, *72*, 301–309. [CrossRef]
44. *GCQES (Grille de Classification Qualité des Eaux de Surface), Arrêté Conjoint du Ministre de l'Equipement et du Ministre Chargé de l'Aménagement du Territoire de l'Urbanisme, de l'Habitat et de l'Environnement*; Bulletin officiel N°1275-02 Du 17 Octobre 2002; de l'Urbanisme, de l'Habitat et de l'Environnement: Paris, France, 2002.
45. Li, F.; Qiu, Z.; Zhang, J.; Liu, C.; Cai, Y.; Xiao, M. Spatial Distribution and Fuzzy Health Risk Assessment of Trace Elements in Surface Water from Honghu Lake. *Int. J. Environ. Res. Public Health* **2017**, *14*, 1011. [CrossRef]
46. Iqbal, J.; Shah, M.H. Health Risk Assessment of Metals in Surface Water from Freshwater Source Lakes, Pakistan. *Hum. Ecol. Risk Assess. Int. J.* **2013**, *19*, 1530–1543. [CrossRef]
47. Kumari, B.; Kumar, V.; Sinha, A.K.; Ahsan, J.; Ghosh, A.K.; Wang, H.; DeBoeck, G. Toxicology of arsenic in fish and aquatic systems. *Environ. Chem. Lett.* **2017**, *15*, 43–64. [CrossRef]
48. Bounouira, H. Etude des Qualités Chimiques et Géochimiques du Bassin Versant de Bouregreg. Ph.D. Thesis, Université Ibn Tofail—Faculté des Sciences, Kenitra, Morocco, 2007; p. 296. Available online: https://tel.archives-ouvertes.fr/tel-00726475 (accessed on 21 October 2024).
49. Rashed, M.N. Cadmium and Lead Levels in Fish (Tilapia Nilotica) Tissues as Biological Indicator for Lake Water Pollution. *Environ. Monit. Assess.* **2001**, *68*, 15. [CrossRef]
50. Malik, N.; Biswas, A.K.; Qureshi, T.A.; Borana, K.; Virha, R. Bioaccumulation of heavy metals in fish tissues of a freshwater lake of Bhopal. *Environ. Monit. Assess.* **2010**, *160*, 267–276. [CrossRef]
51. Malik, R.N.; Nadeem, M. Spatial and temporal characterization of trace elements and nutrients in the Rawal Lake Reservoir, Pakistan using multivariate analysis techniques. *Environ. Geochem. Health* **2011**, *33*, 525–541. [CrossRef] [PubMed]
52. Githaiga, K.B.; Njuguna, S.M.; Gituru, R.W.; Yan, X. Water quality assessment, multivariate analysis and human health risks of heavy metals in eight major lakes in Kenya. *J. Environ. Manag.* **2021**, *297*, 113410. [CrossRef] [PubMed]
53. Crini, G. Chapitre XV. Le sélénium dans les eaux: Une nouvelle substance dangereuse pour demain? In *Eaux Industrielles Contaminées*; Morin-Crini, N., Ed.; Pratiques & Techniques; Presses Universitaires de Franche-Comté: Besançon, France, 2020; pp. 447–466. Available online: https://www.researchgate.net/publication/347197882_Chapitre_XV_Le_selenium_dans_les_eaux_une_nouvelle_substance_dangereuse_pour_demain (accessed on 21 October 2024).
54. Gourcy, L.; Wyns, R.; Garnier-Sereno, C.; Brenot, A. Origine du sélénium dans les eaux souterraines utilisées pour l'alimentation en eau potable dans le bassin de Paris. *Géologues Société Géologique Fr.* **2014**, *179*, 44–49.

55. Lemly, A.D. Symptoms and implications of selenium toxicity in fish: The Belews Lake case example. *Aquat. Toxicol.* **2002**, *57*, 39–49. [CrossRef] [PubMed]
56. Cook, N.J.; Wood, S.A.; Zhang, Y. Transport and fixation of Au, Pt and Pd around the Lac Sheen Cu-Ni-PGE occurrence in Quebec, Canada. *J. Geochem. Explor.* **1992**, *46*, 187–228. [CrossRef]
57. Xiao, J.; Jin, Z.; Wang, J. Geochemistry of trace elements and water quality assessment of natural water within the Tarim River Basin in the extreme arid region, NW China. *J. Geochem. Explor.* **2014**, *136*, 118–126. [CrossRef]
58. dos Santos, M.; Ramires, P.F.; Gironés, M.C.R.; Rubio Armendáriz, M.d.C.; Montelongo, S.P.; Muccillo-Baisch, A.L.; da Silva Junior, F.M.R. Multiple exposure pathways and health risk assessment of selenium for children in a coal mining area. *Environ. Sci. Pollut. Res.* **2021**, *28*, 13562–13569. [CrossRef] [PubMed]
59. Zhang, X.; Wu, Q.; Gao, S.; Wang, Z.; He, S. Distribution, source, water quality and health risk assessment of dissolved heavy metals in major rivers in Wuhan, China. *PeerJ* **2021**, *9*, e11853. [CrossRef] [PubMed]
60. Mohammadi, A.A.; Zarei, A.; Majidi, S.; Ghaderpoury, A.; Hashempour, Y.; Saghi, M.H.; Alinejad, A.; Yousefi, M.; Hosseingholizadeh, N.; Ghaderpoori, M. Carcinogenic and non-carcinogenic health risk assessment of heavy metals in drinking water of Khorramabad, Iran. *MethodsX* **2019**, *6*, 1642–1651. [CrossRef]
61. Myers, R.A.; Gyimah, E.; Gbemadu, K.; Osei, B.; Akoto, O. Appraising groundwater quality and the associated health risks of heavy metal contamination at Suame magazine. *Sci. Afr.* **2023**, *21*, e01794. [CrossRef]

water

MDPI

Article

# Spatiotemporal Variation Characteristics and Source Identification of Nitrogen in the Baiyangdian Lake Water, China

**Qianqian Zhang [1,2], Shimin Xu [3] and Li Yang [3,*]**

[1] Key Laboratory of Coupling Process and Effect of Natural Resources Elements, China Geological Survey, Beijing 100055, China
[2] Hebei and China Geological Survey Key Laboratory of Groundwater Remediation, Institute of Hydrogeology and Environmental Geology, Chinese Academy of Geological Sciences, Shijiazhuang 050061, China
[3] 801 Institute of Hydrogeology and Engineering Geology, Shandong Provincial Bureau of Geology and Mineral Resources (SPBGM), Jinan 250014, China
* Correspondence: sddkj801@yeah.net

**Abstract:** To study the characteristics and sources of nitrogen in the Baiyangdian Lake, this research conducted water quality monitoring during three hydrological periods (normal period, flood period, and dry period), and 165 pieces of routine water quality monitoring data were collected from the three national control sections for Baiyangdian Lake and its inflow rivers. By integrating water chemical analysis with multivariate statistical techniques, the study comprehensively investigated the spatiotemporal variation patterns of nitrogen in Baiyangdian Lake and identified the sources of nitrogen pollution. The results showed that the concentration of total nitrogen (TN) was highest during the dry period, reaching an average of 0.924 mg/L, and 31.3% of the sites exceeded the national Grade III surface water quality standard, reflecting a potential risk of nitrogen pollution. Based on the ion ratio method and principal component analysis (PCA), the main sources of nitrogen pollution in Baiyangdian Lake were identified as manure and domestic sewage, with agricultural fertilizers also having a certain impact on water nitrogen pollution. In addition, the study also compared the nitrogen concentration in Baiyangdian Lake with several important lakes in China. The results showed that the concentrations of TN and ammonium nitrogen ($NH_4^+$-N) in Baiyangdian Lake are lower than those in lakes in areas with similar human activity intensity, indicating that the water quality of Baiyangdian is relatively good. This study can provide a scientific basis for water quality management and pollution prevention for Baiyangdian Lake.

**Keywords:** lakes; nitrogen; water quality; source identification; principal component analysis

## 1. Introduction

Lakes are vital water resources that play an irreplaceable role in regional environmental changes, biogeochemical cycling, ecological function maintenance, safeguarding water supply security, preserving biodiversity, and fostering socio-economic development within their catchment areas [1–3]. However, the increasing intensity of human activities has led to a significant influx of nutrients, such as nitrogen, into lake waters, causing eutrophication and water quality deterioration and severely impacting the ecological functions and value of these lakes [4]. Therefore, studying the spatiotemporal variations of nitrogen in lakes and elucidating its pollution sources is of significant importance for guiding the management of water environmental pollution in lakes [5].

The factors influencing the spatiotemporal variations of nitrogen in lake waters primarily consist of natural and anthropogenic factors [6]. Natural factors, such as seasonal variations in rainfall, can significantly influence the spatiotemporal dynamics of nitrogen within lake waters, specifically manifesting in two aspects: First, rainfall dilutes the concentration of nitrogenous nutrients in lake water, showing a temporal pattern where nitrogen concentrations are lower during the flood season compared to other seasons [7]. Second, the concentration of nitrogen in a lake is affected by the scouring action of rainfall; runoff from rainfall carries nitrogen retained on agricultural and urban surfaces into the lake, leading to higher nitrogen concentrations in the lake water during the flood season than during the dry season [8]. Additionally, the migration and transformation processes of nitrogen in the water body (such as nitrification and denitrification) significantly affect the changes in nitrogen load [9].

Anthropogenic factors affecting the spatiotemporal variations of nitrogen in lakes include various industrial and agricultural activities, domestic sewage, urban development, and land use practices. Yu et al. (2015) found that the spatial differences in land use practices in Xiaoxingkai Lake were the main reasons for the significant differences in nitrate nitrogen ($NO_3^-$-N) and ammonium nitrogen ($NH_4^+$-N) concentrations between the eastern and western parts of the lake, leading to water quality disparities [10]. Mbaye et al. (2016) conducted a study on the spatiotemporal variations of nitrogen in Lake Senegal and found that the concentrations of $NO_3^-$-N and total nitrogen (TN) were significantly higher in the densely populated areas and areas with intensive agricultural development downstream than in the upstream areas of the lake [11]. Zhang et al. (2017) discovered that there was a strong spatial variability in $NO_3^-$-N and $NH_4^+$-N in the Fuyang River, with higher concentrations near urban areas and lower concentrations in areas farther from the city, where the water quality was in a state of severe eutrophication throughout the year [12]. It is evident that the characteristics of spatiotemporal variations of nitrogen in an aquatic environment are influenced by a combination of natural conditions and human activities. Different lakes exhibit similarities and differences in the spatiotemporal characteristics of nitrogen due to their distinct regional rainfall patterns and sources of nitrogen. Therefore, to accurately identify the spatiotemporal patterns of nitrogen in lakes, comprehensive surveys and systematic research are necessary.

Baiyangdian is the largest lake in the North China Plain and one of the few remaining lake-type wetland ecosystems in the region [13]. It is strategically located at the core of the Xiong'an New Area and plays a significant role in flood control, navigation, fisheries, and climate regulation, providing crucial ecological support for the development of the Xiong'an New Area [14]. In recent years, due to the development of industry and agriculture and the increasing demand for water resources in the basin, Baiyangdian Lake has been facing water pollution issues, which have become a critical problem that urgently needs to be addressed for the development of the Xiong'an New Area [15]. Scholars have conducted extensive research on water environmental quality issues [16], the decline in biodiversity of biological communities [17], and heavy metal pollution issues [18] in the Baiyangdian watershed. However, research on the spatiotemporal variation patterns of nitrogen and the identification of pollution sources in Baiyangdian Lake is still relatively weak, which severely restricts the progress of water environment restoration efforts in Baiyangdian Lake.

This study selects Baiyangdian Lake as the research subject. By integrating hydrogeological and geochemical methods with multivariate statistical techniques, this research aims to identify the spatiotemporal variation characteristics of nitrogen in Baiyangdian Lake, assess the water quality, and reveal the main sources of nitrogen pollution. The goal is to provide a scientific basis for the comprehensive management of the Baiyangdian water environment and for the protection of water environments in lakes.

## 2. Materials and Methods

### 2.1. Introduction to the Study Area

Baiyangdian Lake is located between 115°38′ E to 116°07′ E longitude and 38°43′ N to 39°02′ N latitude, primarily situated within the territory of Anxin County, Baoding City, Hebei Province, China (see Figure 1). This area belongs to the Daqing River sub-basin system of the Hai River basin, with a basin area of approximately 366 square kilometers. Within its vast wetland ecosystem, there are 143 lakes of varying sizes, with an average water depth of about 2.3 m. The climate of the region is characterized as a temperate semi-humid continental monsoon climate, with distinct seasons. Spring is usually dry with strong winds and scarce rainfall; summer is relatively hot with abundant rainfall, mainly concentrated from July to September. According to long-term meteorological records, the average temperature of the area is about 12.2 °C, and the average annual precipitation is about 529.7 mm. The soil types in the basin are diverse, including tidal soil, swamp soil, brown soil, and paddy soil, among others. Land use types encompass a variety of categories such as forest land, farmland, construction land, grassland, and water bodies. The water flow direction in the Baiyangdian basin generally exhibits a northeastward flow from west to east. The water source is mainly replenished by nine major rivers entering the lake: the Zhulong River, Xiaoyi River, Tang River, Qingshui River, Bao River, Fu River, Ping River, Cao River, and BaigouYin River. In addition, ecological water replenishment measures, such as the Yellow River Diversion Project and the South-to-North Water Diversion Project, also provide important supplementary water resources for the area.

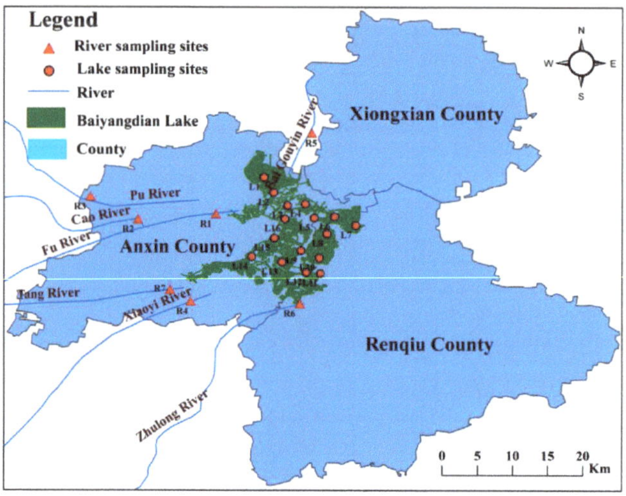


**Figure 1.** Distribution map of sampling points in Baiyangdian Lake.

### 2.2. *Sample Collection and Analysis*

In this study, a total of 16 sampling points were established in Baiyangdian Lake and 7 in the rivers. Sampling was conducted three times: in May 2022 (normal season, spanning from March to May and October to November), in July 2022 (flood season, from June to September), and in December 2022 (dry season, from December to February of the following year, when the lake was under an ice cover). Samples were collected in 2.5 L polyethylene bottles, which were rinsed three times with lake water before sampling. Surface water samplers were used to collect samples at a depth of 0.5 m from the river surface. After collection, the samples were sealed, placed in a refrigerated box, and transported back to the laboratory. They were then filtered through a 0.45 µm glass fiber filter membrane and stored in a 4 °C refrigerator. All samples were tested within 48 h. In addition, 165 pieces of routine water quality monitoring data were collected from the three national control sections (corresponding to sites L1, L5, and L12 in this study) of Baiyangdian Lake from 20 May 2022 to 22 November 2022. These three routine monitoring sites are located in the north, center, and south of the study area, respectively, and can represent the overall water quality situation of Baiyangdian Lake.

The testing indicators mainly included total nitrogen (TN), ammonium nitrogen ($NH_4^+$-N), nitrate nitrogen ($NO_3^-$-N), calcium ($Ca^{2+}$), sodium ($Na^+$), sulfate ($SO_4^{2-}$), and chloride ($Cl^-$). For water chemical analysis: anions ($Cl^-$ and $SO_4^{2-}$) were determined using an ion chromatograph (Perkin-Elmer Lambda 35, San Diego, CA, USA), and bicarbonate ($HCO_3^-$) was measured by standard dilute hydrochloric acid titration; cations ($Ca^{2+}$ and

$Na^+$) were analyzed using an inductively coupled plasma-mass spectrometry (Agilent 7500ce, ICP-MS, Tokyo, Japan). $NH_4^+$-N was measured using the Kjeldahl method with Nessler's reagent; TN was determined using the alkaline potassium persulfate digestion ultraviolet spectrophotometry method, and $NO_3^-$-N was measured using the phenol disulfonic acid spectrophotometric method with an error less than 0.05 mg/L. Sample quality control (QA-QC) involved conducting tests on blank, duplicate, and standard samples. Each sample was tested three times to ensure accuracy. Furthermore, the data on TN concentration from the three national control sites were collected from the Ministry of Ecology and Environment's National Control Section Monitoring Website in China.

### 2.3. *Data Analysis*

#### 2.3.1. Principal Component Analysis

Principal component analysis (PCA) is a mathematical transformation method that converts a given set of correlated variables into another set of uncorrelated variables through linear transformation. These new variables are arranged in descending order of variance. The total variance of the variables remains unchanged during the transformation, with the first variable having the maximum variance, referred to as the first principal component. The second variable has the next highest variance and is uncorrelated with the first, known as the second principal component, and so on, with the last principal component having the smallest variance and being uncorrelated with all previous components.

PCA is a well-established multivariate statistical analysis method that incorporates multidimensional factors into a unified system for quantitative research. In this study, PCA is employed to identify the main pollution sources of nitrogen in the Baiyangdian Lake. The analysis was based on seven selected water quality parameters, namely, TN, $NO_3^-$-N, $NH_4^+$-N, $Na^+$, $Ca^{2+}$, $SO_4^{2-}$, and $Cl^-$. Prior to the application of the data to the PCA model, the study data underwent the Kaiser–Meyer–Olkin (KMO) test and Bartlett's test of sphericity. The results indicated that for the normal, flood, and dry seasons, the KMO values were 0.576, 0.656, and 0.565, respectively, and the Bartlett's test values were 38.4 ($p < 0.01$), 123.5 ($p < 0.001$), and 63.8 ($p < 0.001$), respectively. These findings suggest that the water chemistry data of Baiyangdian Lake for the three hydrological periods meet the requirements for PCA. Based on an eigenvalue >1, two principal components were identified for each of the three hydrological periods, which accounted for 64.1%, 80.6%, and 67.5% of the total variance in the data, respectively.

#### 2.3.2. Cluster Analysis

Cluster analysis is one of the most widely used exploratory methods and can be divided into two categories: variable clustering (R-type clustering) and sample clustering (Q-type clustering). The essence of cluster analysis is to group objects that are most similar based on the degree of closeness between variables or samples, using a stepwise aggregation method until they form a single cluster. In water quality assessment, clustering is often performed based on sampling points and times to analyze the spatiotemporal variation characteristics of water quality or based on evaluation indicators to analyze the similarities between indicators. In this study, the inter-group Euclidean distance method is used to analyze the sampling sites in Baiyangdian Lake through cluster analysis. Additionally, we calculated the silhouette coefficients for the clustering analysis results of three hydrological periods. The results showed that 83%, 92%, and 92% of the silhouette values in the normal flow period, flood season, and dry season clustering analysis results were greater than 0.5, indicating that the clustering results are relatively accurate.

The raw water chemistry data for surface water were organized and statistically analyzed using Excel 2021. Ion concentration ratio plots (Cl vs. $NO_3^-$) were created using Origin 2021 (2019b), and sampling point distribution maps were generated using ArcGIS 10.8. Cluster analysis and principal component analysis plots were created using R 4.4.1.

## 3. Results and Analysis

### 3.1. Temporal Variation Characteristics of Nitrogen Concentration in Baiyangdian Lake

#### 3.1.1. Temporal Variation of Nitrogen Concentration in Baiyangdian Lake

Table 1 summarizes the concentration changes of TN, $NO_3^-$-N, and $NH_4^+$-N in Baiyangdian Lake during different hydrological periods. The temporal variation of TN shows that the average concentration during the dry season (0.924 mg/L) is higher than that during the flood season (0.786 mg/L) and the normal water period (0.759 mg/L). As shown in Figure 2, the temporal variation of TN concentrations at the three routine monitoring sites in the Baiyangdian Lake water body exhibits higher levels during the dry and normal periods compared to the flood season. This is mainly related to the insufficient water replenishment during the dry season, which cannot dilute the pollutants within the lake. Additionally, the concentrations of sites 2, 3, and 5 exceed the Class III standard (1.0 mg/L) of the surface water environmental quality standard of China during the normal, flood, and dry seasons, respectively, indicating that the Baiyangdian Lake still faces a risk of nitrogen pollution. The concentration of $NO_3^-$-N ranges from 0.020 to 1.35 mg/L, 0.130 to 1.07 mg/L, and 0.101 to 2.01 mg/L during the normal, flood, and dry seasons, respectively. Its temporal variation shows that the concentration is highest during the normal water period (0.559 mg/L), followed by the dry season (0.528 mg/L), and lowest during the flood season (0.380 mg/L), which is mainly related to the dilution effect of rainfall during the flood season. Furthermore, the concentration of $NH_4^+$-N is relatively low, with 70% of the sampling sites having $NH_4^+$-N concentrations below the detection limit of 0.025 mg/L. Its temporal variation also shows that the concentration is lowest during the flood season (0.025 mg/L) and the normal water period (0.025 mg/L), and highest during the dry season (0.131 mg/L). Overall, the coefficient of variation for the three indicators is higher during the dry season than during the flood season, indicating that the larger replenishment flow of rivers during the flood season not only promotes adequate mixing of the water in the lake area but also effectively reduces the variability of the indicators in the water body.

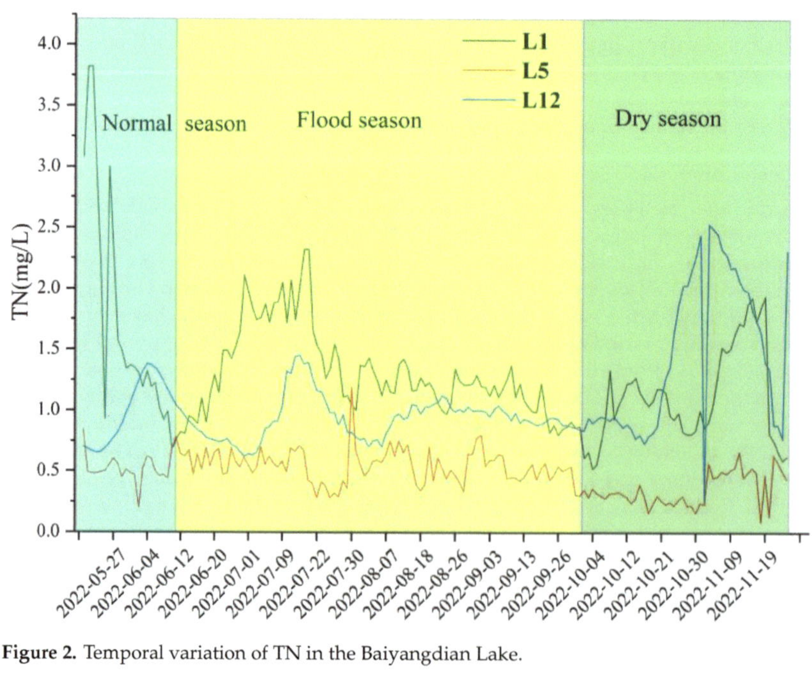


**Figure 2.** Temporal variation of TN in the Baiyangdian Lake.

**Table 1.** Statistics of different nitrogen concentrations in Baiyangdian Lake.

| Parameters | Season | Mean (mg/L) | Min (mg/L) | Max (mg/L) | Standard Deviation | Coefficient of Variation (%) |
|---|---|---|---|---|---|---|
| TN | Normal Season | 0.759 | 0.450 | 1.98 | 0.401 | 52.7 |
| | Flood season | 0.786 | 0.502 | 1.54 | 0.313 | 39.8 |
| | Dry season | 0.924 | 0.532 | 2.51 | 0.536 | 58.0 |
| $NO_3^-$-N | Normal Season | 0.559 | 0.020 | 1.35 | 0.278 | 49.7 |
| | Flood season | 0.380 | 0.130 | 1.07 | 0.256 | 67.4 |
| | Dry season | 0.528 | 0.101 | 2.01 | 0.522 | 98.9 |
| $NH_4^+$-N | Normal Season | 0.025 | 0.025 | 0.032 | 0.002 | 6.9 |
| | Flood season | 0.025 | 0.025 | 0.025 | 0.000 | 0.0 |
| | Dry season | 0.131 | 0.025 | 0.495 | 0.130 | 99.3 |

3.1.2. Temporal Variation of Nitrogen Concentration in Influent Rivers

Table 2 summarizes the concentration changes of TN, $NO_3^-$-N, and $NH_4^+$-N in the influent rivers of Baiyangdian Lake during different hydrological periods. Consistent with the temporal variation of nitrogen concentration in the lake water body, the average concentrations of TN, $NO_3^-$-N, and $NH_4^+$-N show that the flood season (2.12 mg/L, 1.59 mg/L, and 0.074 mg/L, respectively) has lower concentrations compared to the dry season (3.17 mg/L, 2.38 mg/L, and 0.402 mg/L, respectively) and the normal water period (2.37 mg/L, 1.74 mg/L, and 0.553 mg/L, respectively). There are five, five, and six sampling sites in the influent rivers where TN concentration exceeds the Class III standard (1.0 mg/L) of the surface water environmental quality standard GB3838−2002 during the normal, flood, and dry seasons, respectively. This indicates that the pollution of TN in the influent rivers of Baiyangdian Lake is relatively severe. Overall, the coefficients of variation for TN, $NO_3^-$-N, and $NH_4^+$-N in the influent rivers are lower during the normal water period (ice cover period) than during the flood and dry seasons. This suggests that the combined effects of non-point pollution caused by rainfall runoff and point-source pollution from human activities lead to differences in the sources of TN, $NO_3^-$-N, and $NH_4^+$-N in different rivers, resulting in significant variations in their concentrations in the influent rivers during the flood and dry seasons.

**Table 2.** Statistics of different nitrogen concentrations in inflow rivers of Baiyangdian Lake.

| Parameters | Season | Mean (mg/L) | Min (mg/L) | Max (mg/L) | Standard Deviation | Coefficient of Variation (%) |
|---|---|---|---|---|---|---|
| TN | Normal season | 2.37 | 1.46 | 3.10 | 0.654 | 27.6 |
| | Flood season | 2.12 | 0.557 | 4.06 | 1.26 | 59.6 |
| | Dry season | 3.17 | 0.907 | 5.25 | 1.53 | 48.4 |
| $NO_3^-$-N | Normal season | 1.74 | 0.910 | 2.43 | 0.623 | 35.9 |
| | Flood season | 1.59 | 0.247 | 2.52 | 0.958 | 60.1 |
| | Dry season | 2.38 | 0.161 | 4.28 | 1.30 | 54.5 |
| $NH_4^+$-N | Normal season | 0.553 | 0.319 | 0.802 | 0.188 | 34.0 |
| | Flood season | 0.074 | 0.041 | 0.135 | 0.036 | 48.3 |
| | Dry season | 0.402 | 0.191 | 0.750 | 0.218 | 54.3 |

### *3.2. Spatial Variation Characteristics of Nitrogen Concentration in Baiyangdian Lake*

(1) Spatial variation of nitrogen in Baiyangdian Lake during the normal season

The spatial variation of nitrogen in the Baiyangdian watershed generally shows that the concentration of nitrogen in the influent rivers is higher than that at the lake sites (Figure 3). The concentration of nitrogen at the central sites of the lake is lower than that at the surrounding sites, and the concentration at the northern sites of the lake is greater than that at the southern sites (Figures 2 and 3). The spatial variation at the Baiyangdian Lake sites specifically exhibits three distinct spatial characteristics.

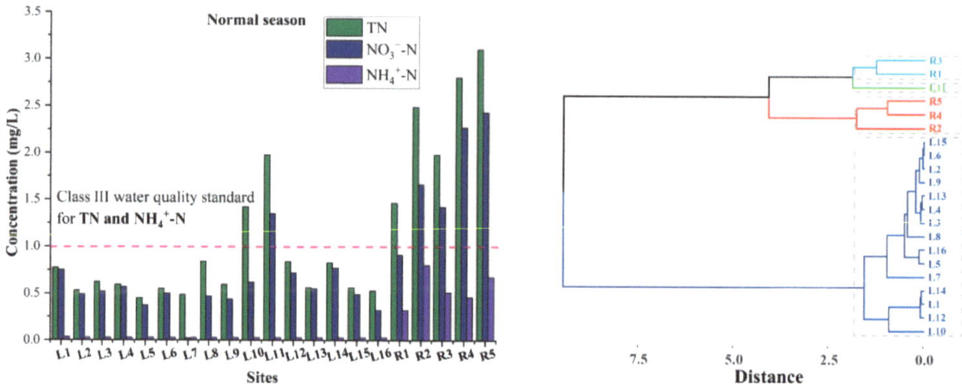


(**a**) (**b**)

**Figure 3.** Spatial variation (**a**) and cluster analysis (**b**) of nitrogen in the Baiyangdian Lake during the normal season.

The first type is that the site is affected by a combination of aquaculture and domestic sewage, resulting in nitrogen concentrations higher than those at other sites. A representative site is L11, with TN and $NO_3{}^-$-N concentrations of 1.98 mg/L and 1.35 mg/L, respectively. The second spatial variation characteristic is that the site is close to the influent rivers and villages, and it is significantly affected by the sewage from these sources, leading to higher water body concentrations compared to other internal lake sites. Representative sites include L1, L12, L10, and L14. The third spatial variation characteristic is for sites within the lake that are farther from villages and thus less affected by human activities, resulting in relatively lower nitrogen concentrations. Regarding the influent rivers, the sites are clustered into two categories. Sites R2, R4, and R5 have higher concentrations of TN (2.79 mg/L), $NO_3{}^-$-N (2.12 mg/L), and $NH_4{}^+$-N (0.645 mg/L) compared to sites R1 and R3. This may be due to the presence of ecological wetlands on the west side of sites R1 and R3, where aquatic plants are abundant and can absorb and assimilate nitrogen from the water body.

(2) Spatial variation of nitrogen in Baiyangdian Lake during the flood season

As shown in Figure 4, during the flood season, the nitrogen in Baiyangdian Lake is significantly affected by rainfall, leading to a decrease in nitrogen concentrations at most lake and influent river sites. For the lake area, there are three main spatial variation patterns.

The first pattern involves sites L14 and L15, which are affected by a combination of non-point and point sources from the Xiaoyi River, resulting in higher nitrogen concentrations compared to other sites. The second pattern is observed at sites L9 to L13, which are located around aquaculture areas and thus have relatively higher nitrogen concentrations due to the impact of aquaculture activities. The third pattern is characterized by sites in the central part of the lake, which are less affected by human activities and therefore have relatively lower nitrogen concentrations.

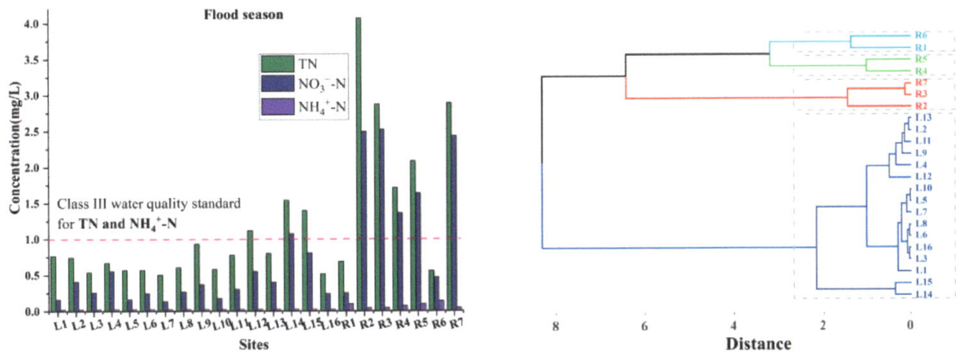


(a) (b)

**Figure 4.** Spatial variation (**a**) and cluster analysis (**b**) of nitrogen in the Baiyangdian Lake during the flood period.

Regarding the influent rivers, they are spatially clustered into three categories. Overall, the average TN concentration at the first category of sites (3.27 mg/L) is higher than that at the second category (1.89 mg/L) and the third category (0.622 mg/L). The first category includes sites R2 (Cao River), R3 (Bao River), and R7 (Tang River), which are significantly affected by rainfall runoff and thus have higher nitrogen concentrations compared to other river sites. The second category includes sites R5 (Baigouyin River) and R4 (Xiaoyi River), where nitrogen concentrations decrease during the flood season due to the dilution effect of rainfall. The third category consists of sites R1 (Fu River) and R6 (Yellow River Diversion inlet). Site R1 has lower nitrogen concentrations due to the combined effects of upstream ecological wetland purification and rainfall dilution. Site R6, connected to the Xiaobai River as part of the Yellow River Diversion Project inlet, has better water quality.

(3) Spatial variation of nitrogen in Baiyangdian Lake during the dry season

As shown in Figure 5, during the dry season, the sites within Baiyangdian Lake are also clustered into three categories: The first category includes sites that are close to the Cao River (R2) and Baigouyin River (R5), such as sites L1–L4 and L16. Sites L1 and L16, with high nitrogen concentrations (exceeding 2.0 mg/L), are clustered with the influent rivers. These two rivers have been contaminated during the dry season (with TN concentrations reaching up to 4.75 mg/L and 5.25 mg/L), significantly impacting the surrounding water bodies. The second category consists of sites located downstream of the first category, influenced by the lake's self-purification capacity, resulting in a decrease in nitrogen concentrations. Representative sites include L5–L8. The third category comprises sites in the southern part of the lake, L9–L15, where human activities such as tourism and aquaculture are minimal or absent during the dry season (ice cover period), leading to relatively lower nitrogen concentrations in the water body.

For the influent rivers of Baiyangdian Lake, they are spatially clustered into three categories as well: the first category includes river sites R2 and R5, which are heavily affected by point-source pollution, with TN concentrations reaching 4.75 mg/L and 5.25 mg/L, respectively, significantly higher than other sites. The second category includes sites R3, R4, and R6, which are relatively more affected by point-source pollution, with nitrogen concentrations lower than those of the first category. The third category consists of sites R1 and R7. Site R1 has relatively lower concentrations due to the purification effects of upstream ecological wetlands, while site R7, being in a frozen state, is less affected by human activities, resulting in lower nitrogen concentrations. These spatial variations during the dry season highlight the differential impacts of various sources of pollution and the lake's self-purification processes on the nitrogen concentrations across different parts of Baiyangdian Lake.

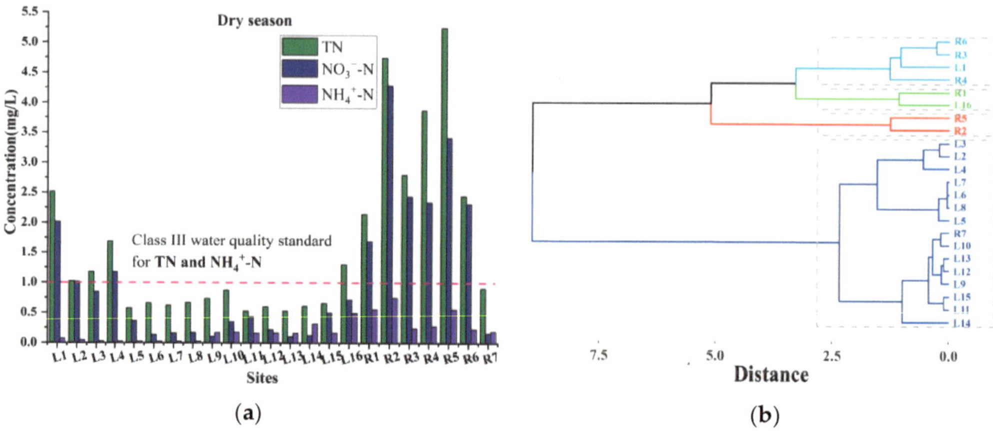


**Figure 5.** Spatial variation (**a**) and cluster analysis (**b**) of nitrogen in the Baiyangdian Lake during the dry period.

In summary, the spatial variation of nitrogen concentration in Baiyangdian Lake is not only influenced by point sources (domestic sewage) but also by the combined effects of agricultural non-point-source pollution caused by rainfall runoff and the replenishment of external water sources. For the overall lake area, the first type of spatial variation is that the nitrogen concentration in the northern part of the lake, which is close to the Baigouyin River replenishment and the agricultural non-point sources of the north, is higher than in the southern part of the lake, except for aquaculture areas. The second type of spatial variation is a gradual decrease in nitrogen concentration from the west to the east side of the lake, reaching the lowest at the outlet (site L7). As for the influent rivers, due to the different environments through which the rivers flow and the differences in upstream water sources, there is no clear regularity in their spatial characteristics.

### 3.3. Water Quality Assessment of Baiyangdian Lake

The single-factor pollution index method is a simple and effective way to determine the pollution status of water quality for a specific use, which can identify the category of water quality and the main pollution indicators. In this study, TN, $NO_3^-$-N, and $NH_4^+$-N were selected, and the content limits specified in the Class III water quality standards of the surface water environmental quality standards were used as the evaluation criteria. The pollution indices of water quality factors at Baiyangdian Lake sites and influent rivers during the three hydrological periods were analyzed. The results (Table 3) indicate that, overall, the pollution indices of $NO_3^-$-N and $NH_4^+$-N in Baiyangdian Lake and influent rivers are both < 1, suggesting that neither has shown pollution characteristics or reached a level that poses a threat to human health. However, the exceedance rates of TN in Baiyangdian Lake during the normal, flood, and dry seasons are 12.5%, 18.8%, and 31.3%, respectively, indicating a state of nitrogen pollution in Baiyangdian Lake. Moreover, the average pollution index of TN in the influent rivers during the three hydrological periods is greater than 2.0, indicating that the pollution of TN in the influent rivers is relatively severe and has a significant impact on the nitrogen pollution of Baiyangdian Lake.

**Table 3.** Pollution index values of different nitrogen forms in Baiyangdian Lake and its influent rivers.

| Sites | Season | Parameters | Standard (mg/L) | Min (mg/L) | Max (mg/L) | Mean (mg/L) | Exceedance Rate (%) |
|---|---|---|---|---|---|---|---|
| Baiyangdian Lake | Normal season | TN | 1.0 | 0.450 | 1.98 | 0.759 | 12.5 |
| | | $NO_3^-$-N | 10 | 0.002 | 0.135 | 0.056 | 0.0 |
| | | $NH_4^+$-N | 1.0 | 0.025 | 0.032 | 0.025 | 0.0 |
| | Flood season | TN | 1.0 | 0.502 | 1.54 | 0.786 | 18.8 |
| | | $NO_3^-$-N | 10 | 0.013 | 0.107 | 0.038 | 0.0 |
| | | $NH_4^+$-N | 1.0 | 0.025 | 0.025 | 0.025 | 0.0 |
| | Dry season | TN | 1.0 | 0.532 | 2.51 | 0.924 | 31.3 |
| | | $NO_3^-$-N | 10 | 0.010 | 0.201 | 0.053 | 0.0 |
| | | $NH_4^+$-N | 1.0 | 0.025 | 0.495 | 0.131 | 0.0 |
| River | Normal season | TN | 1.0 | 1.46 | 3.10 | 2.37 | 100.0 |
| | | $NO_3^-$-N | 10 | 0.090 | 0.240 | 0.174 | 0.0 |
| | | $NH_4^+$-N | 1.0 | 0.320 | 0.800 | 0.553 | 0.0 |
| | Flood season | TN | 1.0 | 0.560 | 4.06 | 2.12 | 71.4 |
| | | $NO_3^-$-N | 10 | 0.020 | 0.250 | 0.159 | 0.0 |
| | | $NH_4^+$-N | 1.0 | 0.040 | 0.130 | 0.074 | 0.0 |
| | Dry season | TN | 1.0 | 0.910 | 5.25 | 3.17 | 85.7 |
| | | $NO_3^-$-N | 10 | 0.020 | 0.430 | 0.238 | 0.0 |
| | | $NH_4^+$-N | 1.0 | 0.190 | 0.750 | 0.402 | 0.0 |

## 4. Discussion

### 4.1. Comparison of Nitrogen Concentrations in Different Types of Lakes

Baiyangdian Lake, the largest shallow lake in the North China Plain, is known as the "Pearl of North China". The quality of its water directly affects the ecological safety of the Xiong'an New Area and also impacts the regional ecological balance and biodiversity of the Beijing–Tianjin–Hebei urban agglomeration [19]. In recent years, national and local governments have carried out comprehensive management of Baiyangdian Lake and its tributaries, and ecological water replenishment measures, such as the Yellow River Diversion Project, have been implemented, leading to continuous improvement in the water quality of Baiyangdian Lake. In this study, we compared the nitrogen levels in Baiyangdian Lake with those in other lakes, including Tai Lake [20], Chao Lake [21], Dian Lake [22], Lugu Lake [23], and Bosten Lake [24] (Figure 6).

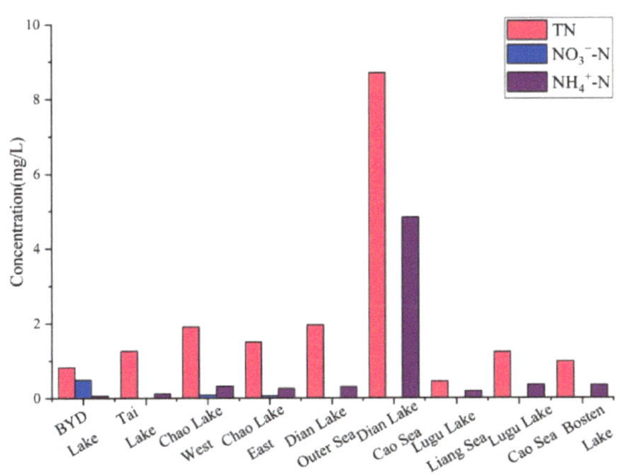


**Figure 6.** Comparison of nitrogen concentrations in lakes from different regions of China.

Baiyangdian Lake is located in the Xiong'an New Area, which has developed rapidly in recent years, and the national governance efforts have been significant, with a population density of about 700 people per square kilometer in the basin. The Tai Lake and Chao Lake basins are in more developed areas with larger population densities, at 650 and 420 people per square kilometer, respectively. Dian Lake and Lugu Lake are both plateau lakes located in the southwestern region of China. Dian Lake is situated to the southwest of Kunming with a population density of 320 people per square kilometer, while Lugu Lake, located at the junction of Sichuan and Yunnan, has a lower population density of 42 people per square kilometer. Bosten Lake is in the northwest region of China, where social and economic development are relatively lagging, with a population density of about 3 people per square kilometer. Compared to Tai Lake and Chao Lake, which have similar population densities, Baiyangdian Lake has lower concentrations of TN and ammonium nitrogen. Compared to the plateau lakes, Baiyangdian Lake has lower TN and ammonium nitrogen concentrations than Dian Lake but higher TN than Lugu Lake, which has a smaller population density. When compared to Bosten Lake, which has a lower population density, the concentration of TN in the Baiyangdian Lake is slightly lower. This differs somewhat from previous research findings. Previous studies have found that nitrogen concentrations in aquatic environments are positively correlated with the population density within the basin [25,26]. In recent years, comprehensive water quality management by national and local governments has led to year-by-year improvements in the water quality of Baiyangdian Lake. Additionally, compared to these other lakes, Baiyangdian Lake is abundant in reeds and other aquatic plants, and the slow water flow in Baiyangdian Lake facilitates the absorption of nitrogen elements by these plants. It is noteworthy that the concentration of ammonium nitrogen in Baiyangdian Lake is lower than that in these other lakes, which may be related to plants' preference for utilizing the more easily absorbed and assimilated $NH_4^+$-N [27]. Furthermore, previous research has found that the dissolved oxygen (DO) concentration in Baiyangdian Lake is relatively high, indicating that strong nitrification may occur in the water body [28].

### *4.2. Identification of Nitrogen Sources in Baiyangdian Lake*

Nitrogen pollution sources in the aquatic environment can be divided into point sources and non-point sources. Point source pollution typically includes industrial wastewater, urban domestic sewage, leachate from municipal solid waste, leakage from septic tanks and sewage pipelines in suburban areas, and seepage from animal farms; non-point-source pollution generally includes surface runoff pollution, irrigation with wastewater, and the application of chemical fertilizers and manure. Researchers commonly categorize nitrogen pollution sources into five types: agricultural fertilizers, animal manure, industrial wastewater and domestic sewage, soil nitrogen, and atmospheric deposition [29].

#### 4.2.1. Ion Ratio Method for Source Tracing

$Cl^-$ serves as an important auxiliary tool for identifying the sources of nitrogen in an aquatic environment. Its inert nature makes it less susceptible to the physical, chemical, and biological processes in the water body, making it a good indicator of sewage input [30]. Many scholars have used the relationship between $[NO_3^-]/[Cl^-]$ and $[Cl^-]$ to identify the sources of $NO_3^-$ in an aquatic environment. Yao et al. (2007) showed that high concentrations of $[Cl^-]$ and low $[NO_3^-]/[Cl^-]$ ratios indicate that $NO_3^-$ primarily comes from domestic sewage and feces [31]; low concentrations of $[Cl^-]$ and high $[NO_3^-]/[Cl^-]$ ratios suggest that $NO_3^-$ mainly originates from fertilizers; low values of both $[NO_3^-]/[Cl^-]$ and $[Cl^-]$ indicate that $NO_3^-$ primarily comes from soil organic nitrogen. To determine the sources of nitrogen in Baiyangdian Lake, this study plotted the relationship between $[NO_3^-]/[Cl^-]$ and $[Cl^-]$ (Figure 7). The $[NO_3^-]/[Cl^-]$ values in Baiyangdian Lake range from 0 to 0.080, with average values of 0.013, 0.017, and 0.019 for the normal, flood, and dry seasons, respectively. The low ratios suggest that fertilizers are not the main source of $NO_3^-$ in Baiyangdian Lake. Overall, the low $[NO_3^-]/[Cl^-]$ and high $[Cl^-]$ values in

Baiyangdian Lake indicate that $NO_3^-$ in the water body is mainly influenced by animal manure and domestic sewage. Previous studies have found that unpolluted water bodies generally have $Cl^-$ concentrations below 49.7 mg/L [12]. In Baiyangdian Lake, 77% of the sampling points have $Cl^-$ concentrations above 49.7 mg/L (the average $Cl^-$ concentrations measured during the three sampling periods for the normal, flood, and dry seasons are 117 mg/L, 72.8 mg/L, and 86.0 mg/L, respectively), further indicating that manure and domestic sewage are potential sources of $NO_3^-$ in Baiyangdian Lake. Seasonally, the $[NO_3^-]/[Cl^-]$ ratio is highest in the dry season, followed by the flood season, and then the normal season. The $[Cl^-]$ values vary significantly across different sampling periods, with the highest concentration in the normal season and the lowest in the flood season, suggesting more severe pollution from domestic sewage and feces during the normal season. Notably, during the flood season, the $[NO_3^-]/[Cl^-]$ values at sites L14 and L15 in the southeast area are relatively high, with $Cl^-$ concentrations below 1.40 mmol/L, possibly due to the predominance of agricultural land use and fewer man-made structures on the southeast shore of Baiyangdian Lake. The abundant rainfall in July may lead to stronger leaching of soil, and surface runoff carries nitrogen from the fields into the lake area. In the dry season, the $[NO_3^-]/[Cl^-]$ values at sites L1, L2, and L4 in the north are relatively high, with $Cl^-$ concentrations ranging from 1.73 to 1.84 mmol/L, possibly due to the concentrated distribution of farmland in the northern area, where agricultural activities have an important impact on the water quality at these sites.

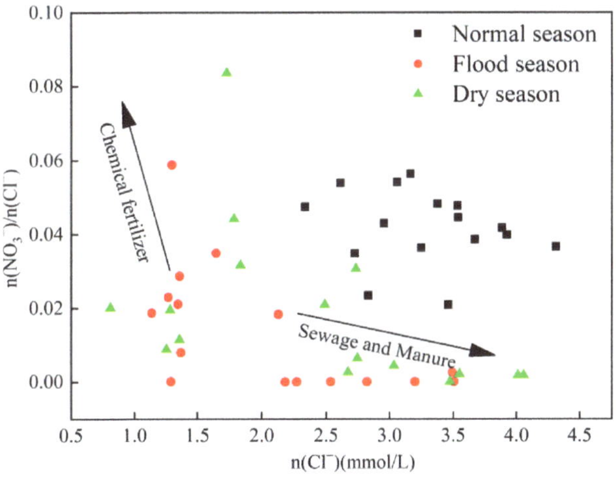


**Figure 7.** $[Cl^-]$ and $[NO_3^-]/[Cl^-]$ relationship for Baiyangdian Lake.

### 4.2.2. Identification of Nitrogen Pollution Sources in Baiyangdian Lake Using PCA

This study utilized PCA to identify the pollution sources in Baiyangdian Lake across three hydrological periods. During the normal water period, the first principal component (PC1) accounted for 37.4% of the total variance (Figure 8), and the variables that showed strong positive correlations with PC1 included TN, $NO_3^-$-N, $Ca^{2+}$ and $SO_4^{2-}$. It is generally believed that TN and $NO_3^-$-N in the aquatic environment primarily originate from human activities, such as the application of agricultural fertilizers, animal manure, soil nitrogen, sewage, and atmospheric deposition [32], while $SO_4^{2-}$ mainly comes from fertilizers, sewage, evaporite dissolution, rainfall, and industrial emissions [33]. Given the high concentration of TN in Baiyangdian Lake and the relative independence of PC1 and PC2, PC1 represents the influence of manure and fertilizer on the water quality of Baiyangdian Lake. Variables that showed strong positive correlations with the second principal component (PC2) were $Cl^-$ and $Na^+$, and a strong negative correlation was observed with $NH_4^+$-N. Typically, $NH_4^+$-N mainly comes from domestic sewage and fertilizers, and

$Cl^-$ primarily originates from domestic sewage, fertilizers, manure, road deicing salt, and the dissolution of natural minerals [34]. In this study, there were no rock salts, and no road deicing salts were used in that season. The high concentration of $Cl^-$ in Baiyangdian Lake is consistent with the characteristics of domestic sewage. Additionally, domestic sewage contains a certain amount of $Na^+$ [35]. This further indicates that the water body of Baiyangdian Lake is affected by domestic sewage, and therefore, PC2 indicates that the water quality of Baiyangdian Lake was impacted by domestic sewage.

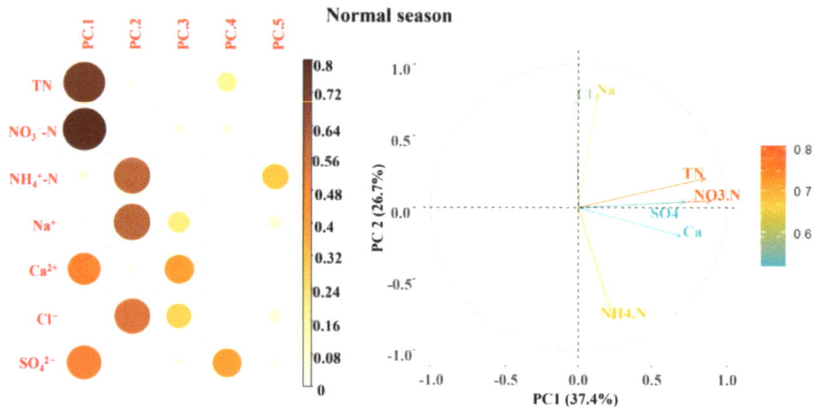


**Figure 8.** Principal component analysis results for Baiyangdian Lake during the normal period.

During the flood season, the first principal component (PC1) accounted for 65.5% of the total variance (Figure 9), with strong positive correlations observed with $Na^+$, $SO_4{}^{2-}$, and $Cl^-$, and strong negative correlations with TN, $NO_3{}^-$-N, and $Ca^{2+}$. As previously mentioned, $Na^+$ and $Cl^-$ in Baiyangdian Lake primarily originate from domestic sewage, while TN and $NO_3{}^-$-N mainly come from manure and fertilizers. Therefore, PC1 represents the influence of domestic sewage, manure, and fertilizer on the Baiyangdian Lake. The second principal component (PC2) showed a strong positive correlation with $NH_4{}^+$-N. However, during the flood season, the concentration of $NH_4{}^+$-N in Baiyangdian Lake is relatively low, close to the detection limit, indicating significant nitrification occurring in the water body. Thus, PC2 represents the impact of nitrification on the nitrogen in Baiyangdian Lake.

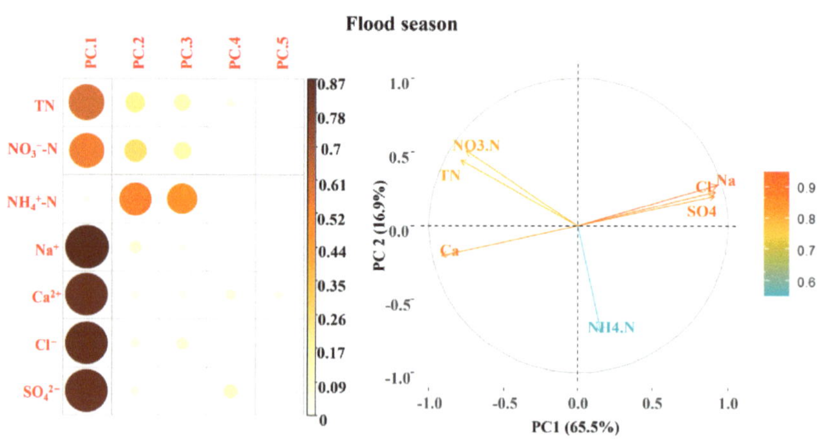


**Figure 9.** Principal component analysis results for Baiyangdian Lake during the flood period.

During the dry season (ice cover period), the first principal component (PC1) accounted for 39.5% of the total variance (Figure 10). There was a strong positive correlation with $Na^+$, a strong negative correlation with TN and $NO_3^-$-N, and a moderate positive correlation with $NH_4^+$-N and $Cl^-$. This suggests that the Baiyangdian Lake was influenced by domestic sewage. Furthermore, the negative correlation between $NH_4^+$-N and both TN and $NO_3^-$-N indicates that nitrification and denitrification processes were occurring in the water body. The second principal component (PC2) showed a strong positive correlation with $SO_4^{2-}$ and $Cl^-$, and a moderate positive correlation with $Ca^{2+}$, TN, and $NO_3^-$-N, indicating that PC2 represents the influence of manure and fertilizer on the Baiyangdian Lake.

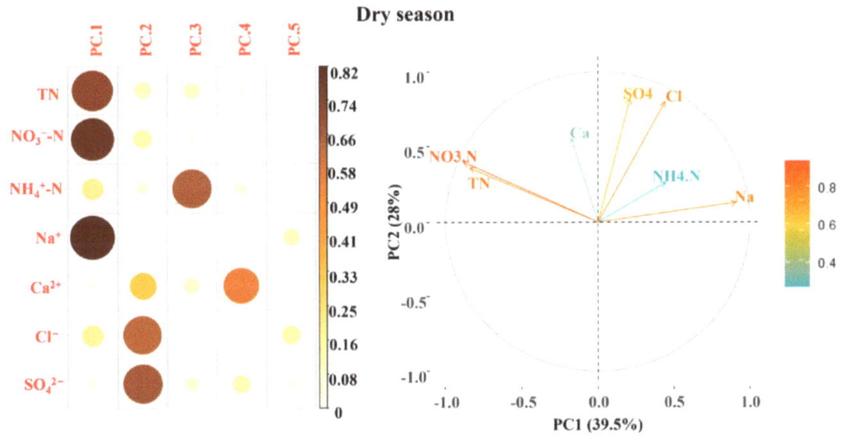


**Figure 10.** Principal component analysis results for Baiyangdian Lake during the dry period.

Overall, the main sources of nitrogen pollution in the Baiyangdian water body during the three hydrological periods are domestic sewage, manure, and fertilizers. However, the contribution of pollution sources to the nitrogen in the water body varies to some extent during different hydrological periods. During the dry season, due to the absence of rainfall runoff to drive the process, it is difficult for residual fertilizers in the fields to enter the water body, hence the main source of pollution during this period is domestic sewage. In contrast, during the flood and normal water periods, rainfall runoff and leaching carry manure and fertilizers from the fields into Baiyangdian Lake, making them the primary sources of nitrogen in the water body during these periods. Therefore, the main measures to control the increase of nitrogen concentration in Baiyangdian Lake are to prevent direct discharge of domestic sewage, implement scientific fertilization methods, and prevent agricultural non-point-source pollution.

## 5. Conclusions

The study provides a comprehensive assessment of nitrogen pollution in Baiyangdian Lake, identifying significant seasonal and spatial variations in nitrogen concentrations. The dry season exhibited the highest average total nitrogen (TN) levels, with 31.3% of sites surpassing national water quality standards, underscoring the lake's vulnerability to nitrogen pollution. Spatial analysis revealed that northern areas and sites proximal to influent rivers are more significantly impacted, likely due to point-source pollution and agricultural non-point-source pollution exacerbated by rainfall runoff. Utilizing the ion ratio method and principal component analysis (PCA), domestic sewage and manure were pinpointed as the predominant sources of nitrogen pollution, with agricultural activities also contributing to nitrogen levels. The study suggests that the water quality of the Baiyangdian Lake is relatively good compared to other lakes with similar human activity intensities, indicating effective management measures. To sustain the water quality and ecological balance, it is recommended to enhance monitoring and implement nitrogen reduction strategies. This includes curbing direct sewage discharge, adopting scientific fertilization strategies that are tailored to the specific conditions of time, location, and crop types, and bolstering ecological restoration efforts such as wetland recovery and vegetation restoration. Our research provides a scientific foundation for ongoing management of and pollution prevention in Baiyangdian Lake.

**Author Contributions:** Q.Z., investigation, methodology, software, data curation, and writing—original draft preparation. S.X., investigation, methodology, and software. L.Y., supervision, methodology, writing—review and editing, and supervision. All authors have read and agreed to the published version of the manuscript.

**Funding:** This work was funded by Open Foundation of the Key Laboratory of Coupling Process and Effect of Natural Resources Elements (No. 2024KFKT015), the Fundamental Research Funds for the Institute of Hydrogeology and Environmental Geology, Chinese Academy of Geological Sciences (No. YK202310), National Natural Science Foundation of China (Grant No. 42377232), Natural Science Foundation of Hebei Province of China (Grant No. D2022504015), and China Geological Survey, China (No. DD20221773).

**Data Availability Statement:** The data are available from the authors upon reasonable request.

**Acknowledgments:** The authors gratefully acknowledge the editor and anonymous reviewers for their valuable comments on this manuscript. The authors also appreciate the financial support from the different organizations.

**Conflicts of Interest:** The authors declare no conflict of interest.

## References

1. Dearig, J.A.; Yang, X.; Dong, X.; Zhang, E.; Chen, X.; Langdon, P.G.; Zhang, K.; Zhang, W.; Dawson, T.; Dawson, T.P. Extending the Timescale and Range of Ecosystem Services through Paleoenvironmental Analyses, Exemplified in the Lower Yangtze Basin. *Proc. Natl. Acad. Sci. USA* **2012**, *109*, 6808–6809.
2. Downing, J.A.; Prairie, Y.T.; Cole, J.J.; Duarte, C.M.; Tranvik, L.J.; Striegl, R.G.; McDowell, W.H.; Kortelainen, P.; Caraco, N.F.; Melack, J.M.; et al. The Global Abundance and Size Distribution of Lakes, Ponds, and Impoundments. *Limnol. Oceanogr.* **2006**, *51*, 2388–2397. [CrossRef]
3. Yang, M.N.; Zhu, L.; Liu, J.T.; Zhang, Y.X.; Zhou, B. Influence of water conservancy project on runoff in the source region of the Yellow River and wetland changes in the Lakeside Zone, China. *J. Groundw. Sci. Eng.* **2023**, *11*, 333–346. [CrossRef]
4. Zhu, Y.; Jin, X.; Tang, W.; Meng, X.; Shan, B. Comprehensive analysis of nitrogen distributions and ammonia nitrogen release fluxes in the sediments of Baiyangdian Lake, China. *J. Environ. Sci.* **2019**, *76*, 319–328. [CrossRef]
5. Zhai, T.L.; Zhang, Q.Q.; Wang, L.; Wang, H.W. Temporal and Spatial Variations in Hydrochemical Components and Driving Factors in Baiyangdian Lake in the Northern Plain of China. *J. Groundw. Sci. Eng.* **2024**, *12*, 293–308. [CrossRef]
6. Galbraith, L.M.; Burns, C.W. Linking Land-use, Water Body Type and Water Quality in Southern New Zealand. *Landscape Ecol.* **2007**, *22*, 231–241. [CrossRef]
7. Li, X.Y.; Li, H.P.; Jiang, S.Y.; Ma, P.; Lai, X.J.; Deng, J.C.; Chen, D.Q. Spatiotemporal Variations in Nutrient Loads in River-Lake System of Changdang Lake Catchment in 2016-2017. *Environ. Sci.* **2020**, *41*, 4042–4052. (In Chinese)
8. Liu, Y.; Zhang, Q. Study on the Spatiotemporal Variation in and Driving Mechanism of Water Quality in Baiyangdian Lake. *Water* **2024**, *16*, 166. [CrossRef]
9. Ding, K.; Zhang, Y.; Zhang, H.; Yu, C.; Li, X.; Zhang, M.; Zhang, Z.; Yang, Y. Tracing Nitrate Origins and Transformation Processes in Groundwater of the Hohhot Basin's Piedmont Strong Runoff Zone through Dual Isotopes and Hydro-Chemical Analysis. *Sci. Total Environ.* **2024**, *919*, 170799. [CrossRef]
10. Yu, S.S. The Water Quality Change and Driving Force Analysis in Xiaoxingkai Lake. Ph.D. Thesis, Northeast Institute of Geography and Agroecology, Chinese Academy of Sciences, Changchun, China, 2015.
11. Mbaye, M.L.; Gaye, A.T.; Spitzy, A.; Dähnke, K.; Afouda, A.; Gaye, B. Seasonal and Spatial Variation in Suspended Matter, Organic Carbon, Nitrogen, and Nutrient Concentrations of the Senegal River in West Africa. *J. Mar. Syst.* **2016**, *57*, 1–13. [CrossRef]
12. Zhang, Q.; Wang, H.; Wang, Y.; Yang, M.; Zhu, L. Groundwater Quality Assessment and Pollution Source Apportionment in an Intensely Exploited Region of Northern China. *Environ. Sci. Pollut. Res.* **2017**, *24*, 16639–16650. [CrossRef] [PubMed]
13. Yang, W.; Yan, J.; Wang, Y.; Zhang, B.; Wang, H. Seasonal variation of aquatic macrophytes and its relationship with environmental factors in baiyangdian lake, China. *Sci. Total Environ.* **2019**, *708*, 135112. [CrossRef] [PubMed]
14. Zhou, L.; Sun, W.; Han, Q.; Chen, H.; Chen, H.; Jin, Y.; Tong, R.; Tian, Z. Assessment of Spatial Variation in River Water Quality of the Baiyangdian Basin (China) during Environmental Water Release Period of Upstream Reservoirs. *Water* **2020**, *12*, 688. [CrossRef]

15. Liu, J.; Zhao, X.; Liu, Y.; Qiao, X.; Wang, X.; Ma, M.; Jin, X.; Liu, C.; Zheng, B.; Shen, J.; et al. High Contamination, Bioaccumulation and Risk Assessment of Perfluoroalkyl Substances in Multiple Environmental Media at the Baiyangdian Lake. *Ecol. Toxicol. Environ. Saf.* **2019**, *182*, 109454. [CrossRef] [PubMed]
16. Li, H.; Shen, H.; Li, S.; Liang, Y.; Zhang, L. Effects of Eutrophication on the Benthic-Pelagic Coupling Food Web in Baiyangdian Lake. *Acta Ecol. Sin.* **2018**, *38*, 2017–2020.
17. Tang, C.; Chen, D.; Yi, Y.; Zhou, Y.; Zhang, S. Effects of Ecological Water Supplement on Vegetation Dynamics in Lake Baiyangdian Wetland. *J. Lake Sci.* **2022**, *34*, 1197–1207. (In Chinese)
18. Xue, P.; Zhao, Q.; Wang, Y.; Geng, L.; Wang, D. Distribution Characteristics of Heavy Metals in Sediment-Submerged Macrophyte-Water Systems of Lake Baiyangdian. *J. Lake Sci.* **2018**, *30*, 1525–1536.
19. Cui, Q.; Pan, Y.; Zhang, H.; Sheng, N.; Dai, J. Elevated Concentrations of Per-Fluorohexanesulfonate and Other Per- and Polyfluoroalkyl Substances in Baiyangdian Lake (China): Source Characterization and Exposure Assessment. *Environ. Pollut.* **2018**, *241*, 684–691. [CrossRef]
20. Hu, Q. Analysis and Comprehensive Evaluation of Water Quality in Taihu Lake from 2011 to 2020. *J. Shantou Univ. (Nat. Sci.)* **2022**, *37*, 65–74. (In Chinese)
21. Wu, L.; Zhang, Y.; Chen, Y.; Geng, M.; Chen, X. Distribution of Water Environmental Factors and Evaluation of Trophic State in the River-Lake System of Chaohu Lake Basin. *J. Hydroecol.* **2023**, *44*, 65–72. (In Chinese)
22. Zhang, Z.; Liu, C.; Zhu, Y.; Li, J. Analysis on Temporal and Spatial Variation of Water Quality in Dianchi Lake from 1990 to 2020. *J. Southwest For. Univ.* **2023**, *43*, 141–151. (In Chinese)
23. Yu, H. Temporal and Spatial Distribution of Nitrogen and Phosphorus and Pollution Loads from Internal and External Sources in Lugu Lake. Ph.D. Thesis, Southwest Jiaotong University, Chengdu, China, 2022. (In Chinese)
24. Saimire, T. Study on the Characteristics and Evaluation of Water Pollution Changes in Lake Bosten. Ph.D. Thesis, Xinjiang Normal University, Urumqi, China, 2023. (In Chinese)
25. Chen, J.; He, D.; Cui, S. The Response of River Water Quality and Quantity to the Development of Irrigated Agriculture in the Last 4 Decades in the Yellow River Basin, China. *Water Resour.* **2003**, *39*, 1047. [CrossRef]
26. Zhang, Q.; Zhang, J.; Wang, H.; Zhai, T.; Liu, L.; Li, G.; Xu, Z. Spatial Patterns in Water Quality and Source Apportionment in a Typical Cascade Development River Southwestern China Using PMF Modeling and Multivariate Statistical Techniques. *Chemosphere* **2023**, *311*, 137–139. [CrossRef] [PubMed]
27. Liu, K.; Greaver, T.L. A Global Perspective on Belowground Carbon Allocation and Dynamics: Patterns and Controls. *Ecol. Lett.* **2010**, *3*, 1459–1474.
28. Han, Q.; Tong, R.; Sun, W.; Zhao, Y.; Yu, J.; Wang, G.; Shrestha, S.; Jin, Y. Anthropogenic Influences on the Water Quality of the Baiyangdian Lake in North China over the Last Decade. *Sci. Total Environ.* **2020**, *701*, 134929. [CrossRef]
29. Chen, X.; Ren, M.; Li, G.; Zhang, J.; Xie, F.; Zheng, L. Identification of nitrate accumulation mechanism of surface water in a mining-rural-urban agglomeration area based on multiple isotopic evidence. *Sci. Total Environ.* **2024**, *912*, 169123. [CrossRef]
30. Wang, W.; Song, X.; Ma, Y. Identification of Nitrate Source Using Isotopic and Geochemical Data in the Lower Reaches of the Yellow River Irrigation District (China). *Environ. Earth Sci.* **2016**, *75*, 936. [CrossRef]
31. Yao, X.; Guo, L.L.; Shih, T.H.; Gavin, S.; He, Z.H. Salinity of Animal Manure and Potential Risk of Secondary Soil Salinization through Successive Manure Application. *Sci. Total Environ.* **2007**, *383*, 106–114.
32. Zhang, Q.; Wang, H.; Liu, L.; Zhai, T.; Zhang, X. Multiple Isotopes Reveal the Driving Mechanism of High $NO_3^-$ Level and Key Processes of Nitrogen Cycling in the Lower Reaches of Yellow River. *J. Environ. Sci.* **2024**, *138*, 597–606. [CrossRef]
33. Torres-Martínez, J.; Mora, A.; Mahlknecht, J.; Kaown, D.; Barceló, D. Determining nitrate and sulfate pollution sources and transformations in a coastal aquifer impacted by seawater intrusion—A multi-isotopic approach combined with self-organizing maps and a bayesian mixing model. *J. Hazard. Mater.* **2021**, *417*, 126103. [CrossRef]
34. Zhang, W.; Jin, X.; Liu, D.; Chao, L.; Shan, B. Temporal and Spatial Variation of Nitrogen and Phosphorus and Eutrophication Assessment for a Typical Arid River—Fuyang River in Northern China. *J. Environ. Sci.* **2017**, *55*, 5. [CrossRef] [PubMed]
35. Liu, C.; Li, S.; Lang, Y.; Xiao, H. Using δ15N- and δ18O-values to identify nitrate sources in Karst ground water, Guiyang, Southwest China. *Environ. Sci. Technol.* **2006**, *40*, 6928–6933. [CrossRef] [PubMed]

water

MDPI

Article

# Integrated Modeling Approach to Assess Freshwater Inflow Impact on Coastal Water Quality

**Shreeya Bhattarai [1], Prem Parajuli [1,*] and Anna Linhoss [2]**

[1] Department of Agricultural and Biological Engineering, Mississippi State University, Starkville, MS 39762, USA
[2] Department of Biosystems Engineering, Auburn University, Auburn, AL 36849, USA
* Correspondence: pparajuli@abe.msstate.edu

**Abstract:** The quality of freshwater input from tributaries of the Western Mississippi Sound (WMSS) impacts the quality of coastal water. Hydrological and hydrodynamic models can be coupled to assess the impact of freshwater inflow from coastal watersheds. This study aims to compare the performance of a hydrodynamic model and a hydrological–hydrodynamic coupled model in detecting the effect of freshwater inflow from the coastal watersheds of the state of Mississippi into the WMSS. A hydrological model, the Soil and Water Assessment Tool (SWAT), and a hydrodynamic model, the visual Environmental Fluid Dynamics Code (vEFDC), were coupled to evaluate the difference between the hydrodynamical modelling approach, which employs an area-weighted approach to define flow and nutrient concentrations, and the more recent coupling model approach, which uses a hydrological model to determine the flow and nutrient load of the model. Furthermore, a nutrient load sensitivity analysis of the effect of freshwater inflow on water quality in the WMSS was conducted in addition to assessing the repercussions of tropical depressions. Hydrological assessments of the major tributaries watersheds of Saint Louis Bay (SLB) at the WMSS were performed using the SWAT model. After calibration/validation of the SWAT model, the streamflow output from the SWAT was incorporated into the vEFDC model. Finally, hydrodynamic simulation of the SWAT-vEFDC model was conducted, and water quality output was compared at different SLB locations. The salinity, dissolved oxygen, total nitrogen (TN), and total phosphorus (TP) were assessed by comparing the vEFDC and SWAT-vEFDC outputs. The results indicated that hydrological input from the SWAT alters the flow and nutrient concentration results as compared to an area-weighted approach. In addition, a major impact on the concentration of TN and TP occurred at the location where the freshwater flows into SLB. This impact diminishes further away from the point of freshwater inflow. Moreover, a 25% nutrient load variation did not demonstrate a difference in water quality at the WMSS besides TN and TP in a post-tropical depression scenario. Therefore, the SWAT-vEFDC coupled approach provided insights into evaluation of the area-weighted method, and of hydrological model output to the hydrodynamical model, the effect of freshwater inflow into coastal waters, and nutrient sensitivity analysis, which are important for integrated coastal ecosystems management.

**Keywords:** coupling models; SWAT; vEFDC; water quality assessment; nutrient sensitivity; extreme weather

## 1. Introduction

The unique environment present in aquatic ecosystems sustains diverse biodiversity and provides environmental benefits by preventing floods and serving as natural filters. Water quality assessment is essential for the sustainable management of aquatic ecosystems [1–3]. A comprehensive understanding of the source, transport, and fate of water contaminants is necessary to develop site-specific management strategies. Numerical modelling techniques offer time- and cost-effective measures to investigate hydrological dynamics under different scales and circumstances [4–9]. Approaches that integrate watershed and receiving water models have gained popularity due to their effective application in understanding water quality hydrodynamics [10–14]. Watershed models simulate hydrology using geomorphologic, land use, and meteorological data and can assess water quality [15]. Receiving water models are implemented in waterbodies, such as large rivers, reservoirs, wetlands, and estuaries, which receive water from the upstream watershed to simulate the hydrodynamics and water quality interactions within the waterbody. Bathymetric, meteorological, hydrological, and water quality data are required for modelling hydrodynamics and water quality in receiving waters [16,17]. Boundary conditions, such as inflow, outflow, tides, winds, water levels, and water quality, define the physical conditions at the boundary and are vital for achieving accurate model outputs [16,17]. Thus, watershed models can provide simulated flow and water quality data for upstream tributaries that lack continuously recorded data.

This study focused on the hydrodynamic and water quality assessment of the Western Mississippi Sound (WMSS), Gulf of Mexico. The WMSS water is designated for shellfish harvesting and recreational activities [18] and thus, water quality is a vital issue. The major sources of freshwater input into the WMSS are the Pearl River, the Jourdan River, the Wolf River, the Bonnet Carré Spillway, and several other smaller streams and bayous including Johnson Bayou, Bayou Portage, Four Dollar Bayou, and Rotten Bayou. In the past, the hydrodynamic model of the WMSS was developed and linked with the hydrodynamic model for the Lower Pearl River (LPR) using the visual Environmental Fluid Dynamic Code (vEFDC) [19,20] tool. In the linked model, the output of the LPR model was input into the WMSS grid, henceforth called the WMSS-LPR_vEFDC model, which represented the riverine flow, estuarine circulation and simulated flow, surface water elevation, temperature, and salinity [19]. The author suggested that the linked WMSS-LPR_vEFDC model was able to be integrated with the water quality and habitat suitability model to study the impact of hydrological variability. Later, the WMSS water quality model, henceforth called the vEFDC model, was developed by Bazgirkhoob et al. [21] to simulate dissolved oxygen at different vertical strata and with the changing seasons. Here, the WMSS-LPR_vEFDC model developed by Armandei et al. [19] was used in conjunction with the water quality model CE-QUAL-ICM to build the vEFDC model. This hydrodynamic water quality model was verified and validated for nitrogen, phosphorus, and chlorophyll-a [21]. Since measured water quality data from the Wolf River and the Jourdan River were unavailable, simulated water quality concentration data from the LPR model were used as inputs for these smaller rivers. It is likely that differences in geomorphology and land use in the smaller rivers compared to the Pearl River lead to differences in water quality flowing out of these rivers. Therefore, to address the research gap in previous studies, where an area-weighted approach was used to determine the flow and nutrient concentration of the Wolf River watershed and the Jourdan River watershed from the proximately located Pear River watershed model, this study aims to incorporate SWAT model flow and nutrient data from the Wolf River and the Jourdan River into the vEFDC model to evaluate the water quality at various locations within WMSS. For this purpose, the coupled hydrological and hydrodynamic models in the present study are henceforth called the SWAT-vEFDC model. The major objective of this study is to compare the output of the vEFDC model developed with an area-weighted approach and the SWAT_vEFDC coupled model to see the variation in the simulated flow and water quality parameters. Moreover, this study assesses the impact of freshwater flows from the coastal watersheds of Mississippi into the WMSS by coupling the SWAT and vEFDC models. This study focuses on coupling two watershed models of the Wolf River and the Jourdan River with the vEFDC model to make a composite river–bay–estuary modelling system for assessing the effect of freshwater introduction on the water quality at various locations in SLB and the WMSS. This study builds upon a previous water quality assessment where the dissolved oxygen level was analyzed throughout the WMSS considering the calculated freshwater inflow and water quality concentration from the Wolf River and the Jourdan River on the basis of the weighted average method obtained from the Pear River model [21]. The findings of this study indicated that a satisfactory trend and correlation between the area-weighted method flow and hydrologically simulated flow was observed. Varying concentrations of total nitrogen (TN) and total phosphorus (TP) were observed at the freshwater inflow area, but were later minimized while moving further towards the WMSS.

## 2. Materials and Methods

### 2.1. Study Area

The location of the study area is shown in Figure 1a–c, illustrating the WMSS, the Jourdan River watershed (JRW), the Wolf River watershed (WRW), and their respective locations in the state of Mississippi, USA. The WMSS is a semi-enclosed estuary in the northern Gulf of Mexico which has a total surface area of 2255 km$^2$ [21]. It is bounded to the east by the Mississippi and Louisiana coastlines, including the city of Gulfport; to the south by Ship Island, Cat Island, and the Biloxi Marsh; and to the west by Lake Borgne. The Jourdan River and the Wolf River are the major sources of freshwater input, draining into the SLB of the WMSS. Forestry land use dominates the JRW, whose estimated area is 538 km$^2$. The land use in JRW consists of 37.9% evergreen forest, 26.6% rangeland, 18.26% pasture, and 17.3% forested wetlands. Likewise, WRW is also a forest-dominated watershed. The WRW area is estimated to be 801 km$^2$, of which 7.7% is pasture, 20.8% is forested wetland areas, 24.9% is range-brush areas, and 46.6% is evergreen forests [22]. The WMSS is designated for shellfish harvesting and recreational activities [18], which aid Mississippi's economy.

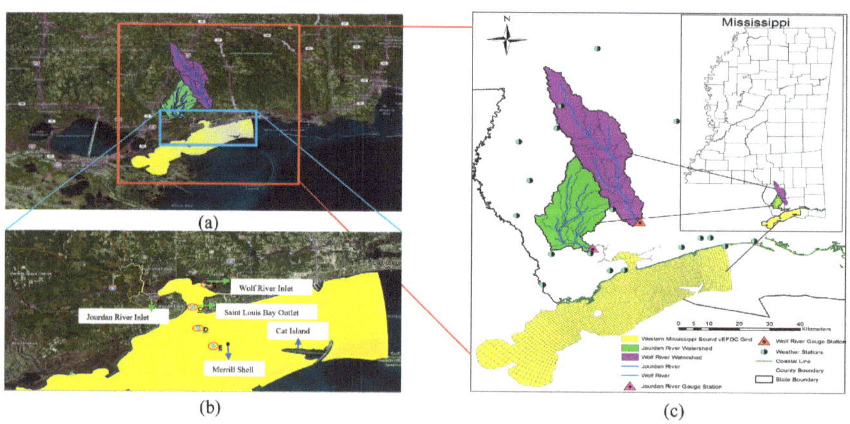


**Figure 1.** (**a**) Wider view of the Western Mississippi Sound showing the vEFDC grid in yellow, the SWAT-delineated upstream Jourdan River watershed in green, and the Wolf River watershed in violet. (**b**) Image showing water quality assessment cell in the Western Mississippi grid at locations A, B, C, D, and E. (**c**) Map of the study area showing the Wolf River watershed, the Jourdan River watershed, and the Western Mississippi Sound grid in the state of Mississippi.

### 2.2. Description of SWAT and EFDC Models and Their Coupling Process

The SWAT [23] is a comprehensive hydrological model which is used to simulate and predict the impact of land management practices and climate change on water quantity and water quality [24]. It is a watershed scale model used in different topographies to simulate a wide range of environmental processes including surface runoff, groundwater flow, erosion, nutrient transport, and crop yield [22,24–27]. The basic input data required to set up, calibrate, and validate the SWAT model are elevation, land use, land cover, soil, climate, management practices, and continuous timestep stream flow data. The SWAT model has been widely used in recent studies for watershed management, climate change assessment, nutrient management, integrated water resource management, and ecosystem services such as carbon sequestration, with basic data and detailed objective-specific data [27–30]. This mathematical model divides the watershed into networks of subbasins that are further divided into unique hydrologic response units (HRUs) based on land use, soil type, and slope class [15]. The model simulates water balance and nutrient movement throughout the watershed.

In this study, the SWAT model setup was carried out by collecting site-specific elevation, land use, soil, climate, and streamflow data as listed in Table 1. Two separate models were set up for the JRW and the WRW. The JRW was divided into 13 subbasins with 233 HRUs and the WRW was divided into 15 subbasins with 489 HRUs. Both models were calibrated and validated according to the parameters listed in Bhattarai et al. [22] using the autocalibration tool SWAT Calibration and Uncertainty Program with Sequential Uncertainty Fitting version 2 (SWAT-CUP SUFI-2) [31]. The JRW was calibrated on a daily timestep from 10 March 2002 to 31 December 2003 and validated from 1 January 2004 to 30 September 2004, utilizing all the observed data available from the single gauge station USGS 02481660 at the JRW. In addition, the WRW was calibrated from January 1997 to December 2003 and validated from January 2004 to December 2010 on a monthly timestep, using the data collected from the USGS 02481510 gauge station. After adjusting the different parameters of both models, acceptable statistical values for comparing the observed and simulated flows were attained. The flowchart in Figure 2 illustrates the process of the SWAT model setup, calibration, and validation. Finally, the simulated streamflow from the calibrated/validated SWAT model was incorporated into the vEFDC model to study the impact of freshwater inflow into the coastal water.

**Table 1.** Basic data and their corresponding sources for the SWAT model setup.

| S. No. | Data | Source |
|---|---|---|
| 1 | Elevation Data:<br>Digital Elevation Model (DEM) (30 m × 30 m) (2020) | United States Geological Survey (USGS)<br>(https://apps.nationalmap.gov/viewer/) (accessed on 4 October 2022) |
| 2 | Land-use and Land-cover Data:<br>Cropland Data Layer (CDL) (2010) | United States Department of Agriculture-National Agricultural Statistics Service (USDA-NASS)<br>(https://nassgeodata.gmu.edu/CropScape/) (accessed on 4 October 2022) |
| 3 | Soil Data:<br>USSURGO (2020) | United States Soil Survey Geographic Database (US-SSURGO)<br>SWAT-USSURGO (https://swat.tamu.edu/data/) (accessed on 4 October 2022) |
| 4 | Weather Data:<br>NOAA (1995–2010)<br>Precipitation,<br>Maximum Temperature,<br>Minimum Temperature | National Oceanic and Atmospheric Administration (NOAA)<br>SWAT—Climate Data (https://swat.tamu.edu/data/) (accessed on 4 October 2022) |
| 5 | Discharge Data:<br>-USGS 02481660 (2002–2005)<br>(Jourdan River Nr Bay St Louis)<br>-USGS 02481510 (1997–2010)<br>(Wolf River Nr Landon) | United States Geological Survey (USGS)<br>(https://waterdata.usgs.gov/ms/nwis/) (accessed on 5 October 2022) |

The EFDC [16,17] is a state-of-the-art hydrodynamical modelling tool used to predict water quality and anthropogenic impacts on rivers, estuaries, lakes and reservoirs, wetlands, and coastal areas. The EFDC can use a three-dimensional hydrodynamic module to simulate the movement and mixing of water and contaminants in the water column and assess the sediment and nutrient transport and fate [16,17]. The EFDC is a numerical modelling tool that can simulate a wide variety of physical and biological phenomena, such as temperature, salinity gradient, dissolved oxygen (DO) level, nutrient load, growth, and decay of phytoplankton. The vEFDC [20] is a graphical user interface of the EFDC tool which offers a convenient way to create meshes, set up and run EFDC simulations, and visualize and analyze outputs. This is a multifunctional tool that can be coupled with watershed models to provide a more comprehensive analysis of the water body [19,21].

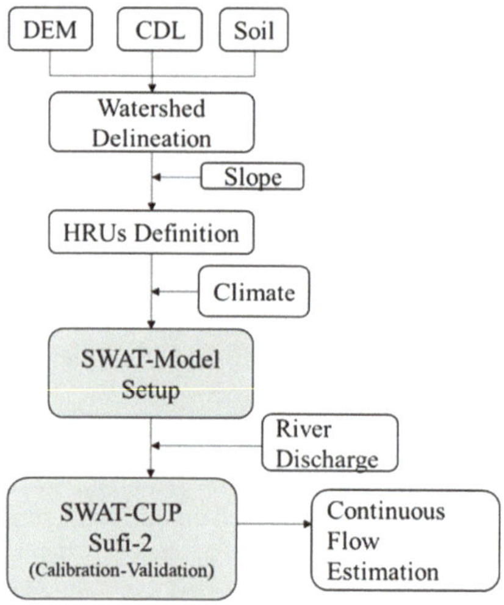


**Figure 2.** Flowchart of the SWAT model setup. (DEM: Digital Elevation Model, CDL: Cropland Data Layer, HRU: Hydrologic Response Units, SWAT-CUP Sufi-2: SWAT Calibration and Uncertainty Programs with Sequential Uncertainty Fitting version 2).

Using the vEFDC tool, an integrated 3D hydrodynamic model of the WMSS and the LPR was developed and calibrated by Armandei et al. [19], but it was not used for the simulation of water quality. Later, Bazgirkhoob et al. [21] assessed DO level, seasonality, and vertical stratification by integrating the vEFDC model and a water quality model named CE-QUAL-ICM. In the present study, the calibrated/validated flow from the SWAT model simulation of the JRW and the WRW are incorporated into the qser.inp file of the vEFDC model to couple the vEFDC model and the SWAT model to assess the impact of freshwater inflow on the concentration of DO, salinity, TN, and TP at different locations in the SLB and the WMSS. Figure 3 illustrates the flowchart of the SWAT-vEFDC coupling process:

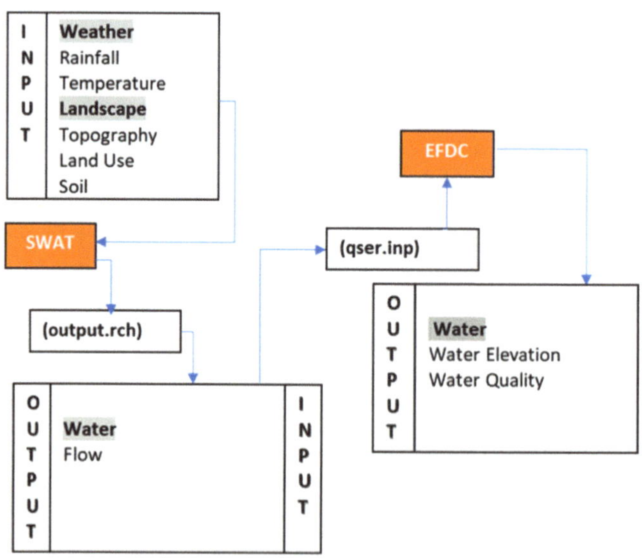


**Figure 3.** Schematic diagram of the SWAT-vEFDC model coupling process.

## 2.3. Water Quality Parameters

Salinity is the amount of dissolved salt in water. Salinity levels vary widely between freshwater, where it measures less than 1000 ppm, and coastal water, where it can exceed 35,000 ppm [32]. Climatic factors and anthropogenic activities can greatly alter salinity [33]. In the EFDC, salinity serves as a hydrodynamic parameter that affects the buoyancy and mixing of water. It can also be used as a conservative tracer that verifies the transport process of the model and assists in the analysis of mass conservation. Salinity is essential for estuarine aquatic life. Additionally, salinity affects the chemical processes [34] influencing the transportation and fate of contaminants in aquatic systems.

Dissolved oxygen (DO) is a critical parameter of aquatic ecosystems, referring to the amount of oxygen dissolved in the water that is available for the survival of aquatic life and for various biogeochemical functions. The amount of dissolved oxygen in a water body affects the way nutrients move through the aquatic ecosystem [35]. It is a major component of water quality models, such as the EFDC, and it is important to monitor and manage DO levels when managing water quality. Nitrogen and phosphorus are critical nutrients [36] in aquatic ecosystems as their availability can limit the growth and production of primary producers. Thus, monitoring TN and TP levels is essential for water quality management [37]. In this assessment, the spatial analysis of the TN and TP parameters in filtered (Filt) and unfiltered (Unfilt) states was also observed. Total nitrogen (TN) refers to the sum of organic nitrogen, ammonia ($NH_4$), nitrate ($NO_3$), and nitrite ($NO_2$). In the vEFDC, TN is simulated as five state variables: three in organic forms (refractory particulate, liable particulate, and dissolved) and two in inorganic forms (ammonium and nitrate) [16,17]. In the model, the nitrate state variable represents the total of nitrate and nitrite.

TN = Dissolved inorganic nitrogen (DIN) + Dissolved organic nitrogen (DON) + Particulate organic nitrogen (PON)

i.e.,

TN = [Nitrate + Nitrite + Ammonia] + [Dissolved organic nitrogen] + [Refractory particulate organic nitrogen (RPON) + Labile particulate organic nitrogen (LPON)]

Therefore,

TN Filt = Dissolved inorganic nitrogen (DIN) + Dissolved organic nitrogen (DON)

TN Unfilt = Dissolved inorganic nitrogen (DIN) + Dissolved organic nitrogen (DON) + Particulate organic nitrogen (PON)

Total phosphorus (TP) refers to the sum of dissolved and particulate phosphorus. In vEFDC, TP is simulated as four state variables: three in organic forms (refractory particulate, liable particulate, and dissolved) and one in an inorganic form (dissolved and particulate phosphate) [16,17].

TP = Dissolved inorganic phosphorus (DIP) + Dissolved organic phosphorus (DOP) + Particulate organic phosphorus (POP)

i.e.,

TP = [Total orthophosphate (PO4)] + [Dissolved organic phosphorus (DOP)] + [Refractory particulate organic phosphorus (RPOP) + Labile particulate organic phosphorus (LPOP)]

Therefore,

TP Filt = Dissolved inorganic phosphorus (DIP) + Dissolved organic phosphorus (DOP)

TP Unfilt = Dissolved inorganic phosphorus (DIP) + Dissolved organic phosphorus (DOP) + Particulate organic phosphorus (POP)

## 3. Results

### 3.1. Flow Assessment: Calibration and Validation

The statistical measures used to calibrate/validate flow and assess the accuracy of the SWAT model included the coefficient of determination ($R^2$), the Nash–Sutcliffe efficiency (NSE), percentage bias (PBAIS), and the Kling–Gupta efficiency (KGE). The KGE was defined as the objective function in the SWAT-CUP. In the JRW, the $R^2$ varied from 0.59 to 0.72, the NSE varied from 0.42 to 0.71, the PBAIS varied from 36.7 to 4.7, and the KGE varied from 0.55 to 0.68, for daily flow. Since the KGE was defined as the objective function, it indicated intermediate performance of the JRW model during calibration and validation, which was in the acceptable range. In the WRW, the $R^2$ varied from 0.82 to 0.75, the NSE varied from 0.81 to 0.73, the PBAIS varied from −4.9 to −3, and the KGE varied from 0.80 to 0.70, for monthly flow. These results also indicated an acceptable performance of the SWAT model during calibration and validation [22,25,38]. For the calibration and validation of the vEFDC model, measured DO data were compared across the simulated data and achieved good statistics in terms of root mean square error, bias and $R^2$ at 11 different locations throughout the WMSS grid [21].

The graphs in Figure 4 show the comparison of the flow data used in the vEFDC model obtained using area-weighted measurement and the SWAT-vEFDC model simulated using a hydrological modelling tool for the Wolf River and the Jourdan River. Time series analysis of these two flow trends showed higher peaks in the area-weighted flow series. Statistical metrices demonstrated $R^2 = 0.66$ and NSE = 0.61 for the Wolf River flow; and $R^2 = 0.80$ and NSE = 0.66 for the Jourdan River. These results signify high goodness of fit for the two datasets.

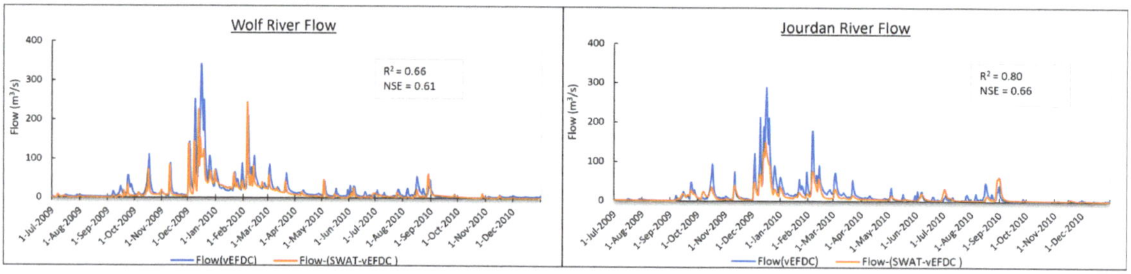


**Figure 4.** Flow comparison between the vEFDC model and the SWAT-vEFDC model.

### 3.2. Water Quality Assessment

The SWAT and vEFDC models were coupled and run from 1 January 2009 to 31 December 2010, skipping the first six months as a warmup period. The temporal variation in the concentration of DO, salinity, TN, and TP as simulated by the vEFDC model and the SWAT-vEFDC coupled model were compared at different locations, as presented in Figure 5a–e. Five locations, namely A, B, C, D, and E, as shown in Figure 1b, were selected to compare the vEFDC and SWAT-vEFDC simulation outputs. Location A (cell I = 62, J = 37, K = 4) is the Jourdan River inlet, location B (cell I = 68, J = 38, K = 4) is the Wolf River inlet, location C (cell I = 66, J = 31, K = 4) is where SLB opens up to the WMSS (approx. radial distance 7 km away from location A and 6 km away from location B), location D (cell I = 65, J = 25, K = 4) is in between the outlet of Saint Louis Bay and Merrill Shell (approx. radial distance 11 km away from location A and 12 km away from location B) and location E (cell I = 66, J = 16, K = 4) lies near Merrill Shell on the WMSS (approx. radial distance 16 km away from location A and 15 km away from location B). The statistical results obtained by comparing the results of the vEFDC and SWAT-vEFDC models are presented in Table 2. The influence of the SWAT-vEFDC model outflow was observed for salinity, dissolved oxygen, and nutrients in the Jourdan inlet (A), the Wolf inlet (B) and the SLB outlet (C) while, in contrast, there was minimal distinction between the models' outputs at location D and location E.

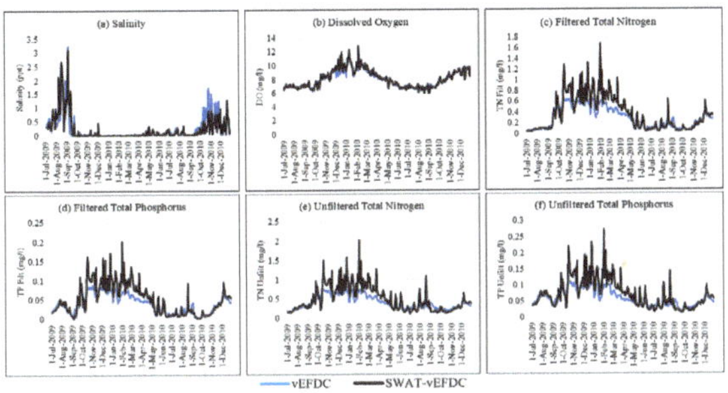


**(a)**

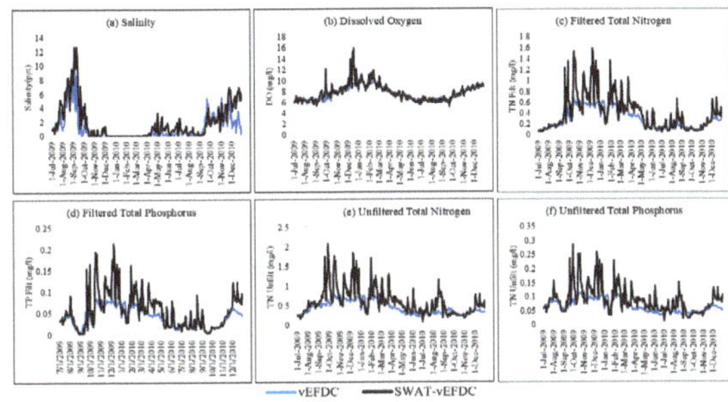


**(b)**

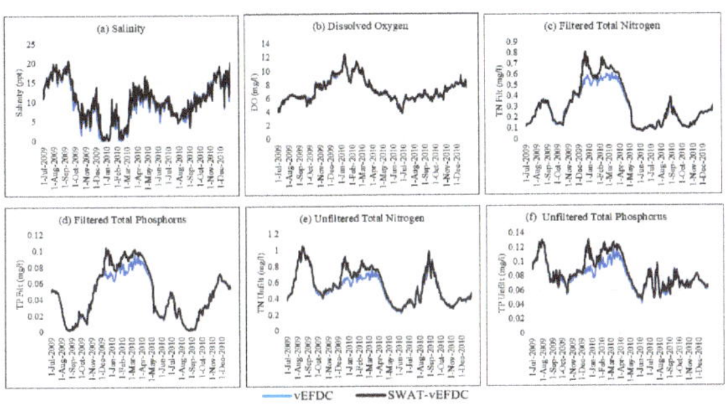


**(c)**

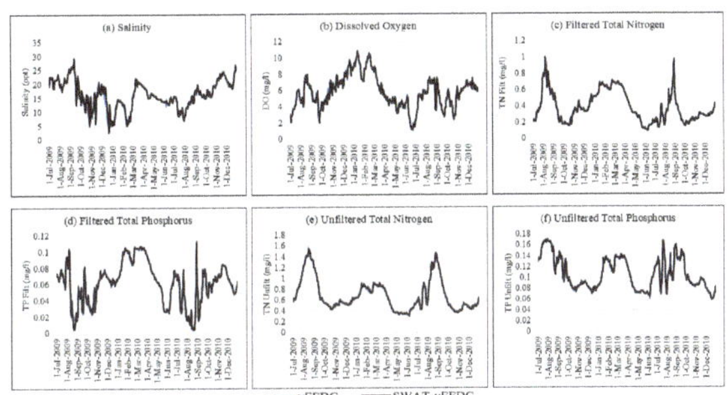


**(d)**

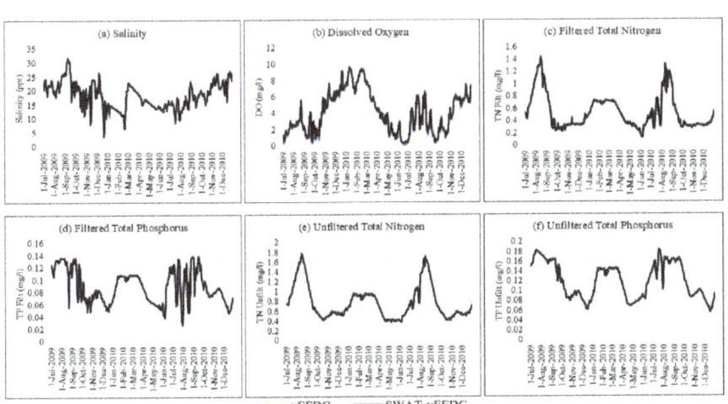


**(e)**

**Figure 5.** Temporal variation of salinity, dissolved oxygen, filtered total nitrogen, filtered total phosphorus, unfiltered total nitrogen, and unfiltered total phosphorus at the inlet of the Jourdan River at (**a**) location A, (**b**) location B, (**c**) location C, (**d**) location D and (**e**) location E. In (**d**,**e**), the vEFDC graph and the SWAT-vEFDC graph are overlapped.

**Table 2.** Summary of the statistical comparison between the vEFDC and SWAT-vEFDC models.

| Location | Salt | | DO | | TN Filt | | TP Filt | | TN Unfilt | | TP Unfilt | |
|---|---|---|---|---|---|---|---|---|---|---|---|---|
| | $R^2$ | NSE | $R^2$ | NSE | $R^2$ | NSE | $R^2$ | NSE | $R^2$ | NSE | $R^2$ | NSE |
| A | 0.83 | 0.83 | 0.92 | 0.88 | 0.84 | 0.32 | 0.82 | 0.23 | 0.76 | −0.3 | 0.75 | −0.3 |
| B | 0.76 | 0.39 | 0.76 | 0.57 | 0.75 | − 0.35 | 0.71 | −0.6 | 0.54 | −2.4 | 0.50 | −2.3 |
| C | 0.98 | 0.97 | 1 | 1 | 0.97 | 0.9 | 0.97 | 0.93 | 0.94 | 0.9 | 0.89 | 0.76 |
| D | 0.99 | 0.99 | 1 | 1 | 1 | 1 | 0.99 | 0.99 | 1 | 1 | 1 | 1 |
| E | 0.99 | 0.99 | 0.99 | 0.98 | 0.98 | 0.98 | 1 | 1 | 1 | 1 | 1 | 1 |

Increased differences were observed in the water quality parameters, especially TN Filt, TP Filt, TN Unfilt, and TP Unfilt, along the transects where freshwater flows into SLB, as shown in Figure 6. The graphs in Figure 6 compare the average daily concentration from 1 July 2009 to 31 December 2010 of the vEFDC and SWAT-vEFDC models plotted together.

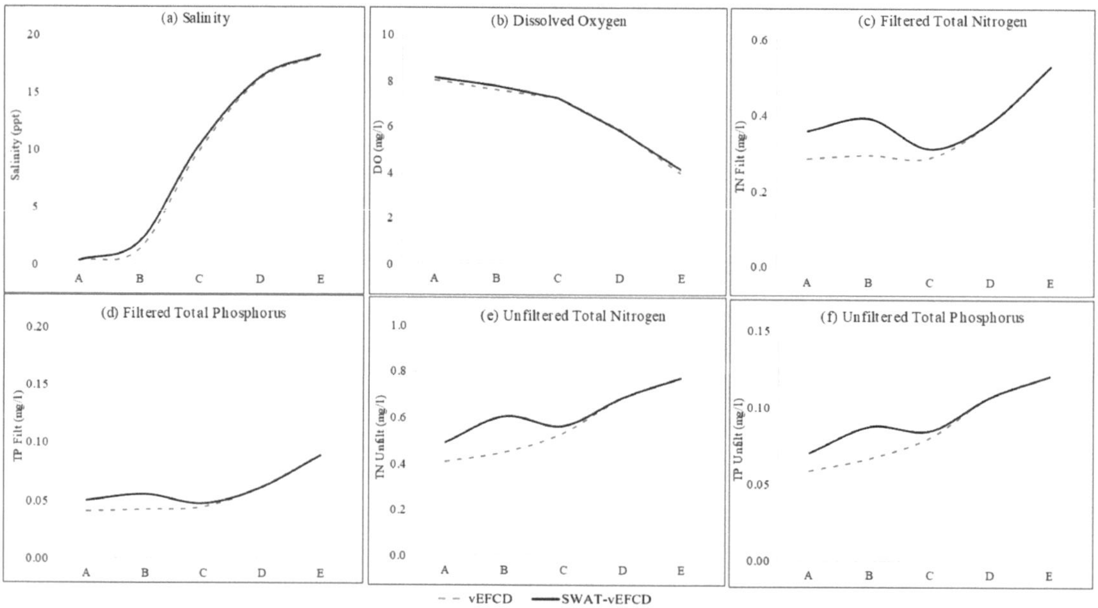


**Figure 6.** Average daily concentration of (**a**) salinity, (**b**) dissolved oxygen, (**c**) filtered total nitrogen, (**d**) filtered total phosphorus, (**e**) unfiltered total nitrogen, and (**f**) unfiltered total phosphorus from 1 July 2009 to 31 December 2010 simulated by the vEFDC and SWAT-vEFDC models at locations A, B, C, D, and E.

The box plot in Figure 7a–f shows the distribution of the concentration of DO, salinity, TN Filt, TP Filt, TN Unfilt and TP Unfilt at different locations. These plots are discussed individually in the following sections.

Salinity: Statistically comparing the performance of both models in terms of salinity, $R^2 = 0.83$ and NSE = 0.83 for location A. The average salinity concentration was 0.25 ppt in the vEFDC model and 0.23 ppt in the SWAT-vEFDC model. Similarly, for location B, $R^2 = 0.76$ and NSE = 0.39. Here, the concentration of salinity changed from 1.23 ppt to 1.88 ppt between the vEFDC and SWAT-vEFDC models. At location C, where $R^2 = 0.98$ and NSE = 0.97, the average daily simulated salinity concentration increased from 9.83 ppt in the vEFDC model to 10.27 ppt in the SWAT- vEFDC model. At location D, the results obtained were 0.99 for both $R^2$ and NSE. Here, the daily average salinity concentration changed from 16.19 ppt to 16.26 ppt. Further down, at location E, the average daily salinity was 18.24 ppt in both the models, resulting in a 0.99 $R^2$ and 0.99 NSE. Both models showed lower salinity near the river outlets and higher salinity further away from the river outlets and toward the models' southern boundary. The concentration of salinity was the highest at location E. This shows that differences in the estimation of salinity concentration resulted from the two models.

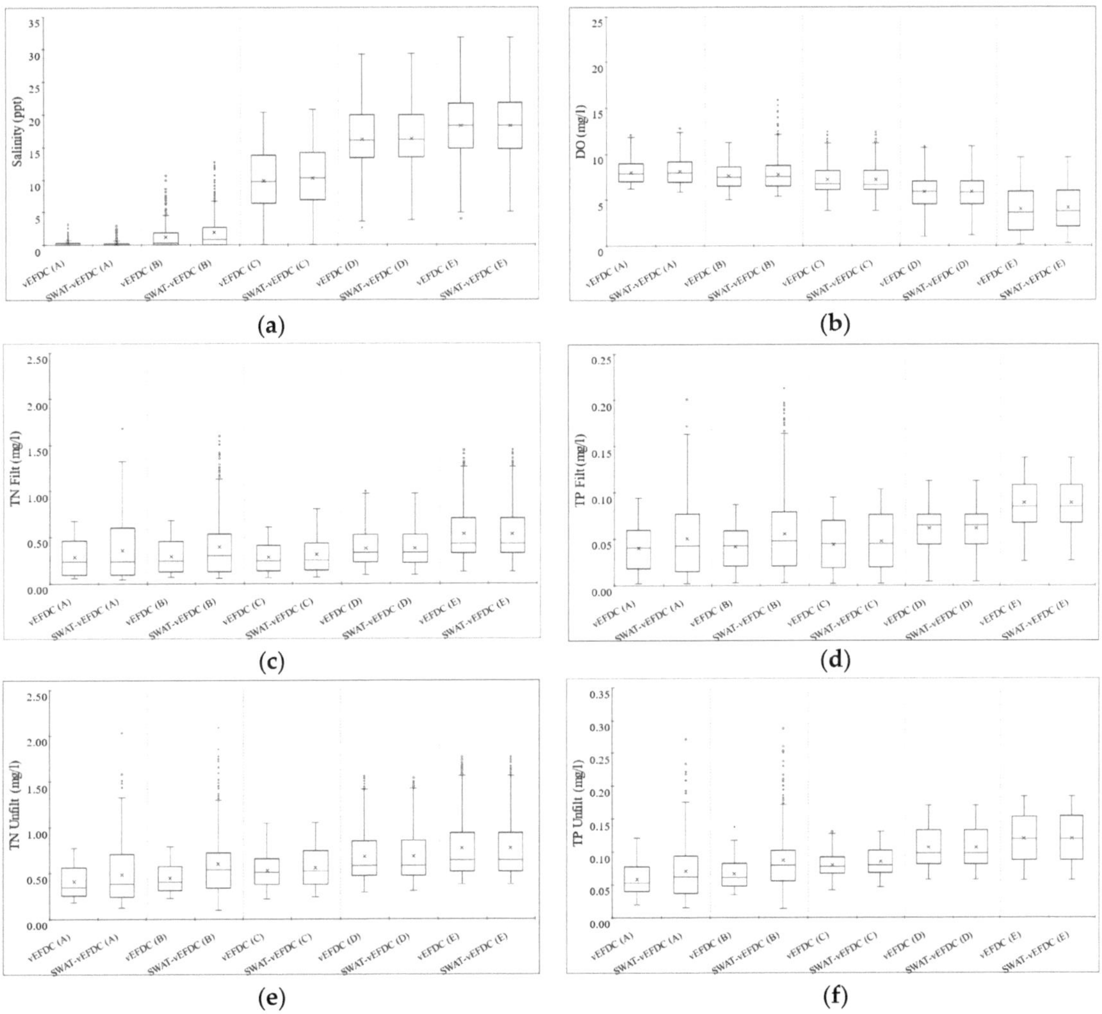


**Figure 7.** Distribution of (**a**) salinity, (**b**) dissolved oxygen (DO), (**c**) filtered total nitrogen, (**d**) filtered total phosphorus, (**e**) unfiltered total nitrogen and (**f**) unfiltered total phosphorus simulated by the vEFDC and SWAT-vEFDC models at locations A, B, C, D, and E across the Western Mississippi Sound.

Dissolved oxygen: Evaluating the performance of both models with respect to DO, $R^2 = 0.92$ and NSE = 0.88 were obtained for location A, with an average DO level increasing from 8.01 mg/L to 8.12 mg/L. At location B, the average daily DO was $R^2 = 0.76$ and NSE = 0.57, with the DO level changing from 7.62 mg/L to 7.81 mg/L. At location C, the average daily DO was $R^2 = 1.00$ and NSE = 1.00, with the DO level decreasing from 7.24 mg/L to 7.22 mg/L. At location D, $R^2$ and NSE were both 1.00, with the DO level changing from 5.85 mg/L to 5.84 mg/L. At location E, the average daily DO was $R^2$ = 0.99 and NSE = 0.98, with the DO level increasing from 3.95 mg/L to 4.12 mg/L. The trends of simulated DO from both the models show that DO was highest at location A and location B, which are freshwater inlets. DO at location C was 7.24 mg/L and 7.22 mg/L as simulated by the vEFDC model and the SWAT-EFDC models, respectively. The DO decreased as it moved along the Western Mississippi sound near Merrill Shell, and was observed to be lowest at location E. These results demonstrate that the SWAT-vEFDC model estimated an increased difference in DO compared to that estimated by the vEFDC model.

TN Filt: Assessing vEFDC and SWAT-vEFDC model performance at estimating TN Filt, $R^2$ = 0.84 and NSE = 0.32 results were obtained for location A, with the average TN Filt level increasing from 0.28 mg/L to 0.36 mg/L. At location B, the average daily TN Filt was $R^2$ = 0.75 and NSE = −0.35, with the TN Filt level changing from 0.29 mg/L to 0.39 mg/L. At location C, the average daily TN Filt was $R^2$ = 0.97 and NSE = 0.90, with the TN Filt level changing from 0.29 mg/L to 0.31 mg/L. At location D, $R^2$ and NSE were both 1.00, with the TN Filt level remaining unchanged at 0.38 mg/L. At location E, $R^2$ and NSE were both 0.98, with the TN Filt level remaining constant at 0.53 mg/L. The trends of simulated TN Filt show that TN Filt was low at location A and location B, which are freshwater inlets. TN Filt level at location C was 0.29 mg/L and 0.31 mg/L as simulated by the vEFDC model and the SWAT-vEFDC integrated models, respectively. The TN Filt level increased nearer to Merrill Shell and was observed to be highest at location E. These results demonstrated that there were increases difference between the concentrations of TN Filt estimated by the vEFDC model alone and by the SWAT-vEFDC model.

TP Filt: Similarly, comparing the TP Filt obtained from the vEFDC model and the SWAT-vEFDC model, $R^2$ = 0.82 and NSE = 0.23 results were obtained at location A, with the average TP Filt level increasing from 0.04 mg/L to 0 mg/L. At location B, $R^2$ = 0.71 and NSE = −0.56, with the TP Filt level changing from 0.04 mg/L to 0.06 mg/L. At location C, $R^2$ = 0.97 and NSE = 0.93, with the TP Filt level changing from 0.04 mg/L to 0.05 mg/L. At Location d, $R^2$ and NSE were both 0.999, with the TP Filt level remaining unchanged at 0.06 mg/L. At location E, $R^2$ and NSE were both 1, with the TP Filt level being constant at 0.09 mg/L. The trends of simulated TP Filt level showed that TP Filt was low at location A, location B, and location C. TP Filt level at location C was 0.04 mg/L and 0.05 mg/L as simulated by the vEFDC model and the SWAT-vEFDC models, respectively. The TP Filt level increased nearer to Merrill Shell and was observed to be highest at location E. Again, these results demonstrated that there were increased differences between the estimation of TP Filt from the vEFDC model and the SWAT-vEFDC model.

TN Unfilt: Likewise, comparing the performance of TN Unfilt between the vEFDC model and the SWAT-vEFDC model, $R^2$ = 0.76 and NSE = −0.25 results were obtained at location A, with the average TN Unfilt level increasing from 0.41 mg/L to 0.49 mg/L. At location B, $R^2$ = 0.54 and NSE = −2.37, with the TN Unfilt level changing from 0.45 mg/L to 0.60 mg/L. At location C, $R^2$ = 0.94 and NSE = 0.89, with the TN Unfilt level increasing from 0.53 mg/L to 0.56 mg/L. At location D, $R^2$ and NSE were both 1.00, with the TN Unfilt level remaining constant at 0.68 mg/L. Similarly, at location E, the average daily $R^2$ and NSE were both 1.00, with the TN Unfilt remaining unchanged at 0.77 mg/L. The trends of simulated TN Unfilt show that TN Unfilt was lower at locations A, B, and C as compared to locations D and E. The TN Unfilt level at location C was 0.53 mg/L and 0.56 mg/L as simulated by the vEFDC model and the SWAT-vEFDC models, respectively. The TN Unfilt level increased as it moved along the Western Mississippi Sound near Merrill Shell, and was observed to be highest at location E. These results demonstrate considerable difference between the estimation of TN Unfilt from the vEFDC model and the SWAT-vEFDC model.

TP Unfilt: Evaluating model performance for TP Unfilt between the vEFDC model and the SWAT-vEFDC model, $R^2$ = 0.75 and NSE = −0.26 results were obtained at location A, with the average TP Unfilt level increasing from 0.06 mg/L to 0.07 mg/L. At location B, $R^2$ = 0.50 and NSE = −2.26, with the TP Unfilt level increasing from 0.07 mg/L to 0.09 mg/L. At location C, $R^2$ = 0.89 and NSE = 0.76, with the TP Unfilt level remaining constant at

0.08 mg/L. At location D, $R^2$ and NSE were both 1.00, with the TP Unfilt level remaining constant at 0.11 mg/L. Likewise, at location E, $R^2$ and NSE were both 1.00, with the TP Unfilt level remaining constant at 0.12 mg/L. The trends of simulated TP Unfilt show that TP Unfilt was lowest at locations A, B, and C. The TP Unfilt level at location C remained constant at 0.008 mg/L as simulated by both the vEFDC model and the SWAT-vEFDC coupled model. The TP Unfilt level increased as it moved along the Western Mississippi Sound near Merrill Shell, and was observed to be highest at location E. These results demonstrate that levels of TP Unfilt varied between the vEFDC and SWAT-vEFDC models.

### 3.3. Sensitivity Analysis

The impact of changing riverine inflow nutrient concentrations on water quality within the WMSS was studied during an extreme weather event. Various water quality parameters from the Jourdan River and the Wolf River inflows were adjusted. A 25% increase and 25% decrease in RPOP, LPOP, DOP, $PO_4$, RPON, LPON, DON, $NH_4$, and $NO_2$ + $NO_3$ were made one at a time in the model to perform the nutrient sensitivity analysis [39]. From 10 to 18 August 2010, a tropical depression named Five 2010 (https://oceanservice.noaa.gov/news/historical-hurricanes/) (accessed on 3 August 2023) passed through the watershed area. Therefore, 5 August, 15 August, and 25 August were selected as the representative dates on which to observe the changes in nutrition distribution throughout the grid in order to account for 25% increase, baseline, and 25% decrease scenarios. In Figure 8a–e, the changes in water quality concentration during the tropical depression are shown for DO, TN Filt, TP Filt, TN Unfilt, and TP Unfilt at selected locations in the WMSS.

The impact of nutrient concentration changes on DO levels throughout the selected locations in the WMSS during the tropical depression is shown in Figure 8a. Before the tropical depression, in the scenario with a 25% nutrient load increase, the DO level was high closer to the freshwater inflow area, the adjacent SLB area, and the area towards the end of the WMSS grid. The DO level was observed reducing with the occurrence of the tropical depression. After the tropical depression was over, the number of grids with lower DO levels increased in the midsection of the WMSS. Nutrient variation shows notable change in the 25% increment scenario before the tropical depression, whereas notable variation in DO level in the inflow from the JRW and the WRW was not observed in the nutrient load reduction scenario throughout the WMSS grid. All the images in Figure 8a show that the DO level was higher at the boundary and lower in the inside section of the WMSS grid.

Figure 8b shows TN Filt sensitivity analysis due to the nutrient load variation during the tropical depression throughout the WMSS. Notable effects of the nutrient load variation were not observed near the inflow zone and throughout the WMSS grid. The major impact due to the tropical depression was detected as represented by a higher concentration of TN Filt in Figure 8b, and its impact was observed prevailing through 25 August.

Similarly, in Figure 8c, the impact of nutrient load variation from a freshwater inflow in TP Filt concentration during the tropical depression throughout the WMSS is presented. Notable effects of nutrient load variation were not seen near the inflow zone and throughout the WMSS; a rather more notable impact was observed because of the extreme weather. Higher concentrations of TP Filt were observed before and after the tropical depression, and a decrease in the concentration of TP Filt was observed during the tropical depression. The reason for this could be a tropical depression-induced flash flood draining a larger volume of freshwater into the WMSS.

Figure 8d represents the TN Unfilt sensitivity analysis due to the nutrient load variation in the freshwater during Five throughout the WMSS. Before the tropical depression, freshwater inflow did not show notable variations in TN Unfilt concentration near the inflow zone. However, during the tropical depression, more influence from the coastal water was observed in the WMSS reaching SLB. The TN Unfilt concentration at the WRW inflow zone was observed more in the baseline and 25% increase nutrient load scenarios, surpassing the 25% reduced nutrient load scenario. An aftereffect of Five was observed in the post-tropical depression scenario, where TN Unfilt concentration was higher in all nutrient load scenarios.

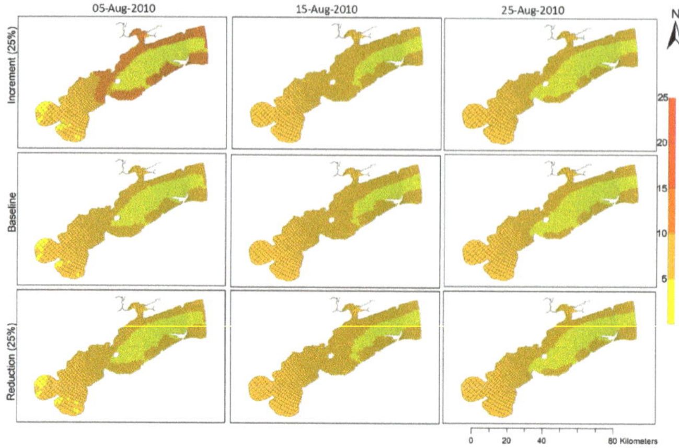


(a)

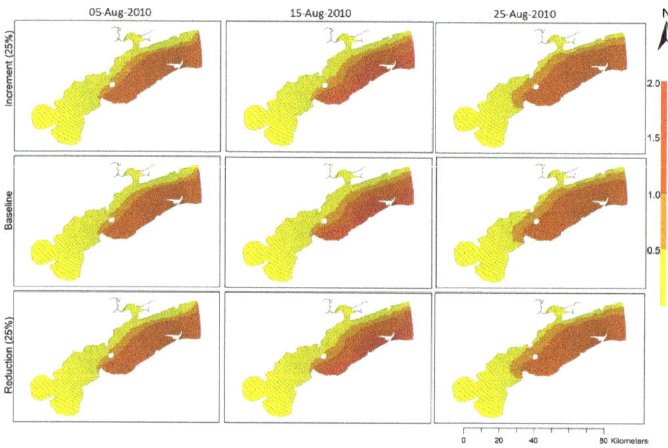


(b)

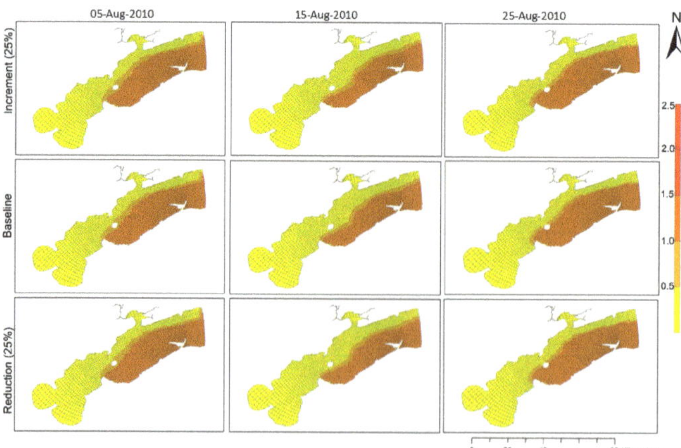


(c)

**Figure 8.** *Cont.*

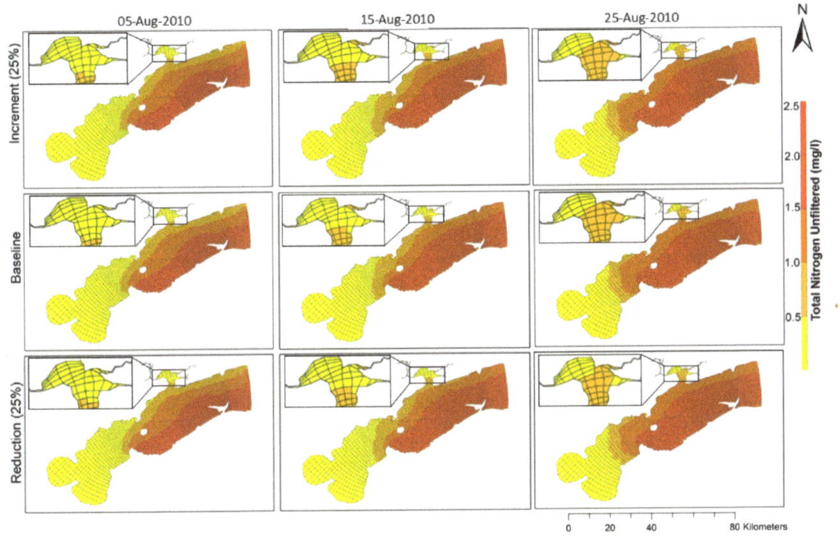


(d)

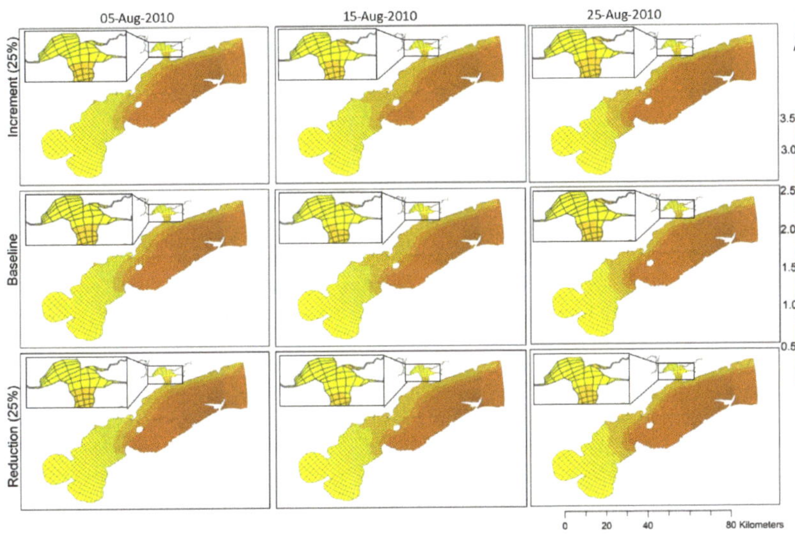


(e)

**Figure 8.** (**a**) Sensitivity analysis of nutrient load variation in DO level throughout the WMSS grid. (**b**) Sensitivity analysis of nutrient load variation in total nitrogen filtered concentration throughout the WMSS grid. (**c**) Sensitivity analysis of nutrient load variation in total phosphorus filtered concentration throughout the WMSS grid. (**d**) Sensitivity analysis of nutrient load variation in total nitrogen unfiltered concentration throughout the WMSS grid. (**e**) Sensitivity analysis of nutrient load variation in total phosphorus unfiltered concentration throughout the WMSS grid.

Similarly, Figure 8e shows the sensitivity analysis of nutrient load variation on TP Unfilt concentration during the tropical depression, throughout the WMSS grid. An extended impact on TP Unfilt concentration was seen reaching SLB in all scenarios. The image shows that during and after the occurrence of Five, the concentration of TP Unfilt increased at the WRW inflow zone as compared to that before the tropical depression in all nutrient load scenarios.

## 4. Discussion

In this study, the concept of coupling hydrological and hydrodynamical models for a better understanding of hydrodynamical processes in the WMSS has been demonstrated. Graphical comparisons of flow simulated by the SWAT model and that obtained through the weighted-area model showed similar trends, where the weighted-area flow showed higher peaks and the statistical metrices demonstrated goodness of fit. In addition, the introduction of freshwater simulated by the coupled model showed a substantial change in the water quality parameters. A pronounced impact on TN and TP was observed at the point where freshwater from the Wolf River and the Jourdan River flows into the bay. This effect decreased towards the end of bay and further diminished as the water moved away from the bay. The weakening effect was due to the increasing mixture of freshwater in the coastal water. Ultimately, the impact of the freshwater was greatly reduced by the time it reached the Western Mississippi Sound. Nutrient sensitivity analysis during extreme weather conditions demonstrated that a 25% variation in the nutrient load of freshwater inflow had lower influence in the water quality of the WMSS in comparison to the influence of wind and storm surges during a tropical depression.

The results of this study are consistent with previous studies which assessed water quality by coupling the hydrodynamics model with the hydrological model. Hwang et al. [10] evaluated different water quality improvement scenarios in an estuarine reservoir, with a levee in middle, which was affected by point and non-point source pollution from the upstream basin and thus concluded that SWAT-EFDC coupling could improve water quality results. Wu et al. [13] stated that the combination of the SWAT and EFDC models addressed the limitations in measured data of upstream watersheds, and it was feasible to identify the association between sources of pollution and water quality response. Kim et al. [12] developed a watershed–estuary linkage model with the HSPF-EFDC-WASP tool to access the impact of the drainage gate operation on water quality in the Ganwol Reservoir, Korea. The study reported that the outcome of the simulations for different water quality parameters across various locations of the reservoir was different and a pronounced effect was observed at the location near to the drainage gate operation, the inlet point [12], similar to the findings of our study. Additionally, Shin et al. [40] demonstrated the influence of wind during hurricane events in the variation of water quality parameters at Lake Okeechobee, corresponding to the outcomes of our analysis.

This study assessed the water quality at five different locations of SLB and observed variations in the effects of freshwater inflow. This study also highlights the diminishing effect of freshwater inflow when the nutrient parameters were observed further along the coastal water body. The study attempted to fill the gap in previous hydrodynamical modelling assessments of the Western Mississippi Sound by using hydrologically simulated flow data instead of using area-weighted flow data from an adjacent watershed in the hydrodynamical model. The results of this model-coupling approach are consistent with previous studies at different sites around the globe, which were able to discuss the relationship between the spatial variation of water quality parameters within coastal waters.

## 5. Conclusions

This study focused on the coupling of two watershed-scale SWAT models developed for the Wolf River and the Jourdan River with the vEFDC model of the estuary and bay linking the Western Mississippi Sound to make an integrated river–bay–estuary model. The calibrated/validated streamflow and nutrients data from the hydrological and water quality SWAT model were provided as an input to the calibrated/validated hydrodynamic vEFDC model to understand the execution performance of the hydrodynamic model with area-weighted flow and hydrologically simulated flow. The trend of area-weighted flow and hydrological-model simulated flow was studied. Additionally, the impact of freshwater inflow on the water of Saint Louis Bay and the Western Mississippi Sound was also assessed along with the impact of nutrient load variation and extreme weather. Evaluation of salinity, dissolved oxygen, total nitrogen, and total phosphorus at various locations in Saint Louis Bay and the Western Mississippi Sound was conducted with the results obtained from the vEFDC and SWAT-vEFDC model simulations. After comparing the concentrations of water quality parameters obtained from the vEFDC and SWAT-vEFDC models, this study revealed that the freshwater inflow had a substantial impact on the levels of TN and TP in the bay, which gradually decreased as the water diluted when moving further along into the Western Mississippi Sound. Nutrient sensitivity analysis during the tropical depression demonstrated that a 25% variation in the nutrient load inflow contributed less than the influence of wind and storm surges in changing the water quality. In conclusion, coupling hydrological models with hydrodynamical models can be useful in assessing the impact of freshwater inflows into estuarine systems. The SWAT-vEFDC coupled approach thus provided insight into the variations in the calculated and simulated flow and water quality parameters. This study also provides insights into approaches to assessing the impact of freshwater inflows into coastal waters, understanding the model coupling method to couple a watershed model and a receiving water model can develop an integrated river–bay–estuary modelling system which can aid in integrated coastal management.

In conclusion, this study provides an insight into water quality management and aquatic ecosystem response during hurricanes. The results can be helpful in the prediction of nutrient surges in ecosystems due to runoff during extreme weather. Sediment and nutrient loading can be minimized from the coastal watershed into estuaries, wetlands, and bays by scientific result-driven policies. Moreover, habitat restoration efforts can be optimized which in turn will aid the economy of the state of Mississippi. The interaction between storm surges, watershed runoff, and sediment–nutrient transport and extreme weather conditions helps support coastal planning and emergence preparedness, and enhances understanding of ecosystem resiliency and extreme event impacts on coastal areas. This dynamic modelling approach will eventually aid in integrated coastal zone management, formulating long-term adaptation strategies to mitigate extreme weather. This study improves our understanding of applying models to the Wolf and Jourdan Rivers, which are two major tributaries feeding the Saint Louis Bay. However, future application of such models to other small bayous, such as Johnson, Portage, Four Dollar, and Rotten, may help to provide additional information on the hydrodynamics of the Saint Louis Bay. In addition, the availability and use of longer-time observed data for validating the model's performance might further improve model calibration and validation processes including sensitivity analysis of the model's parameters.

**Author Contributions:** S.B.: methodology, model development, data analysis, and writing original draft. P.P.: conceptual, supervision, funding acquisition, reviewing, and editing. A.L.: reviewing and editing. All authors have read and agreed to the published version of the manuscript.

**Funding:** This project was paid for in part with federal funding from the U.S. Department of the Treasury, the Mississippi Department of Environmental Quality, and the Mississippi-Based RESTORE Act Center of Excellence under the Resources and Ecosystems Sustainability, Tourist Opportunities, and Revived Economies of the Gulf Coast States Act of 2012 (RESTORE Act). The statements, findings, conclusions, and recommendations are those of the authors and do not necessarily reflect the views of the Department of the Treasury, the Mississippi Department of Environmental Quality, or the Mississippi-Based RESTORE Act Center of Excellence. We would like to acknowledge the partial support from the Bagley College of Engineering, Mississippi State University.

**Data Availability Statement:** The original contributions presented in the study are included in the article, further inquiries can be directed to the corresponding author.

**Acknowledgments:** We would like to acknowledge the partial support of the Bagley College of Engineering, Mississippi Agricultural and Forestry Experiment Services (MAFES), and the College of Agriculture and Life Sciences at Mississippi State University, the United States Geological Survey (USGS), and all our collaborators in this study.

**Conflicts of Interest:** The authors declare that they do not have any possible financial interests or personal relationships that could have influenced the work. No conflicts of interest are declared by the authors.

## References


1. Jordan, S.J.; Benson, W.H. Introduction Sustainable Watersheds: Integrating Ecosystem Services and Public Health Supplementary Issue: Ecosystem Services and Environmental Health. *Environ. Health Insights* **2015**, *9*, EHI-S19586. [CrossRef]
2. Meshesha, T.W.; Wang, J.; Melaku, N.D. Modelling Spatiotemporal Patterns of Water Quality and Its Impacts on Aquatic Ecosystem in the Cold Climate Region of Alberta, Canada. *J. Hydrol.* **2020**, *587*, 124952. [CrossRef]
3. Prasad, M.B.K.; Long, W.; Zhang, X.; Wood, R.J.; Murtugudde, R. Predicting Dissolved Oxygen in the Chesapeake Bay: Applications and Implications. *Aquat. Sci.* **2011**, *73*, 437–451. [CrossRef]
4. Bhusal, A.; Ghimire, A.B.; Thakur, B.; Kalra, A. Evaluating the Hydrological Performance of Integrating PCSWMM and NEXRAD Precipitation Product at Different Spatial Scales of Watersheds. *Model. Earth Syst. Env.* **2023**, *9*, 4251–4264. [CrossRef]
5. Fabian, P.S.; Kwon, H.-H.; Vithanage, M.; Lee, J.-H. Modeling, Challenges, and Strategies for Understanding Impacts of Climate Extremes (Droughts and Floods) on Water Quality in Asia: A Review. *Env. Res.* **2023**, *225*, 115617. [CrossRef]
6. Flipo, N.; Gallois, N.; Schuite, J. Regional Coupled Surface–Subsurface Hydrological Model Fitting Based on a Spatially Distributed Minimalist Reduction of Frequency Domain Discharge Data. *Geosci. Model. Dev.* **2023**, *16*, 353–381. [CrossRef]
7. Parajuli, U.; Bhusal, A.; Babu Ghimire, A.; Shin, S. Comparing HEC-HMS, PCSWMM, and Random Forest Models for Rainfall-Runoff Evaluation to Extreme Flooding Events. In Proceedings of the World Environmental and Water Resources Congress 2023: Adaptive Planning and Design in an Age of Risk and Uncertainty–Selected Papers from World Environmental and Water Resources Congress, Henderson, NV, USA, 21–24 May 2023; pp. 1250–1262. [CrossRef]
8. Sun, G.; Wei, X.; Hao, L.; Sanchis, M.G.; Hou, Y.; Yousefpour, R.; Tang, R.; Zhang, Z. Forest Hydrology Modeling Tools for Watershed Management: A Review. *Ecol. Manag.* **2023**, *530*, 120755. [CrossRef]
9. Tarpanelli, A.; Paris, A.; Sichangi, A.W.; O'Loughlin, F.; Papa, F. Water Resources in Africa: The Role of Earth Obser vation Data and Hydrodynamic Modeling to Derive River Discharge. *Surv. Geophys.* **2022**, *44*, 97–122. [CrossRef]
10. Hwang, S.; Jun, S.M.; Song, J.H.; Kim, K.; Kim, H.; Kang, M.S. Application of the SWAT-EFDC Linkage Model for as sessing Water Quality Management in an Estuarine Reservoir Separated by Levees. *Appl. Sci.* **2021**, *11*, 3911. [CrossRef]
11. Hwang, S.; Jun, S.M.; Kim, K.; Kim, S.H.; Lee, H.; Kwak, J.; Kang, M.S. Development of a Framework for Evaluating Water Quality in Estuarine Reservoir Based on a Resilience Analysis Method. *J. Korean Soc. Agric. Eng.* **2020**, *62*, 105–119. [CrossRef]
12. Kim, S.; Kim, S.; Hwang, S.; Lee, H.; Kwak, J.; Song, J.H.; Jun, S.M.; Kang, M.S. Impact Assessment of Water-Level Management on Water Quality in an Estuary Reservoir Using a Watershed-Reservoir Linkage Model. *Agric. Water Manag.* **2023**, *280*, 108234. [CrossRef]
13. Wu, L.; Chen, Z.; Ding, X.; Liu, H.Y.; Wang, D.Q. Research on Water Environmental Capacity Accounting of the Yongzhou Section of Xiangjiang River Basin Based on the SWAT-EFDC Coupling Model. *J. Water Clim. Chang.* **2022**, *13*, 1106–1122. [CrossRef]
14. Zhang, C.; Huang, Y.; Javed, A.; Arhonditsis, G.B. An Ensemble Modeling Framework to Study the Effects of Climate Change on the Trophic State of Shallow Reservoirs. *Sci. Total Environ.* **2019**, *697*, 134078. [CrossRef] [PubMed]
15. Neitsch, S.L.; Arnold, J.G.; Kiniry, J.R.; Srinivasan, R.; Williams, J.R. SWAT2000 User's Manual VERSION 2000. 2005. Available online: https://www.scribd.com/document/145722143/SWAT2000-User-s-Manual (accessed on 7 March 2023).
16. Tetra Tech Inc. *The Environmental Fluid Dynamics Code User Manual US EPA*, Version 1.01; Tetra Tech Inc.: Fairfax, VA, USA, 2007.
17. Tetra Tech Inc. *The Environmental Fluid Dynamics Code Theory and Computation Volume 3: Water Quality Module*; Tetra Tech Inc.: Fairfax, VA, USA, 2007; Volume 2, p. 96.
18. Mississippi Commission on Environmental Quality. *Regulations for Surface Water Quality Criteria for Intrastate, Interstate, and Coastal Waters*; MDEQ: Jackson, MI, USA, 2021; pp. 39289–40385.
19. Armandei, M.; Linhoss, A.C.; Camacho, R.A. Hydrodynamic Modeling of the Western Mississippi Sound Using a Linked Model System. *Reg. Stud. Mar. Sci.* **2021**, *44*, 101685. [CrossRef]
20. Tetra Tech Inc. *Visual EFDC 2.0 User's Guide, A Grid Generation, Editing, Pre- and Post-Processing Tool for EFDC Hydrodynamic and Water Quality Modeling 2018*; Tetra Tech Inc.: Fairfax, VA, USA, 2018.
21. Bazgirkhoob, H.; Linhoss, A.; Armandei, M. A Numerical Tool for Dissolved Oxygen Simulation in the Western Mississippi Sound. *J. Coast. Res.* **2022**, *38*, 699–711. [CrossRef]
22. Bhattarai, S.; Parajuli, P.B. Best Management Practices Affect Water Quality in Coastal Watersheds. *Sustainability* **2023**, *15*, 4045. [CrossRef]
23. Arnold, J.G.; Srinivasan, R.; Muttiah, R.S.; Williams, J.R. Large Area Hydrologic Modeling and Assessment Part 1: Model Development1. *J. Am. Water Resour. Assoc.* **1998**, *34*, 73–89. [CrossRef]
24. Wang, Y.; Jiang, R.; Xie, J.; Zhao, Y.; Yan, D.; Yang, S. Soil and Water Assessment Tool (SWAT) Model: A Systemic Review. *J. Coast. Res.* **2019**, *93*, 22–30. [CrossRef]
25. Bhattarai, S.; Parajuli, P.B.; To, F. Comparison of Flood Frequency at Different Climatic Scenarios in Forested Coastal Watersheds. *Climate* **2023**, *11*, 41. [CrossRef]
26. Venishetty, V.; Parajuli, P.B.; To, F. Assessing the Effect of Spatial Variation in Soils on Sediment Loads in Yazoo River Watershed. *Hydrology* **2023**, *10*, 62. [CrossRef]

27. Venishetty, V.; Parajuli, P.B. Assessment of BMPs by Estimating Hydrologic and Water Quality Outputs Using SWAT in Yazoo River Watershed. *Agriculture* **2022**, *12*, 477. [CrossRef]
28. Francesconi, W.; Srinivasan, R.; Pérez-Miñana, E.; Willcock, S.P.; Quintero, M. Using the Soil and Water Assessment Tool (SWAT) to Model Ecosystem Services: A Systematic Review. *J. Hydrol.* **2016**, *535*, 625–636. [CrossRef]
29. Liu, M.; Chen, X.; Yao, H.; Chen, Y. A Coupled Modeling Approach to Evaluate Nitrogen Retention within the Shanmei Reservoir Watershed, China. *Estuar. Coast. Shelf Sci.* **2015**, *166*, 189–198. [CrossRef]
30. Parajuli, P.B.; Risal, A. Evaluation of Climate Change on Streamflow, Sediment, and Nutrient Load at Watershed Scale. *Climate* **2021**, *9*, 165. [CrossRef]
31. Abbaspour, K.C. *SWAT-CUP 2012: SWAT Calibration and Uncertainty Programs—A User Manual*; Eawag: Dübendorf, Switzerland, 2014; Volume 106.
32. Saline Water and Salinity | U.S. Geological Survey. Available online: https://www.usgs.gov/special-topics/water-science-school/science/saline-water-and-salinity (accessed on 7 March 2023).
33. Tang, X.; Xie, G.; Deng, J.; Shao, K.; Hu, Y.; He, J.; Zhang, J.; Gao, G. Effects of Climate Change and Anthropogenic Activities on Lake Environmental Dynamics: A Case Study in Lake Bosten Catchment, NW China. *J. Env. Manag.* **2022**, *319*, 115764. [CrossRef]
34. Telesh, I.V.; Khlebovich, V.V. Principal Processes within the Estuarine Salinity Gradient: A Review. *Mar. Pollut. Bull.* **2010**, *61*, 149–155. [CrossRef]
35. Kang, M.; Peng, S.; Tian, Y.; Zhang, H. Effects of Dissolved Oxygen and Nutrient Loading on Phosphorus Fluxes at the Sediment–Water Interface in the Hai River Estuary, China. *Mar. Pollut. Bull.* **2018**, *130*, 132–139. [CrossRef]
36. Zhou, J.; Han, X.; Brookes, J.D.; Qin, B. High Probability of Nitrogen and Phosphorus Co-Limitation Occurring in Eutrophic Lakes. *Environ. Pollut.* **2022**, *292*, 118276. [CrossRef]
37. Park, J.; Kim, K.T.; Lee, W.H. Recent Advances in Information and Communications Technology (ICT) and Sensor Technology for Monitoring Water Quality. *Water* **2020**, *12*, 510. [CrossRef]
38. Moriasi, D.N.; Gitau, M.W.; Pai, N.; Daggupati, P.; Gitau, M.W.; Member, A.; Moriasi, D.N. Hydrologic and Water Quality Models: Performance Measures and Evaluation Criteria. *Trans. ASABE* **2015**, *58*, 1763–1785. [CrossRef]
39. Luo, C.; Li, Z.; Wu, M.; Jiang, K.; Chen, X.; Li, H. Comprehensive Study on Parameter Sensitivity for Flow and Nutrient Modeling in the Hydrological Simulation Program Fortran Model. *Environ. Sci. Pollut. Res.* **2017**, *24*, 20982–20994. [CrossRef] [PubMed]
40. Shin, S.; Her, Y.; Muñoz-Carpena, R.; Yu, X. Quantifying the Contribution of External Loadings and Internal Hydrodynamic Processes to the Water Quality of Lake Okeechobee. *Sci. Total Environ.* **2023**, *883*, 163713. [CrossRef] [PubMed]

*water*

MDPI

*Article*

# Assessment of Ecological Hazards in the Inaouen Wadi and Its Tributaries Using the Presence of Potentially Toxic Elements in Its Sediments

**Sanae Rezouki [1,2], Tarik Moubchir [2,3,*], Laila El Hanafi [4], Rachid Flouchi [5], Ilham Zahir [3], Mashail N. Alzain [6,*], Bouchra El Guerrouj [2,7], Omar Noman [8], Abdelaaty A. Shahat [8] and Aimad Allali [1,2,5]**

1. Laboratory of Plant, Animal and Agro-Industry Production Laboratory, Faculty of Sciences, Ibn Tofail University, Kenitra 14000, Morocco; rezoukisanae@gmail.com
2. Centre de l'Oriental des Sciences et Technologies de l'Eau et de l'Environnement (COSTEE), Université Mohammed Premier, Oujda 60000, Morocco; elguerroujb@gmail.com
3. Polyvalent Team in Research and Development, Polydisciplinary Faculty, Sultan Moulay Slimane University, Beni Mellal 23000, Morocco; ilham.zahir@usms.ma
4. Laboratory of Functional Ecology and Environmental Engineering Faculty of Sciences and Technology, Sidi Mohamed Ben Abdellah University, Fez P.O. Box 1796, Morocco; lailaelhanafi4@gmail.com
5. High Institute of Nursing Professions and Health Techniques, Fez, Morocco; rachid.flouchi@usmba.ac.ma (R.F.); aimad.allali@uit.ac.ma (A.A.)
6. Department of Biology, College of Science, Princess Nourah bint Abdulrahman University, P.O. Box 84428, Riyadh 11761, Saudi Arabia
7. Laboratoire d'Amélioration des Productions Agricoles, Biotechnologie et Environnement (LAPABE), Faculté des Sciences, Université Mohammed Premier, Oujda 60000, Morocco
8. Department of Pharmacognosy, College of Pharmacy, King Saud University, P.O. Box 2457, Riyadh 11451, Saudi Arabia; onoman@ksu.edu.sa (O.N.); ashahat@ksu.edu.sa (A.A.S.)

* Correspondence: tarik.moubchir@usms.ac.ma (T.M.); mnalzain@pnu.edu.sa (M.N.A.)

**Abstract:** Inaouen wadi is the second largest tributary of the Sebou river, one of Morocco's major rivers, which holds significant economic and social importance. Unfortunately, this watercourse is severely impacted by pollution from various human activities, particularly industrial sources. However, available data on the presence of potentially toxic elements (PTEs) that could harm human health in this region remain limited. PTEs pose major environmental risks due to their toxicity, persistence, and bioaccumulation. This study aimed to assess the concentrations of PTEs in the sediments of Inaouen wadi and its main tributaries based on sediment samples collected from 12 locations in 2019. The concentrations of Cd, Pb, Cr, Ag, Al, Cu, Fe, and Zn were measured using inductively coupled plasma atomic emission spectroscopy (ICP–AES), and sediment contamination levels were evaluated using multiple indices: the enrichment factor (EF), the geo-accumulation index (Igeo), the potential ecological hazard index (RI), and the modified ecological risk index (MRI). The results indicate that concentrations of Pb, Cd, Cr, Cu, Fe, and Zn are significantly influenced by urban discharges, particularly at sites S1, S3, and S5 near the cities of Taza and Oued-Amlil. The maximum values recorded were 7.01 g/kg for Pb, 0.9 g/kg for Cd, 0.1 g/kg for Cr, 19.9 g/kg for Fe and 1.9 g/kg for Zn. The enrichment factor (EF) revealed anthropogenic sources of Fe and Pb, confirming the human origin of these elements. The geo-accumulation index (Igeo) showed that the areas around stations S1, S3, and S5 are highly contaminated by Pb, Cd, and Fe, a finding also supported by the MRI. The study identified potential ecological risks at stations S1, S3, and S5, highlighting the urgent need for improved pollution management practices to mitigate environmental risks.

**Keywords:** Inaouen wadi; potentially toxic elements; sediment contamination; ecological risks; anthropogenic activities; pollution

## 1. Introduction

The contamination of water, soil, and sediment by potentially toxic metals (PTMs) has become a serious environmental concern in many developing countries, including Morocco [1–3]. Human activities, such as agriculture, mining, industry, artisanal work, and the production of domestic waste, have impacted the quality of aquatic environments [4]. The severity of these effects increases when effluents are released into the environment without pretreatment. Aquatic ecosystems, particularly rivers and oceans, are the primary recipients of toxic metals, and even slight changes in environmental conditions can disrupt their normal functioning. The presence of heavy metals in rivers poses a threat to biodiversity and human health due to heavy metals' toxicity, persistence, and ability to accumulate in the food chain [5–9].

Sediment is a vital component of river systems, formed through the transport and deposition of materials by watercourses [10]. It plays a crucial role in river and lake ecosystems. During various physicochemical processes, suspended solids and ions present in the water are adsorbed and accumulate in riverbed sediment [11]. However, when external conditions change, heavy metals can be released into the water, increasing the potential ecological risks to aquatic systems and human health [12].

Due to their persistence, toxicity, and ability to bioaccumulate in food webs, contaminant inputs, particularly those of potentially toxic elements (PTEs), can affect aquatic species directly and have indirect effects on the health of humans [5,7–9]. An effective method for evaluating the presence of PTEs in rivers that originate from both natural and human activities is to analyze the sediment on the riverbed [13–15]. The reason for this is because sediment possesses a significant ability to retain PTEs and functions as a reservoir for contaminants [16,17].

The geochemical cycling of elements, such as manganese oxides, iron, clays, carbonates, organic matter, and residual fractions, is predominantly controlled by sediment components [18]. Nonetheless, the behavior of PTEs in sediment is contingent upon many biogeochemical processes [19]. In fact, under certain physicochemical circumstances, such as differences in pH, dissolved oxygen, or redox potential, these substances may be released into the dissolved phase and become more easily absorbed by living organisms [20,21]. Moreover, semi-arid basins, which experience extended periods of drought and get little rainfall, are especially vulnerable to pollution by PTEs [22,23], as is the case for the Inaouene wadi. Thus, the purpose of this investigation is, firstly, to study the concentrations of PTEs in sediments of the Inaouene wadi and its tributaries. Secondly, we will determine the degree of contamination in the sediments studied, the potential sources of contamination, and the hazards of adverse effects on the aquatic environment through the calculation of indices and enrichment factors. Finally, we will perform a statistical analysis of the obtained data using principal component analysis (PCA) to identify associations between the PTEs and sampling stations [24].

## 2. Materials and Methods

### 2.1. Study Area

To obtain a more accurate sample of the watercourses, we chose eleven specific locations along the Inaouene wadi. Out of these, six locations (S1, S2, S5, S7, S9, and S10) represented the primary tributaries of the Inaouene wadi. The remaining five locations (S3, S4, S6, S8, and S11) were situated on the bed of the wadi. Additionally, the twelfth location (S12) corresponds to the Idriss the first dam (Idriss I dam), also known as Barrage Idriss I, as depicted in Figure 1. The sampling sites (Figure 1) were selected to include a diverse range of activities in the research region, including wastewater discharge and household, industrial, and agricultural activities, as shown in Appendix A. The stations were chosen based on their accessibility and their ability to highlight the key features of the wadi and its tributaries.

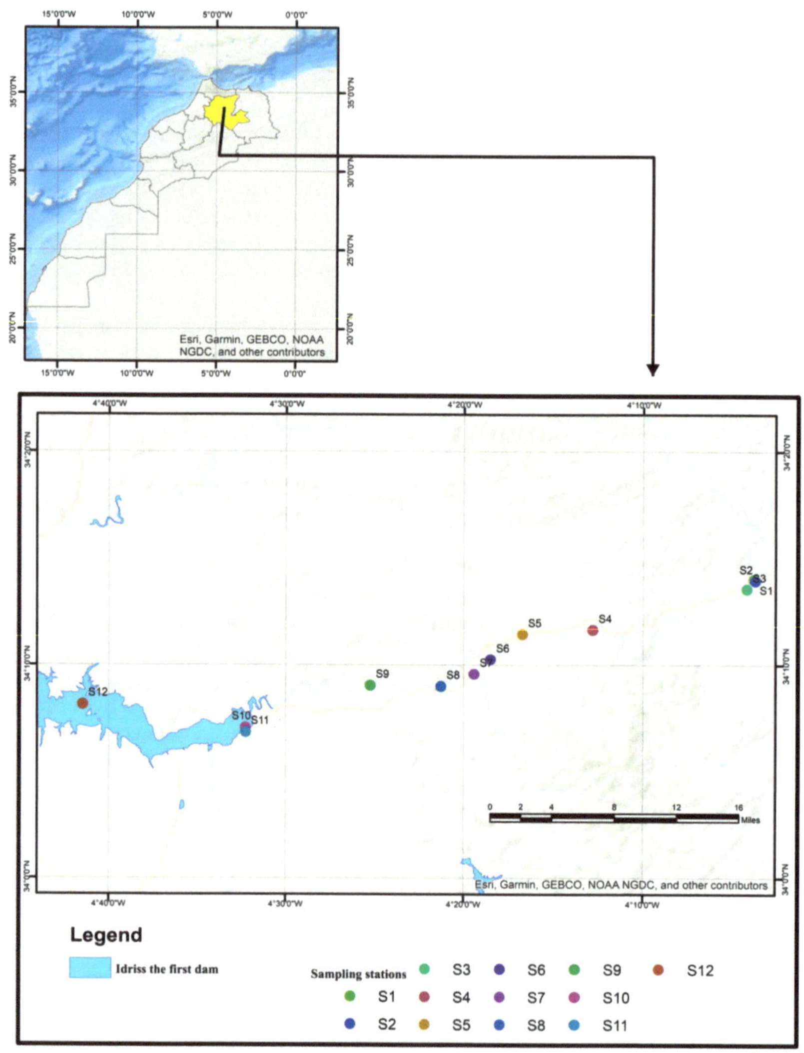


**Figure 1.** Location of the sampling stations in the studied area.

## 2.2. Sampling and Pre-Processing of Samples

The sediment samples were collected in 2019, with careful attention given to precisely determining the time of sampling. Surface sediment samples, taken from the top 0 to 5 cm deep, were collected following the methodologies described in studies conducted by Branchu et al., and Lionard and Coquery in 2013 [25,26]. A modified Ekman dredge was used to sample sediment at the 12 stations along the main course of the Inaouene Wadi, taking into account variations observed after the confluence with the tributaries (Figure 1). These samples were collected underwater from the riverbanks and were carefully stored in polyethylene bottles to ensure their integrity. In the field, the samples from each site were thoroughly mixed to ensure a homogeneity that was representative of the local conditions. Each set of samples was sealed, labeled, and transported in coolers maintained at a constant temperature of 4 °C to preserve their physicochemical characteristics. Once in the laboratory, the sediments were air-dried to remove residual moisture without affecting their chemical composition. The dried material was then divided into four equal parts and carefully ground using an agate mortar to avoid external contamination. The parts were subsequently sieved into three distinct granulometric fractions: fine particles (diameter $\leq 63$ µm), medium-sized particles (63–200 µm), and coarse particles (200–2000 µm). This granulometric separation enabled the study of variations in heavy metal concentrations across different grain sizes, thus providing detailed information on the distribution of pollutants within the sediment.

### 2.3. Physicochemical Treatments and Analyses

The main physicochemical analyses of the samples were focused on characterizing concentrations of Cadmium (Cd), Lead (Pb), Chromium (Cr), Silver (Ag), Aluminum (Al), Copper (Cu), Iron (Fe), and Zinc (Zn) and the geochemical processes that affect their dynamics. Concentrations of the elements were measured using the inductively coupled plasma atomic emission spectroscopy method (ICP–AES) [27].

The sediment digestion process was performed according to the method outlined by da Silva et al. in 2014 [28]. Subsequently, metals were assessed using ICP–AES in a laboratory at the Innovation Campus of the Mohammed Ben Abdallah University in Fez. This instrument operated in simultaneous mode. The experimental conditions were as follows: power: 1.3 kW; plasma gas flow: 15 L/min; auxiliary gas flow rate: 1.5 L/min; nebulization pressure: 200 kPa; stabilization time: 20 s; reading time per replicate: 5 s; pump speed: 15 rpm; and end flush time: 30 s. The standard solution employed was Merck's premium Certipur®, available in both single- and multi-element forms (1000 mg/L). The Certipur® standards are traceable to NIST SRM.

### 2.4. Quality Indexes

#### 2.4.1. The Enrichment Factor (EF) and Geo-Accumulation Index (Igeo)

The enrichment factor (EF) is a calculation method introduced in the early 1970s to investigate the source of metallic elements in the atmosphere [29]. It has since been applied to sediment analysis [30]. This tool helps to assess the extent of enrichment of each element in the examined soils and sediments and to determine whether this enrichment is natural or the result of human activities. Aluminum, considered an inert and immobile detrital element, is used as a reference in the EF formula (Equation (1)) for calculation purposes [30,31].

$$\mathrm{EF} = \frac{(\mathrm{C_n}/\mathrm{X_n})\mathrm{sample}}{(\mathrm{C_r}/\mathrm{X_r})\mathrm{refence}} \quad (1)$$

In the formula, $C_n$ and Cr represent the concentrations of the specific element in the soil under study and in the reference local geochemical background (LGB), respectively, while $X_n$ and $X_r$ represent the concentrations of aluminum (the occurring element's concentration in the Earth's crust, also known as Clarke values) in the soil under study and in the LBG, respectively.

The geo-accumulation index (Igeo) was determined to assess the anthropogenic nature of sediment contamination by PTEs and to determine the difference in concentrations between elements found in the sediments and the mean concentrations that naturally exist in the continental plate [25,32] (Equation (2)).

$$\mathrm{Igeo} = \log_2 \frac{\mathrm{C_n}}{1.5\mathrm{XB_n}} \quad (2)$$

where $C_n$ represents the concentration of the PTE in the sediment and $B_n$ represents the average value of the PTE in the upper continental plate [33]. Igeo has a categorization system consisting of seven levels. A value of Igeo less than or equal to zero indicates that the contamination level is at level 0 and the sample is basically uncontaminated. A value of $0 < \mathrm{Igeo} < 1$ represents level 1 and indicates that the sample is in an uncontaminated to moderately contaminated condition. A value of $1 < \mathrm{Igeo} < 2$ corresponds to level 2 and indicates that the sample is in a moderately contaminated condition. A value of 2 <Igeo< 3 represents level 3 and indicates that the sample is in a condition of moderate to heavy contamination. A value of 3 < Igeo < 4 corresponds to level 4 and indicates that the sample is in a heavily contaminated condition. A value of 4 < Igeo < 5 represents level 5 and indicates that the sample is in a condition of heavy to extreme contamination. An Igeo value greater than 5 corresponds to level 6 and indicates that the sample is in an extremely contaminated condition.

#### 2.4.2. Ecological Hazard Factor (Eh), Potential Ecological Hazard Index (RI) and Modified Potential Ecological Risk Index (MRI)

As described by Solomon et al. (2000) [34], the ecological hazard assessment index, also called the potential ecological risk index (RI), was performed by determining the ecological hazard factor (Eh), using Equation (3), according to the method established by Hakanson (1980) [31] to estimate the hypothetical effect of PTEs in sediments on organisms in the aquatic ecosystem.

$$\mathrm{Eh} = \mathrm{Tr} \times \mathrm{Cf} \tag{3}$$

where "Tr" is a coefficient that indicates a specific toxicity index for each PTE, and Cf is the contamination coefficient calculated by dividing Cd (the present concentration of heavy metals in sediments) by Cr (the pre-industrial record of heavy metal concentrations in sediments).

The potential ecological hazard index (RI) was calculated by the sum of the Eh evaluated at every single site using Equation (4).

$$\mathrm{RI} = \sum \mathrm{Eh} \tag{4}$$

The composition of local bedrock was used as a reference.

The classification of Eh is as follows: an Eh < 40 indicates a low potential for ecological hazard; 40 < Eh < 80 indicates a moderate potential for ecological hazard; 80 < Eh < 160 indicates a considerable potential for ecological hazard; an Eh between 160 and 320 indicates a high potential for ecological hazard; and an Eh value above 320 indicates a very high potential for ecological hazard. The ecological hazard potential index (RI) is categorized into the following ranges: an RI value of 150 is categorized as a minor ecological danger, a value between 150 and 300 is considered a moderate ecological hazard, a value between 300 and 600 is defined as a large ecological hazard, and a value equal to or more than 600 is considered a very high ecological risk.

Modified potential ecological risk index.

The potential for heavy metal contamination is determined by both the overall concentration of heavy metals and their specific chemical forms. A study on the correlation between the speciation of heavy metals and their bioavailability has shown that the speciation of heavy metals may significantly impact their bioavailability. However, up until now, risk assessment models have not considered the chemical speciation of heavy metals, focusing only on their overall concentrations [35].

The formulae to calculate the MRI are as follows [34]:

$$\hat{\mathrm{C}}\mathrm{d} = \mathrm{Cd} \times \Omega \tag{5}$$

where $\Omega = \mathbf{A}\partial + \mathbf{B}$

$$\hat{\mathrm{C}}\mathrm{f} = \hat{\mathrm{C}}\mathrm{d}/\mathrm{Cr} \tag{6}$$

$$\hat{\mathrm{E}}\mathrm{h} = \mathrm{Tr} \times \hat{\mathrm{C}}\mathrm{f} \tag{7}$$

$$\mathrm{MRI} = \sum \hat{\mathrm{E}}\mathrm{h} \tag{8}$$

where $\hat{\mathrm{C}}\mathrm{f}$, $\hat{\mathrm{C}}\mathrm{d}$, $\hat{\mathrm{E}}\mathrm{h}$, and MRI are the modified forms of Cf, Cd, Eh, and RI, respectively; $\Omega$ is the modified index of heavy metal concentration; **A** is the percentage of Metal in carbonate and exchangeable fractions presented in the Table 1; **B** is the value of 1-**A**; $\partial$ is the toxic index corresponding to different ratios of exchangeable and carbonate fractions. The ∂ value listed in Table 2 was obtained according to Zhu et al. (2012) [35].

**Table 1.** Percentage of metal in the carbonate and exchangeable fractions of the sediment samples.

| | Site | S1 | S2 | S3 | S4 | S5 | S6 | S7 | S8 | S9 | S10 | S11 | S12 |
|---|---|---|---|---|---|---|---|---|---|---|---|---|---|
| % of Metal in Carbonate and Exchangeable Fractions (A) | % Pb | 50 | 65 | 50 | 59 | 67 | 46 | 64 | 57 | 48 | 55 | 66 | 45 |
| | % Cd | 83 | 50 | 63 | 67 | 67 | 60 | 40 | 71 | 50 | 40 | 69 | 67 |

**Table 2.** Values of ∂ referring to metal in the carbonate and exchangeable fractions [31,35].

| % of Metal in the Carbonate and Exchangeable Fraction (A) | ∂ |
|---|---|
| >50 | 1.60 |
| 31–50 | 1.40 |
| 11–30 | 1.20 |
| 1–10 | 1.00 |
| <1 | 1.00 |

### 2.5. Statistical Methods

The statistical data analysis was conducted through the following steps using SPSS (V21) and R Studio software (Version 2023.3.0.386). Initially, we carried out a normality test through the Shapiro–Wilk test at a 5% significance to prove the homogeneity of variances. The results were standardized by applying Z-score normalization; this step was crucial because it centered the variables around zero, which could help reveal similarities based on patterns rather than absolute values. This is important because HAC, the test applied afterwards, relies on the distance between observations to perform clustering, thus offering insights into the similarities and differences between variables and facilitating the reduction of results into clusters [35]. Subsequently, principal component analysis (PCA) was undertaken for all samples. PCA was selected as a powerful analytical tool in this research article due to its ability to uncover latent patterns within complex datasets. By reducing the dimensionality of integrated water chemistry variables, PCA facilitated the identification of dominant chemical sources, the detection outliers, and the visualization of relationships among variables.

## 3. Results

### 3.1. Sediment Grain Size

Granulometry allowed us to determine the distribution of sedimentary particles along the Inaouene wadi. Sedimentary particle categories were classified according to their size, provided the relative proportions of these categories as a percentage of the total sediment mass, and subsequently, were used to deduce the texture of the soil studied (Figure 2).

Particle sizes in the sediment varied among locations (Figure 2) such that sediments at each location could be assigned to classes. Gravels were the most dominant at the upstream level of the Inaouene (S1, S2, S3, S4, S5 and S6). However, near the Idriss I dam, at locations S10, S11, and S12, fine elements appeared to dominate, with proportions up to 86%. Similarly, clay and silt dominated at site S8 with 34%. Conversely, at sites S7 and S9, coarse sand had the greatest percentages of 31% and 37%, respectively.

The geology of the Inaouene watershed [36] is characterized by limestone and the dolomitic outcrops of the northern Middle Atlas as well as the carbonate terrains of the pre-rific zone, where eroding agents (wind, rain, frost) tear off fine particles or whole fragments of carbonate rocks. Then, these elements are moved towards the lowest point of the Inaouene wadi.

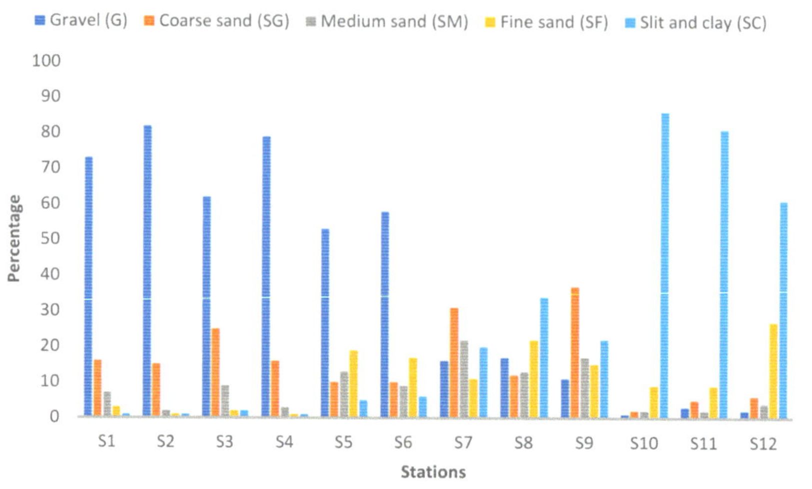


**Figure 2.** Distribution of sedimentary particles at each station along the Inaouene wadi and its tributaries.

### 3.2. Trace Elements in Sediments

Concentrations (g/kg) were determined in the sediment samples. Except for Cu, which remained at the same concentration, a value below 0.011 g/kg, at all the stations studied, the concentrations of all the other PTEs changed from one station to another. For instance, the Al concentration oscillated between 0.05 and 2.08 g/kg (Figure 3).

Based on these results, it can be concluded that, overall, the sediment in the Inaouene wadi contains relatively high concentrations of the elements studied. Concentrations of PTEs in the main stem of the Inaouene wadi (S3, S4, S6, S8 and S11), Larbaa wadi (S1), and Amlil wadi (S5) were greater than those found in sediments at Lahdar wadi (S2), Zirg wadi (S7), Bouhlou wadi (S9), and Matmata wadi (S10). Also, the stations where the concentrations are very great are those close to urban agglomerations; these concentrations decrease progressively from upstream to downstream.

This decrease in the Inaouene wadi and its main tributaries could be linked to the distance from the point of discharge of treated water (S1) or from the points of discharge of gray water from riparian sources. At the same time, this evolution is triggered by several phenomena, including the adsorption, desorption, suspension, and absorption of these trace elements [37,38].

These phenomena are influenced by several factors, such as the pH of the water, the content of organic matter in the sediment and the suspended matter in the water [39,40]. An alkaline pH in water favors the adsorption of metals on colloids and sediments [41]. Alternatively, the slight decrease in water pH observed downstream in this river (with values ranging from 8.02 at the upstream stations to a minimum of 7.23 at the downstream stations) could reduce the adsorption of several metallic elements, which would then be carried through the water in this environment [42].

The assimilation and concentration of metals by aquatic plants and macrophytes, which are widely known to occur [43,44], could also contribute to a decrease in concentrations of trace metals in sediments due to the absorption of metals directly from the water column or from the pore water of sediments.

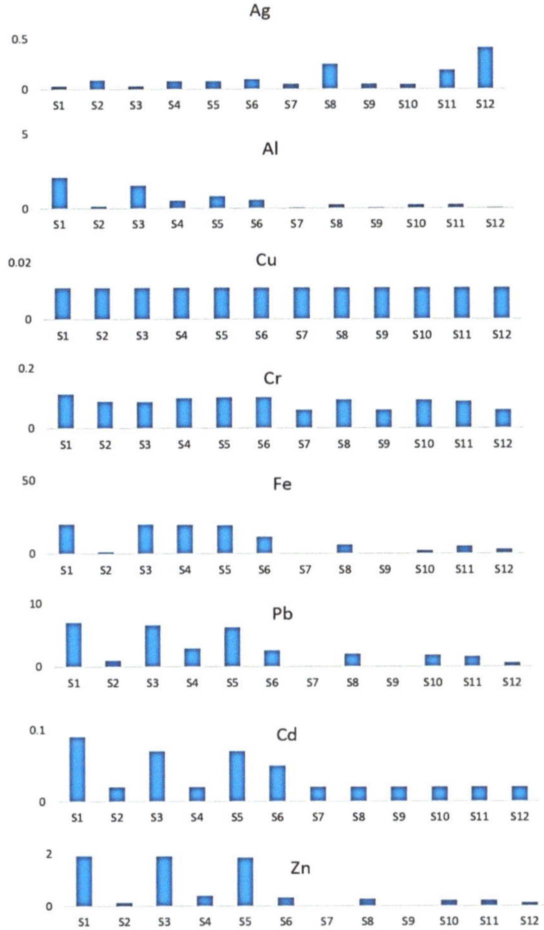


**Figure 3.** Concentrations of PTEs in sediment samples (g/kg) from the Inaouene wadi and its main tributaries.

Concentrations of PTEs obtained in the sediment samples of the present study were compared with those of other rivers in Morocco (Table 3).

**Table 3.** Comparison of PTEs concentrations in sediments.

| | Potential Toxic Elements (mg kg$^{-1}$) | | | | | | |
|---|---|---|---|---|---|---|---|
| | **Cd** | **Cr** | **Cu** | **Pb** | **Zn** | **Fe** | **Reference** |
| World's rivers | 0.0003 | 0.12 | 0.05 | 0.04 | 0.11 | - | [42] |
| Sebou wadi | 0.00001–0.011 | 0.47–77.5 | 12.5–625 | 0.5–19 | 30–1250 | 19–30.300 | [43] |
| Moulouya wadi | 0.19–0.8 | - | 19.4–27.88 | 19.2–710 | 81–949 | - | [44] |
| Khmiss wadi | 0.05–0.67 | 4.15–67.5 | 31–87 | 10–57 | 41.25–200 | 2087–7800 | [45] |
| Rdom wadi | 0–0.01 | 1–1.9 | 0.2–2.1 | 1.2–2.7 | 1–3.8 | 178–468 | [46] |
| Tislit-Talsint wadi | 0.35 | 33.30 | 58.54 | 85.84 | 142.65 | 39.390 | [47] |
| Sebou wadi | - | 77.93–119 | - | - | 79.27–121 | 33.330–44.190 | [48] |
| Hassar wadi | 1.5–17.4 | 107.7–258.9 | 165.4–326.6 | 15.03–148.2 | 458.02–1768.9 | 1700–2500 | [49] |
| Inaouene wadi | 90–0.1 | 10–113 | 1.1–11 | 1–7012 | 2.1–1900 | 82–19.905 | Present study |

### 3.3. Statistical Analysis

Results of the hierarchical analysis suggest two main groups of sediments according to concentrations of PTEs (Figure 4). The number of clusters was chosen according to the minimum number of clusters that explain most of the variation in PTE concentrations in the sediment samples. Thus, group 1 is composed of soil samples from the upstream region of the Inaouene wadi (S1, S3, S4, S5 and S6), which is characterized by the presence of a great pollution load due to domestic and industrial discharge (mainly at its margins), agricultural discharge from the city of Taza for S1 and S3, and discharge with larger concentrations of Fe, Pb and Zn in the Amlil wadi for S5. Group 2 includes samples from the downstream region of the Inaouene wadi and its left bank tributaries (S7, S8, S9, S10, S11 and S12). This is mainly due to the low levels of PTEs in the water of these stations, which are characterized by good water quality [45]. It is noteworthy that S2, which is located upstream of the Inaouene wadi, also belongs to group 2 since this station is characterized by a lesser concentration of PTEs according to Figure 3.

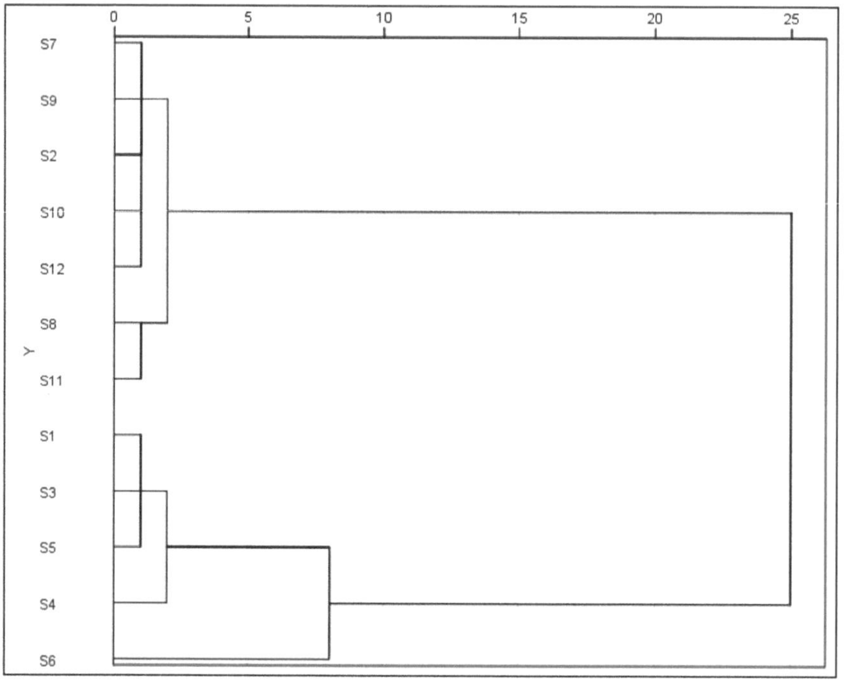


**Figure 4.** Hierarchical ascending classification (HAC) of the studied stations, where Y axis represent the sample sites and X axis represent the resized class combination distance.

Analyses of the heavy metal concentrations and granulometric composition of bottom sediments from the river at each station revealed a significant level of water pollution by these toxic elements, with high values of Cd, Fe, Al, Zn, and Pb. To determine the relationship between trace metal elements, sediment granulometry, and the distribution of stations based on these characteristics, 12 environmental variables were evaluated using principal component analysis (PCA). The PCA analysis identified two significant variables, Component 1 and Component 2, with eigenvalues over 1.0. Component 1 accounted for 57.52% of the variation and showed a substantial and positive correlation with the amounts of Cd, Cu, Al, Zn, and Pb (Figure 5). Component 2, which accounted for 18.83% of the variation, had a substantial negative correlation with concentrations of Al. By projecting the sample sites perpendicular to the individual PTE vectors or extending them beyond the origin of the coordinate system, we may estimate the background sediment concentration (Figure 6). The highest levels of Cd, Cu, Al, Zn, and Pb were found at stations S1 and S3, whereas the lowest levels were seen at sample locations S7, S8, S9, S10, S11, and S12. Component 2 had a positive correlation with gravel content and a negative correlation with fine sand content. The angle formed by the vectors of the different metals give insights into their interrelationships. The connection between variables is larger as the angle decreases but, when the angle reaches 90°, there is no association. The investigation revealed a significant link between the concentrations of PTEs in the bottom sediments and the amount of gravel present. However, no correlation was found between PTE concentrations and the levels of clay, silt, and sand.

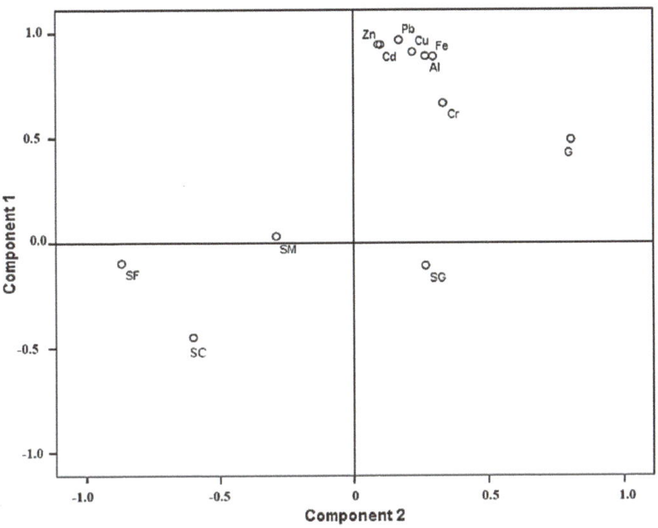


**Figure 5.** PTE concentrations as a function of background sediment sample texture.

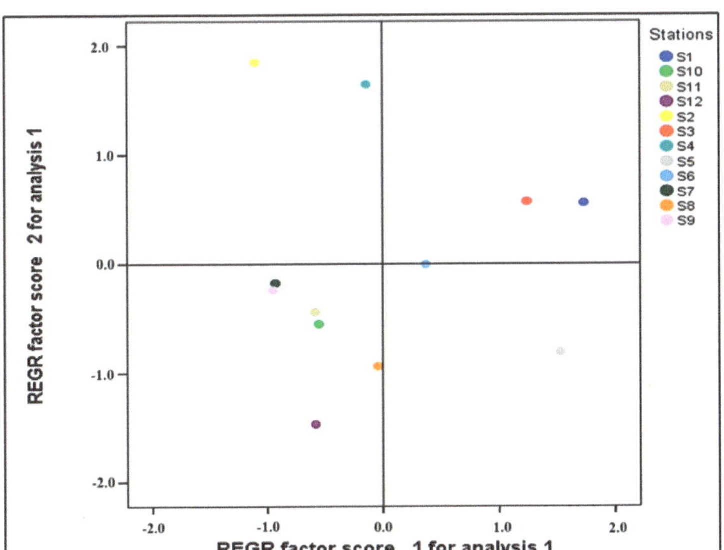


**Figure 6.** The stations in the F1 × F2 factorial plane.

### 3.4. Pollution Indexes and Ecological Hazards

#### 3.4.1. The Geo-Accumulation Index (Igeo)

The geo-accumulation index (Igeo) was used to evaluate the degree of contamination by PTEs in the sediment at the bottom. The Igeo values for Cu and Cr were found to be below zero, indicating that they may be categorized as uncontaminated with these elements. Igeo values for Al were less than 1 except at four locations, S1, S3, S5 and S6. At these locations, the Igeo values were slightly greater than 1.0, which indicated that these stations can be classified as highly to extremely contaminated with Al. The greatest contamination was with Fe, Cd and Pb, and was noted at stations S1, S3, S4, S5 and S6; these stations were considered extremely polluted with these elements. Meanwhile, the others were classified as polluted to heavily polluted with these elements. For Zn at all the other locations—except for S7 and S9, which showed Igeo values less than 0—Igeo values greater than 0 were observed. Thus, according to Igeo classification, the background sediment of the sampling sites S7 and S9 are practically uncontaminated with Zn. Meanwhile, the sediment samples from S10, S11, S4, S6, and S8 showed moderate contamination, with Igeo values greater than 1.0. Whereas those of S1, S3, and S5 displayed heavy contamination, with Igeo values greater than 4.0, which demonstrates severe pollution of the background sediments (Figure 7).

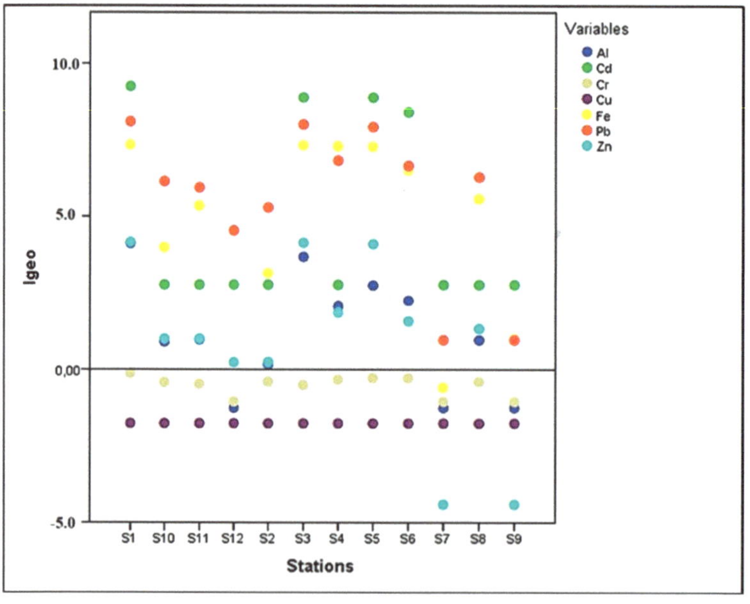


**Figure 7.** Spatial distribution of the geo-accumulation index values (Igeo).

#### 3.4.2. The Enrichment Factor (EF)

The enrichment factor (EF) was used to identify potential sources of PTEs. For Cr, Cd, Cu, and Zn, EF values were generally below 1.0 (Figure 8), indicating that these PTEs might have a natural origin resulting from weathering processes. For Pb, EF values ranged from 0.98 to 2.73, with values exceeding 1.0 observed in the sediment of ten sites. Additionally, EF values for Fe fluctuated between 0.38 and 8.91, with values greater than 5 at stations S4 and S8. An enrichment factor above 1.5 suggests that the PTEs likely originated from external sources, such as anthropogenic or diffuse pollution sources [46,47].

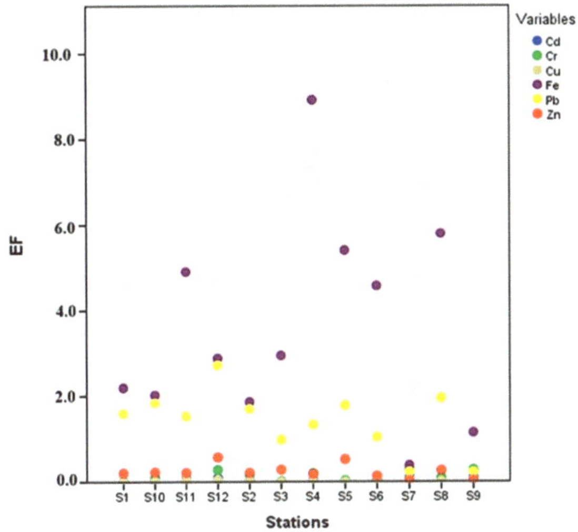


**Figure 8.** Spatial distribution of trace element enrichment factor values in the sediment samples of the Inaouene wadi.

3.4.3. Ecological Hazard Assessment by Potential Ecological Hazard Index and Modified Potential Ecological Hazard Index

Values for the hazard index (RI), which is based on the contamination factor of each PTE [31], ranged from 5.93 to 241.91. In general, RI values were less than 150 at nine locations, which indicated a low ecological hazard associated with the concentration of PTEs in the background sediment. Values over 150 were observed at three locations in the upper Inaouene wadi (S1, S3, and S5), indicating a moderate ecological hazard to benthic organisms. The greatest contribution to the RI was Pb, with an average percentage of 37% (Figure 9).

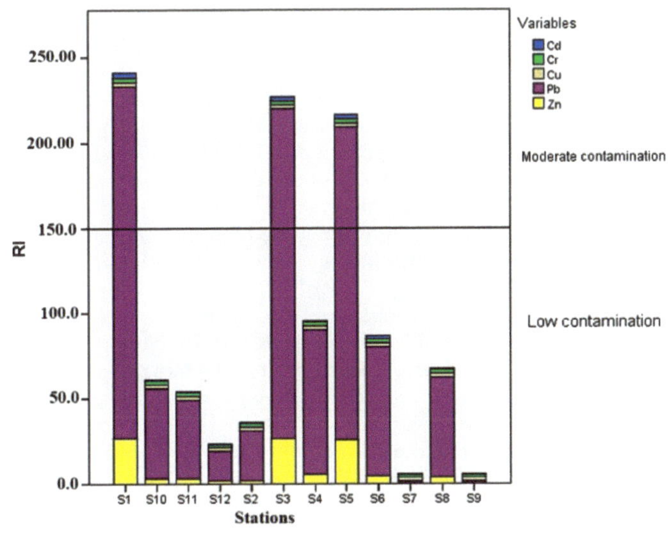


**Figure 9.** Spatial variation in the potential ecological hazard index (RI) along the Inaouene wadi.

All MRI values were below 40 (Figure 10) in all sites, indicating a low ecological risk in the region.

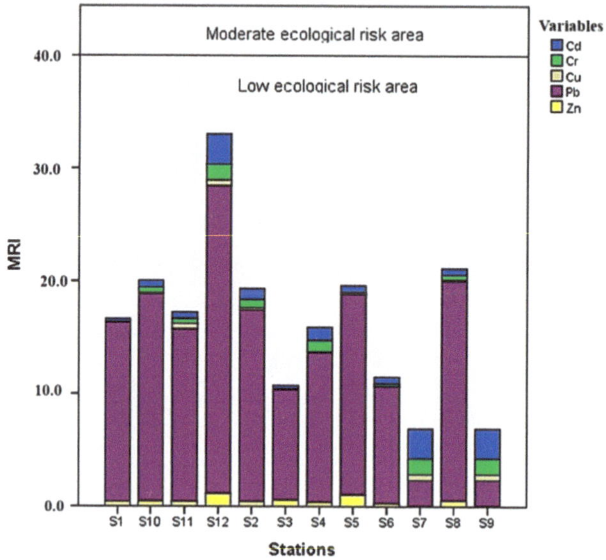


**Figure 10.** Spatial variation in the modified potential ecological hazard index (MRI) along the Inaouene wadi.

## 4. Discussion

The concentrations of potentially toxic elements (PTEs) in sediments and their interrelationships are primarily influenced by human sources, sediment texture, and organic matter content [48,49]. The composition of sand, silt, and clay is influenced by the meteorological, hydrological, and geological characteristics of the watershed. These variables impact the transportation of PTEs in riverine systems. Globally, background sediments are characterized by relatively great variability in concentrations of metals. The sediments analyzed in this study are in a watershed affected by pollution upstream of the Inaouene wadi; therefore, concentrations of PTEs in background sediments are greater upstream and lesser downstream. Elevated levels of PTEs in sediments may occur when residential wastewater is introduced, surpassing the national criteria for sediment quality. Our findings corroborated the results obtained by Rezouki et al. (2021) [45], which demonstrated the existence of chemical and biological pollution linked to human activities (such as agricultural runoff and domestic wastewater). Furthermore, the study suggested that local anthropogenic activities have the potential to intensify the contamination of certain metals in the sediment at the bottom [45].

The principal component analysis (PCA) confirmed our obtained results and interpretations. The PC1 and PC2 axes explained 76.35% of the total information. PC1 was positively influenced by Al, Cd, Cu, Fe, Pb, and Zn, suggesting that these elements may originate from common sources [50]. Stations S1, S2, and S3 were negatively associated with the PC1 axis, indicating that they were significantly impacted by heavy metals due to their upstream locations close to the contamination sources of the city of Taza and the Oued Amlil, which are characterized by wastewater discharge [8,51]. PC2 showed a strong association with the variables of fine sand (FS) and Al. The graphical representation of the twelve studied stations on the PC1 $\times$ PC2 plot clearly illustrates the differences between these stations. Component 2 exhibited a positive correlation with gravel content and a negative correlation with fine sand and Al, characterizing the stations that were associated with low levels of trace metal elements. The study revealed a significant link between the concentrations of PTEs in bottom sediments and gravel content. However, no correlation was found between PTE concentrations and the levels of clay, silt, or sand.

Relatively great values of the Igeo, EF, and RI at locations upstream suggest that the origin of metals is anthropogenic. The greater concentrations of trace metals at upstream locations were from contributions of wastewater from the city of Taza and Amlil wadi. Saleem et al. (2018) [52] found that the levels of PTEs were particularly high in locations near urban and semi-urban regions. PTEs may also originate from untreated urban/industrial waste discharge, agricultural runoff, and vehicular emissions [52]. The small increase in concentrations of PTEs observed near the Idriss I dam could be a consequence of changing hydrologic conditions in the reservoirs during situations of flooding and increased water flow [53].

Sediment quality indices (RI and MRI) were used as indicators to evaluate the toxicological hazards presented by PTEs, including Pb, Cd, Zn, Cr, and Cu, to the benthic environment. The results for areas S1, S3, S4, S5, and S6 showed that there is still a significant toxicity danger and that the concentrations of Cr, Cu, Cd, Fe, Pb, and Zn have recently risen. Inferring conclusions only from the overall quantities of PTEs in sediment, without considering naturally occurring amounts arising from the change of bedrock, might sometimes result in inaccurate findings [54].

Polymetallic pollution poses a possible ecological hazard (RI). The hazard level was low (ER < 150) for all areas except for S1, S3, and S5. In these places, the RI suggests a moderate ecological hazard (40 < RI < 80), specifically for Pb. These discrepancies are associated with the calculation technique for the individual risk (RI) for each element, which considers the toxicity weight (30 for Cd, 5 for Cu and Pb, and 2 for Cr). The overall concentration was inadequate to evaluate the possible risk of metal availability to living organisms [22,55]. The partitioning of metals into residual and non-residual fractions was assessed with the use of common EDTA and HCl extractions, which are frequently employed to recover the largest proportion of trace elements that are loosely bound to organic and inorganic complexes [56,57]. HCl is capable of extracting metals from exchangeable and carbonate phases, but EDTA is more effective at extracting metals that are tightly bonded to oxides and organics with a recognized affinity for metals [18,58,59].

## 5. Conclusions

This study assessed the concentrations of potentially toxic elements (PTEs) in the sediments of the Inaouene Wadi across 12 sampling stations, revealing significant levels of contamination, particularly near urban and agricultural discharge points in the upstream region of the river (stations S1, S3, and S5). Correlation analyses using PCA and HAC demonstrated a positive association among most of these elements (Cd, Pb, Cr, Ag, Al, Cu, Fe, and Zn), indicating a common source for these pollutants. This connection was especially pronounced at stations S1, S3, and S5, which were directly impacted by domestic and industrial discharges from the cities of Taza and Oued Amlil. The use of geochemical indicators, such as the geo-accumulation index (Igeo), enrichment factor (EF), ecological hazard index (RI), and the modified ecological risk index (MRI) provided a comprehensive assessment of the environmental impact. The PTE contamination in the Inaouene Wadi, particularly for Cd, Pb, Cr, Ag, Al, Cu, Fe, and Zn, is deeply concerning. It poses a significant threat to sediment quality, endangering local biodiversity and raising serious concerns for human health. These findings underscore the urgent need for targeted pollution control measures to protect the aquatic ecosystem and public health. Effective strategies should include improvements in wastewater treatment facilities, the implementation of better agricultural practices, and the enhancement of waste management systems. Continuous monitoring and further studies are essential to evaluate the effectiveness of these measures and ensure sustainable management of the Inaouene Wadi's water resources.

**Author Contributions:** Conceptualization, S.R. and T.M.; data curation, S.R. and T.M.; formal analysis, S.R. and T.M.; funding acquisition, M.N.A., B.E.G., O.N. and A.A.S.; investigation, S.R. and T.M.; methodology, S.R., T.M., I.Z. and A.A.; project administration, I.Z.; resources, L.E.H. and

R.F.; software, S.R. and T.M.; supervision, I.Z., M.N.A., B.E.G. and A.A.; validation, L.E.H., R.F., I.Z., M.N.A., B.E.G., O.N., A.A.S. and A.A.; visualization, M.N.A., B.E.G., O.N. and A.A.S.; writing—original draft, S.R., T.M., and A.A.; writing—review & editing, L.E.H., R.F., I.Z. and A.A. All authors have read and agreed to the published version of the manuscript.

**Funding:** This study was funded by the Researchers Supporting Project number PNURSP2024R103 at the Princess Nourah bint Abdulrahman University, Riyadh, Saudi Arabia.

**Data Availability Statement:** The datasets presented in this study are available upon reasonable request from the corresponding author.

**Acknowledgments:** The authors extend their appreciation to the Researchers Supporting Project number PNURSP2024R103 of the Princess Nourah bint Abdulrahman University, Riyadh, Saudi Arabia. The authors are also thankful to the Researchers Supporting Project number RSPD2024R1057 from the King Saud University, Riyadh, Saudi Arabia.

**Conflicts of Interest:** The authors declare no conflicts of interest.

## Appendix A. Description of the Sampling Stations and Their Characteristics

| Presentation of the Stations | | Geographic Coordinates | | Description of the Stations |
|---|---|---|---|---|
| | | LATITUDE N | LONGITUDE W | |
| S1 | Larbaa wadi | 34°13′58.841″ | 4°3′42.36″ | The blackish color of the water, along with its unpleasant odor and visual nuisance, indicated the influence of wastewater discharge. Additionally, the presence of solid waste accumulated at the edge of the watercourse further exacerbates environmental impact. |
| S2 | Lahdar wadi | 34°14′3.659″ | 4°3′50.099″ | The right bank is fed by tributaries that collect runoff from the pre-Rifian hills and are characterized by Jurassic marl and limestone formations. |
| S3 | Upstream of Inaouene wadi | 34°13′40.942″ | 4° 4′7.528″ | Upstream of the Inaouene wadi, the waters exhibit a blackish color, accompanied by olfactory and visual nuisances. |
| S4 | Inaouene wadi located in Ghiata Al Gharbia | 34°11′57.611″ | 4°12′24.865″ | The presence of agricultural activity was used to estimate the degree of self-purification of the watercourses. |
| S5 | Amlil wadi | 34°11′17.293″ | 4°16′39.879″ | The right bank of the Inaouene wadi, characterized by a blackish water color and olfactory and visual nuisance, helped to detect the influence of wastewater discharge and was affected by uncontrolled dumping. |
| S6 | Inaouene wadi | 34°11′16.967″ | 4°16′39.88″ | The detection of wastewater discharge helped estimate the degree of self-purification of the watercourses. |
| S7 | Zirg wadi | 34°09′54.2″ | 4°20′24.5″ | Presence of agricultural activity. |
| S8 | Inaouene wadi after Zirg | 34°09′54.5″ | 4°20′24.4″ | The degree of self-purification of the watercourses was estimated. |
| S9 | Bouhlou wadi | 34°07′56.4″ | 4°24′33.8″ | The tributary of the left bank of the Inaouene wadi, fed by the Tazekka massif and the limestone of the Middle Atlas, is little affected by anthropic and agricultural factors and has a bottom made up of coarse gravel. |
| S10 | Metmata wadi | 34°07′00.6″ | 4°32′17.4″ | This region is a tributary of the left bank of the Inaouene wadi and was little affected by agricultural and anthropic activity. |
| S11 | Downstream of Inaouene wadi sidi abd jlil | 34°08′16.2″ | 4°30′09.6″ | The color and odor of the water seemed to be acceptable. |
| S12 | Idriss 1st dam | 34°7′31.989″ | 4°39′50.032″ | All parameters seemed to be acceptable. |

## References

1. Chedadi, M.; Bassouya, M.; Agour, A.; Asmi, H.E.; Merzouki, M.; Bari, A. Assessment of sediments contamination by potential toxic elements, in relationships with the physicochemical characteristics of Moroccan Oued Fez river. *Sci. Afr.* **2023**, *22*, e01949. [CrossRef]
2. Pobi, K.K.; Nayek, S.; Gope, M.; Rai, A.K.; Saha, R. Sources evaluation, ecological and health risk assessment of potential toxic metals (PTMs) in surface soils of an industrial area, India. *Env. Geochem. Health* **2020**, *42*, 4159–4180. [CrossRef] [PubMed]

3. Tholkappian, M.; Ravisankar, R.; Chandrasekaran, A.; Jebakumar, J.P.P.; Kanagasabapathy, K.; Prasad, M.; Satapathy, K. Assessing heavy metal toxicity in sediments of Chennai Coast of Tamil Nadu using Energy Dispersive X-ray Fluorescence Spectroscopy (EDXRF) with statistical approach. *Toxicol. Rep.* **2018**, *5*, 173–182. [CrossRef] [PubMed]
4. Hayzoun, H.; Garnier, C.; Durrieu, G.; Lenoble, V.; Le Poupon, C.; Angeletti, B.; Ouammou, A.; Mounier, S. Organic carbon, and major and trace element dynamic and fate in a large river subjected to poorly-regulated urban and industrial pressures (Sebou River, Morocco). *Sci. Total Environ.* **2015**, *502*, 296–308. [CrossRef]
5. Faouzi, J.; Bedoui, I.; Rezouki, S.; Moubchir, T.; Allali, A.; EloutassI, N.; Lahkimi, A. Vertical transfer of bacteriological and parasitological pollutants from irrigation water to soil and crops. *Ecol. Eng. Environ. Technol.* **2023**, *24*, 93–103. Available online: https://yadda.icm.edu.pl/yadda/element/bwmeta1.element.baztech-0ef0576d-5dd7-4d24-ab67-4db7b4207fa7 (accessed on 10 July 2024). [CrossRef]
6. Moubchir, T.; Eloutassi, N.; Bendaoud, A.; Belkhiri, A.; Maai, M.; Moubchir, M.; Zahir, I. Heavy metals analysis and quality evaluation in drinking groundwater around an abandoned mine Area of Ouichane (Nador's Province, Morocco). *J. Ecol. Eng.* **2023**, *24*, 118–127. Available online: https://bibliotekanauki.pl/articles/24201772.pdf (accessed on 10 July 2024).
7. Qiu, Y.-W. Bioaccumulation of heavy metals both in wild and mariculture food chains in Daya Bay, South China. *Estuar. Coast. Shelf Sci.* **2015**, *163*, 7–14. [CrossRef]
8. Rezouki, S.; El-Haji, S.; Moubchir, T.; Faouzi, J.; Bendaoud, A.; Beniaich, G.; Allali, A.; Eloutassi, N. Evaluation of the impact of anthropic activities on the physico-chemical and microbiological parameters of the water of the wadi inaouene and its tributaries. *Ecol. Eng. Environ. Technol.* **2023**, *24*, 282–294. Available online: https://yadda.icm.edu.pl/yadda/element/bwmeta1.element.baztech-688757fd-6e59-42cf-8af8-e6108b4e6956 (accessed on 10 July 2024).
9. Volpe, M.; La Cara, F.; Volpe, F.; De Mattia, A.; Serino, V.; Petitto, F.; Zavalloni, C.; Limone, F.; Pellecchia, R.; De Prisco, P.; et al. Heavy metal uptake in the enological food chain. *Food Chem.* **2009**, *117*, 553–560. [CrossRef]
10. Parsa, M.; Shirazy, A.; Shirazi, A.; Pour, A.B. Processing and interpretation of geochemical data for mineral exploration. In *Geospatial Analysis Applied to Mineral Exploration*; Elsevier: Amsterdam, The Netherlands, 2023; pp. 171–188. Available online: https://www.sciencedirect.com/science/article/pii/B9780323956086000044 (accessed on 3 October 2024).
11. Zhang, S.; Chen, B.; Du, J.; Wang, T.; Shi, H.; Wang, F. Distribution, assessment, and source of heavy metals in sediments of the Qinjiang River, China. *Int. J. Environ. Res. Public Health* **2022**, *19*, 9140. [CrossRef]
12. Liu, X.; Sheng, Y.; Liu, Q.; Li, Z. Ecological and environmental risks of heavy metals in sediments in Dingzi Bay, South Yellow Sea. *Mar. Pollut. Bull.* **2023**, *188*, 114683. [CrossRef] [PubMed]
13. N'guessan, Y.M.; Probst, J.L.; Bur, T.; Probst, A. Trace elements in stream bed sediments from agricultural catchments (Gascogne region, SW France): Where do they come from? *Sci. Total Environ.* **2009**, *407*, 2939–2952. [CrossRef] [PubMed]
14. Cabral Pinto, M.M.S.; Silva, M.M.V.G.; Ferreira da Silva, E.A.; Dinis, P.A.; Rocha, F. Transfer processes of potentially toxic elements (PTE) from rocks to soils and the origin of PTE in soils: A case study on the island of Santiago (Cape Verde). *J. Geochem. Explor.* **2017**, *183*, 140–151. [CrossRef]
15. Rodríguez, J.; Tueros, I.; Borja, A.; Belzunce, M.; Franco, J.; Solaun, O.; Valencia, V.; Zuazo, A. Maximum likelihood mixture estimation to determine metal background values in estuarine and coastal sediments within the European Water Framework Directive. *Sci. Total Environ.* **2006**, *370*, 278–293. [CrossRef]
16. Birch, G. Determination of sediment metal background concentrations and enrichment in marine environments–a critical review. *Sci. Total Environ.* **2017**, *580*, 813–831. [CrossRef]
17. Singh, K.P.; Mohan, D.; Singh, V.K.; Malik, A. Studies on distribution and fractionation of heavy metals in Gomti river sediments—A tributary of the Ganges, India. *J. Hydrol.* **2005**, *312*, 14–27. [CrossRef]
18. Leleyter, L.; Probst, J.-L. A New Sequential Extraction Procedure for the Speciation of Particulate Trace Elements in River Sediments. *Int. J. Environ. Anal. Chem.* **1999**, *73*, 109–128. [CrossRef]
19. Charriau, A.; Lesven, L.; Gao, Y.; Leermakers, M.; Baeyens, W.; Ouddane, B.; Billon, G. Trace metal behaviour in riverine sediments: Role of organic matter and sulfides. *Appl. Geochem.* **2011**, *26*, 80–90. [CrossRef]
20. Lao, C.; Casas, J.M.; Rosas, H. Heavy metals and metalloids in sediments from the Llobregat basin, Spain. *Environ. Geol.* **2003**, *44*, 325–332. [CrossRef]
21. Hassimi, H.; Taleb, A.; Bouezmarni, M.; Karzazi, O.; Taleb, M.; Kherbeche, A.; Debbaut, V. The effect of the physicochemical conditions variations on the behavior of heavy metals trapped in polluted fluvial system sediments: The case of Oued Sebou, Morocco. *Appl. Water Sci.* **2019**, *9*, 1–8. [CrossRef]
22. Benabdelkader, A.; Taleb, A.; Probst, J.; Belaidi, N.; Probst, A. Anthropogenic contribution and influencing factors on metal features in fluvial sediments from a semi-arid Mediterranean river basin (Tafna River, Algeria): A multi-indices approach. *Sci. Total Environ.* **2018**, *626*, 899–914. [CrossRef] [PubMed]
23. Neto, J.R.; Andrade, E.M.; Palácio, H.A.; Sales, M.M.; Maia, A.R. Influence of land use/occupation on water quality in the Trussu river valley, Ceará, Brazil. *Rev. Ciência Agronômica* **2017**, *48*, 59–69.
24. Moubchir, T.; Faouzi, J.; Loukili, E.H.; Bendaoud, A.; Belkhiri, A.; Rezouki, S.; Eloutassi, N.; Zahir, I. Investigation of the Hydrochemistry Quality of the Ouichane Groundwater (Morocco) Using Multivariate Statistical Methods and Diagram Analysis. *Moroc. J. Chem.* **2024**, *12*, 734–746.

25. Branchu, P.; Badin, A.L.; Bechet, B.; Eisenlohr, L.; Priol, T.L.; Marseille, F.; Trielli, E. Pollution D'origine Routière et Environnement de Proximité. *VertigO-La Rev. Électronique En Sci. De L'environnement* **2013**, *15*. Available online: https://journals.openedition.org/vertigo/12775 (accessed on 10 July 2024). [CrossRef]
26. Lionard, E.; Coquery, M.; Lyon, I. *Échantillonnage des Sédiments: Préparation d'un Essai Collaborative*; Hal Inrae: Paris, France, 2013; p. 24.
27. Montarges-Pelletier, E.; Jeanneau, L.; Faure, P.; Bihannic, I.; Barres, O.; Lartiges, B.S. The junction of Fensch and Moselle rivers, France; mineralogy and composition of river materials. *Environ. Geol.* **2007**, *53*, 85–102. [CrossRef]
28. Da Silva, Y.J.A.B.; Nascimento, C.W.A.D.; Biondi, C.M. Comparison of USEPA digestion methods to heavy metals in soil samples. *Environ. Monit. Assess.* **2014**, *186*, 47–53. [CrossRef]
29. Lepel, E.A.; Stefansson, K.M.; Zoller, W.H. The enrichment of volatile elements in the atmosphere by volcanic activity: Augustine volcano 1976. *J. Geophys. Res. Ocean.* **1978**, *83*, 6213–6220. [CrossRef]
30. Sinex, S.A.; Helz, G.R. Regional geochemistry of trace elements in Chesapeake Bay sediments. *Environ. Geol.* **1981**, *3*, 315–323. [CrossRef]
31. Hakanson, L. An ecological risk index for aquatic pollution control.a sedimentological approach. *Water Res.* **1980**, *14*, 975–1001. [CrossRef]
32. Muller, G. Index of geoaccumulation in sediments of the Rhine River. *Geo J.* **1969**, *2*, 108–118.
33. McLennan, S.M. Chapter 7. Rare Earth Elements in Sedimentary Rocks: Influence of Provenance and Sedimentary Processes. In *Geochemistry and Mineralogy of Rare Earth Elements*; De Gruyter: Berlin, Germany, 2018; Volume 189, pp. 169–200. [CrossRef]
34. Solomon, K.; Giesy, J.; Jones, P. Probabilistic risk assessment of agrochemicals in the environment. *Crop. Prot.* **2000**, *19*, 649–655. [CrossRef]
35. Zhu, H.-N.; Yuan, X.-Z.; Zeng, G.-M.; Jiang, M.; Liang, J.; Zhang, C.; Yin, J.; Huang, H.-J.; Liu, Z.-F.; Jiang, H.-W. Ecological risk assessment of heavy metals in sediments of Xiawan Port based on modified potential ecological risk index. *Trans. Nonferrous Met. Soc. China* **2012**, *22*, 1470–1477. [CrossRef]
36. Benzougagh, B.; Meshram, S.G.; Baamar, B.; Dridri, A.; Boudad, L.; Sadkaoui, D.; Mimich, K. Relationship between landslide and morpho-structural analysis: A case study in Northeast of Morocco. *Appl. Water Sci.* **2020**, *10*, 175. [CrossRef]
37. Deschamps, T.; Benzaazoua, M.; Bussière, B.; Belem, T.; Mbonimpa, M. Mécanismes de rétention des métaux lourds en phase solide: Cas de la stabilisation des sols contaminés et des déchets industriels. *VertigO-La Rev. Électronique En Sci. De L'environnement* **2006**, *7*. [CrossRef]
38. Kimball, B.A.; Callender, E.; Axtmann, E.V. Effects of colloids on metal transport in a river receiving acid mine drainage, upper Arkansas River, Colorado, U.S.A. *Appl. Geochem.* **1995**, *10*, 285–306. [CrossRef]
39. Elliott, H.A.; Liberati, M.R.; Huang, C.P. Competitive Adsorption of Heavy Metals by Soils. *J. Environ. Qual.* **1986**, *15*, 214–219. [CrossRef]
40. Geesey, G.; Borstad, L.; Chapman, P. Influence of flow-related events on concentration and phase distribution of metals in the lower Fraser River and a small tributary stream in British Columbia, Canada. *Water Res.* **1984**, *18*, 233–238. [CrossRef]
41. Basta, N.T.; Tabatabai, M.A. Effect of cropping systems on adsorption of metals by soils: II. Effect of pH. *Soil Sci.* **1992**, *153*, 195–204. [CrossRef]
42. Benbouih, H.; Nassali, H.; Ei Abdellaoui, A.; Srhiri, A. Speciation chimique du zinc et du cuivre dans les sediments du lac Fouarat au Maroc. *J. Soc. Chim. De Tunis.* **2001**, *4*, 1279–1286.
43. Campbell, P.G.C. *Associate Committee on Scientific. Biologically Available Metals in Sediments*; National Research Council of Canada: Ottawa, ON, Canada, 1988; Volume 27694.
44. Kovacs, M.; Nyary, I.; Toth, L. The microelement content of some submerged and floating aquatic plants. *Acta. Bot. Hung.* **1984**, *30*, 173–185.
45. Rezouki, S.; Allali, A.; Touati, N.; Mansouri, D.; Eloutassi, N.; Fadli, M. Spatio-temporal evolution of the physico-chemical parameters of the Inaouen wadi and its tributaries. *J. Chem.* **2021**, *9*, 3–9.
46. Yongming, H.; Peixuan, D.; Junji, C.; Posmentier, E.S. Multivariate analysis of heavy metal contamination in urban dusts of Xi'an, Central China. *Sci. Total Environ.* **2006**, *355*, 176–186. [CrossRef] [PubMed]
47. Zhang, J.; Liu, C. Riverine Composition and Estuarine Geochemistry of Particulate Metals in China—Weathering Features, Anthropogenic Impact and Chemical Fluxes. *Estuar. Coast. Shelf Sci.* **2002**, *54*, 1051–1070. [CrossRef]
48. Lin, Q.; Liu, E.; Zhang, E.; Li, K.; Shen, J. Spatial distribution, contamination and ecological risk assessment of heavy metals in surface sediments of Erhai Lake, a large eutrophic plateau lake in southwest China. *CATENA* **2016**, *145*, 193–203. [CrossRef]
49. Wang, X.; Zhang, L.; Zhao, Z.; Cai, Y. Heavy metal pollution in reservoirs in the hilly area of southern China: Distribution, source apportionment and health risk assessment. *Sci. Total Environ.* **2018**, *634*, 158–169. [CrossRef]
50. El Hajjami, S.; Abriak, N.-E.; Souabi, S.; El Alami, M. +Study of Metallic Contamination of Oued Sebou Sediments, Morocco. *Environ. Technol. Innov.* **2021**, *23*, 101680. [CrossRef]
51. Laaraj, M.; Benaabidate, L.; Mesnage, V. Assessment of inaouene river pollution for potable water supply, Northern Morocc. *J. Ecol. Eng.* **2020**, *21*, 68–80. Available online: https://yadda.icm.edu.pl/baztech/element/bwmeta1.element.baztech-7a14f851-1dc6-4bf7-afbc-0d81c8bf73f9 (accessed on 3 October 2024). [CrossRef]
52. Saleem, M.; Iqbal, J.; Akhter, G.; Shah, M.H. Fractionation, bioavailability, contamination and environmental risk of heavy metals in the sediments from a freshwater reservoir, Pakistan. *J. Geochem. Explor.* **2018**, *184*, 199–208. [CrossRef]

53. Bing, H.; Wu, Y.; Zhou, J.; Sun, H.; Wang, X.; Zhu, H. Spatial variation of heavy metal contamination in the riparian sediments after two-year flow regulation in the Three Gorges Reservoir, China. *Sci. Total Environ.* **2019**, *649*, 1004–1016. [CrossRef]
54. Burton, G.A., Jr. Sediment quality criteria in use around the world. *Limnology* **2002**, *3*, 65–76. [CrossRef]
55. Badassan, T.E.; Avumadi, A.M.; Ouro-Sama, K.; Gnandi, K.; Jean-Dupuy, S.; Probst, J.L. Geochemical composition of the Lomé lagoon sediments, Togo: Seasonal and spatial variations of major, trace and rare earth element concentrations. *Water* **2020**, *12*, 3026. [CrossRef]
56. Cuvier, A.; Leleyter, L.; Probst, A.; Probst, J.-L.; Prunier, J.; Pourcelot, L.; Le Roux, G.; Lemoine, M.; Reinert, M.; Baraud, F. Why comparison between different chemical extraction procedures is necessary to better assess the metals availability in sediments. *J. Geochem. Explor.* **2021**, *225*, 106762. [CrossRef]
57. Leleyter, L.; Rousseau, C.; Biree, L.; Baraud, F. Comparison of EDTA, HCl and sequential extraction procedures, for selected metals (Cu, Mn, Pb, Zn), in soils, riverine and marine sediments. *J. Geochem. Explor.* **2012**, *116–117*, 51–59. [CrossRef]
58. Baize, D. Teneurs totales en éléments traces métalliques dans les sols (France): Références et stratégies d'interprétation. *Programme ASPITET* **1997**, 1–410.
59. Kuo, S.; Lai, M.; Lin, C. Influence of solution acidity and $CaCl_2$ concentration on the removal of heavy metals from metal-contaminated rice soils. *Environ. Pollut.* **2006**, *144*, 918–925. [CrossRef] [PubMed]

*water*

MDPI

*Article*

# Under the Strong Influence of Human Activities: The Patterns and Controlling Factors of River Water Chemistry Changes—A Case Study of the Lower Yellow River

**Chaobin Ren [1] and Lu Liu [2,*]**

[1] School of Civil Engineering, Nanyang Institute of Technology, Nanyang 473004, China; robin.reen@nyist.edu.cn
[2] Hebei Geological Environment Monitoring Institute, Shijiazhuang 050021, China
* Correspondence: liulu0011@163.com

**Abstract:** This study provides an in-depth analysis of the hydrochemical characteristics and their controlling factors in the lower reaches of the Yellow River. Through water quality sampling and analysis over two hydrological periods within a year, combined with hydrochemical methods and machine learning techniques, the study reveals the joint impact of natural factors and human activities on the spatiotemporal variations in hydrochemical constituents. The findings indicate that the water in the lower reaches of the Yellow River exhibits weak alkalinity (the pH is between 7 and 8), with the primary hydrochemical type being $HCO_3 \cdot SO_4$—Ca·Na·Mg. The temporal variation in the hydrochemical constituents is mainly influenced by rainfall, where nitrate levels are higher during the flood season due to the flushing effect of rainfall, whereas other hydrochemical constituents show an opposite temporal pattern due to the dilution effect of rainfall. The spatial variation in the Yellow River's hydrochemistry is primarily controlled by a combination of human activities and rainfall. Using Gibbs diagram analysis, it is identified that rock weathering is the main source of ionic constituents, while agricultural fertilization, industrial emissions, and domestic wastewater discharge have significant impacts on the hydrochemical constituents. Compared to other rivers worldwide, the concentration of hydrochemical constituents in the lower reaches of the Yellow River is relatively high, especially nitrate and sulfate, which is closely related to the geological characteristics of the Yellow River basin and intense human activities in the middle and lower reaches. Principal component analysis reveals that the main controlling factors for hydrochemical constituents during the dry season in the lower reaches of the Yellow River are rock weathering dissolution and industrial activities, followed by domestic wastewater; during the flood season, the main controlling factors are rock weathering dissolution and industrial activities, followed by agricultural activities and domestic wastewater. The research findings provide theoretical support for water resource management and water quality protection in the lower reaches of the Yellow River.

**Keywords:** yellow river; hydrochemistry; ion sources; controlling factors; machine learning

## 1. Introduction

Rivers serve as a critical link between terrestrial ecosystems and the oceans, not only facilitating the transfer of materials and the exchange of energy but also playing a vital role in the socio-economic development within their basins [1,2]. However, in the wake of rapid industrial and urban growth, human activities have exerted a significant impact on the hydrochemical composition of rivers [3], which severely affects the ecological attributes and functions of rivers. Therefore, studying the patterns of river hydrochemical changes and their controlling factors can help us better understand the health of river ecosystems and predict and assess the potential impact of environmental changes on river ecosystems [4,5].

The patterns of variation in river hydrochemical components are influenced by a multitude of factors, including natural factors such as climatic conditions, topography, hydrological cycles, and geological background, as well as anthropogenic factors like land use changes, industrial discharges, agricultural activities, and urbanization processes. These factors interact through physical, chemical, and biological processes, collectively determining the chemical composition and evolutionary trends of river waters [6–9]. Previous studies have found that the temporal variation in surface water hydrochemical components is affected by the dilution and flushing effects of rainfall, exhibiting different seasonal patterns. For instance, rivers dominated by point source pollution are primarily influenced by the dilution effect of rainfall, showing that hydrochemical components have lower concentrations during the flood season than during the dry season [10]. In contrast, water bodies dominated by non-point source pollution have their hydrochemical components influenced by the flushing effect of rainfall, with higher concentrations during the flood season compared to the dry season [11]. The spatial variation in hydrochemical components in surface water is mainly affected by the intensity of human activities within the watershed, showing spatial heterogeneity. For example, Mbaye et al. [12] conducted a study on the spatiotemporal variation in nitrogen in Lake Senegal and found that the concentrations of $NO_3^-$-N and TN were significantly higher in densely populated areas and downstream river sections with intensive agricultural development than in the upstream areas of the lake. Zhang et al. [13] found strong spatial differences in $NO_3^-$-N and $NH_4^+$-N in the Fuyang River, with higher concentrations near urban areas and lower concentrations further away from the city. It can be seen that the spatiotemporal characteristics of surface water hydrochemical components are influenced by a combination of natural conditions and human activities. Due to the varying intensities of natural and human activities within different regions, there are differences in the spatiotemporal variations in the hydrochemical components. Therefore, to accurately identify the spatiotemporal variation patterns of hydrochemical components in rivers, comprehensive surveys and systematic research efforts are required.

The Yellow River Basin, as China's second-longest river, has significant implications for regional water resource management and ecological and environmental protection [14]. Currently, hydrochemical research in the Yellow River Basin mainly focuses on water quality monitoring [15], identification of pollution sources [16], and the spatiotemporal variations in nutrients [17]. However, due to the Yellow River Basin spanning multiple climatic types and geological units, especially in the lower reaches of the Yellow River (the section from Henan Province to Shandong Province), where human activity is intense (with population densities as high as 577 and 638 people/km$^2$) [14], the impact on the Yellow River's hydrochemistry is profound, and its hydrochemical characteristics and evolutionary laws are more complex. In-depth research is needed to reveal the underlying control mechanisms.

This study focuses on the downstream section of the Yellow River from Zhengzhou City in Henan Province to the estuary area in Dongying City, Shandong Province, as the focus of the research. By integrating geochemical methods and machine learning techniques, it aims to deeply analyze the patterns and controlling factors of hydrochemical component variations in the downstream section of the Yellow River. The specific research content includes (1) analyzing the spatiotemporal variation characteristics of the hydrochemical composition of the main stream water in the lower reaches of the Yellow River; (2) identifying the mechanisms by which natural factors and human activities influence the spatiotemporal variations in hydrochemistry in the lower Yellow River; and (3) revealing the controlling factors of hydrochemical variation during different periods. The results of this study will provide a scientific basis for water resource management and protection in the downstream area of the Yellow River.

## 2. Materials and Methods

### 2.1. Overview of the Study Area

The Yellow River originates from the northern foothills of the Bayan Har Mountains on the Qinghai–Tibet Plateau, with a total length of 5464 km, making it the second-longest river in China [18]. This study primarily focuses on the downstream section of the Yellow River, starting from Zhengzhou city in Henan Province in the west to the estuary in Dongying City, Shandong Province, in the east (Figure 1), with a total length of approximately 750 km. The study area has a typical monsoon climate with distinct seasons, characterized by dry and windy conditions in spring with frequent sandstorms, hot and rainy summers, and uneven distribution of precipitation throughout the year, with rainfall concentrated in the months of July to October. The average annual rainfall of more than 900 mm, and mean annual temperature of 12–14 °C. The main soil types in the study area include fluvic soils, loess, and saline soils, while land use types comprise arable land (72.2%), construction land (16.1%), bare land (0.4%), lawn (2.5%), forest land (4.1%), and water bodies (4.7%).

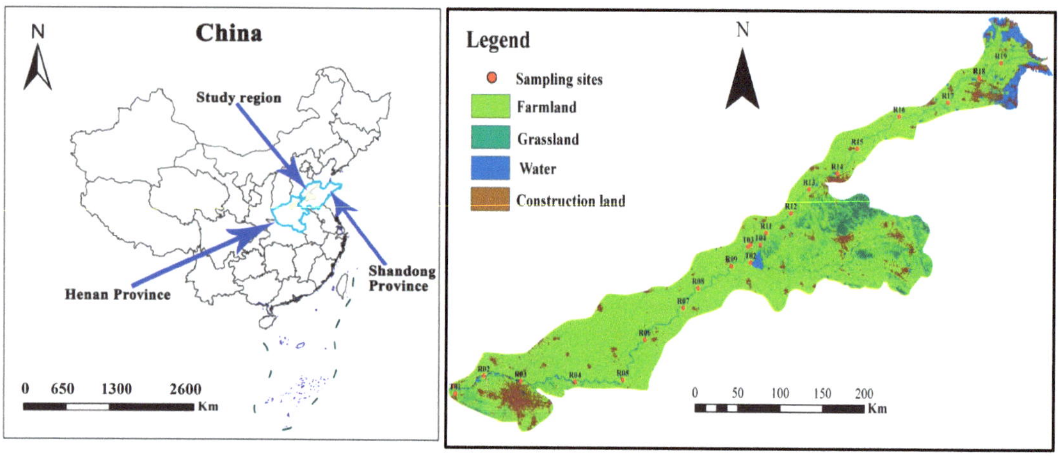


**Figure 1.** Distribution map of river sample sites.

### 2.2. Sample Collection and Testing

This study conducted water sampling during two hydrological periods in April 2021 (low flow season) and July 2021 (high flow season), with a total of 23 sampling sites, including 19 main stream sampling sites and 4 tributary sampling sites. The positions of the sampling sites are shown in Figure 1. A polyethylene surface water sampler was chosen for this study. Before sampling, polyethylene plastic bottles were cleaned with ultra-pure water, and then rinsed three times with river water samples. Water specimens were typically procured from the river's core channel, specifically at depths exceeding 50 cm below the water's surface. Each sample was divided into two portions: one for the determination of cations ($Na^+$, $K^+$, $Ca^{2+}$, and $Mg^{2+}$), acidified with reagent-grade nitric acid on-site to a pH < 2; the other for the determination of anions ($HCO_3^-$, $SO_4^{2-}$, $Cl^-$ and $NO_3^-$), without adding any reagents. All samples were filled to the top of the container without air bubbles and stored in a refrigerator, protected from light. Sample analysis and testing were completed by the Testing Center of the Institute of Hydrogeology and Environmental Geology, Chinese Academy of Geological Sciences. All samples were tested within one week after collection. pH was measured on-site using a Hach portable multi-parameter water quality meter (HACH HQ40d, Loveland, CO, USA). The concentration of total dissolved solids (TDS) was ascertained through the evaporative technique. For the quantification of sodium ($Na^+$), potassium ($K^+$), calcium ($Ca^{2+}$), and magnesium ($Mg^{2+}$), we employed atomic absorption spectrophotometry (Agilent7500ce, ICP-MS, Tokyo, Japan).

Sulfate ($SO_4^{2-}$), chloride ($Cl^-$), and nitrate ($NO_3^-$) were quantified using ion chromatography (Perkin-ElmerLambda 35, Waltham, MA, USA), and $HCO_3^-$ was determined by titration with hydrochloric acid.

*2.3. Data Analysis*

This study utilized SPSS 25.0 software for statistical calculations and analysis of the data, including mean, standard deviation, and principal component analysis. Origin 2022 was employed for plotting the ion ratio diagrams, and ArcGIS 10.5 was used for mapping the sampling sites in the study area.

## 3. Results and Analysis

*3.1. Characteristics of the Hydrochemical Components in the Lower Yellow River*

The statistical analysis results of the hydrochemical components in the lower Yellow River are presented in Table 1. The results were checked using the ion balance method, and it was found that the inorganic ion balance constant [NICB = $(TZ^+ - TZ^-)/TZ^+ \times 100\%$] for all water samples was within ±5%. This suggests that the concentrations of dissolved cations and anions within the water are in equilibrium, validating the authenticity and dependability of the water quality analysis outcomes.

**Table 1.** Statistical characteristics of the hydrochemical composition of the lower Yellow River.

| Parameters | | pH | $K^+$ | $Na^+$ | $Ca^{2+}$ | $Mg^{2+}$ | $Cl^-$ | $SO_4^{2-}$ | $HCO_3^-$ | $NO_3^-$ | TDS |
|---|---|---|---|---|---|---|---|---|---|---|---|
| Mainstream | Min (mg/L) | 7.09 | 3.74 | 50.1 | 62.1 | 24.1 | 43.8 | 124 | 198 | 11.1 | 437 |
| | Max (mg/L) | 8.10 | 4.29 | 98.1 | 71.5 | 31.9 | 94.0 | 163 | 288 | 18.2 | 608 |
| | Mean (mg/L) | 7.42 | 3.96 | 74.3 | 66.8 | 28.0 | 69.2 | 141 | 230 | 15.1 | 529 |
| | SD (mg/L) | 0.313 | 0.148 | 20.7 | 1.69 | 3.33 | 22.6 | 14.5 | 27.0 | 2.15 | 70.8 |
| | CV (%) | 4.22 | 3.75 | 27.8 | 2.53 | 11.9 | 32.7 | 10.3 | 11.7 | 14.2 | 13.4 |
| Tributary | Min (mg/L) | 7.09 | 3.80 | 24.9 | 29.3 | 19.6 | 29.8 | 92.6 | 52.8 | 7.94 | 371 |
| | Max (mg/L) | 8.14 | 14.21 | 137 | 91.3 | 34.8 | 153 | 186 | 320 | 23.2 | 731 |
| | Mean (mg/L) | 7.65 | 6.74 | 70.9 | 62.5 | 26.1 | 83.0 | 140 | 180 | 15.7 | 497 |
| | SD (mg/L) | 0.35 | 3.11 | 36.7 | 20.2 | 5.47 | 42.2 | 31.9 | 74.6 | 5.18 | 103 |
| | CV (%) | 4.57 | 46.1 | 51.8 | 32.4 | 20.9 | 50.9 | 22.8 | 41.5 | 33.1 | 20.8 |

Note: pH, dimensionless; Min = Minimum; Max= Maximum; SD = Standard deviation; CV = Coefficient of variation.

The pH range of the main stream of the Yellow River is between 7.09 and 8.10, with an average of 7.42 ± 0.313, showing weak alkalinity. The main cations are $Na^+$ and $Ca^{2+}$, with average values of 74.3 ± 20.7 mg/L and 66.8 ± 1.69 mg/L, respectively, while the main anions are $HCO_3^-$ and $SO_4^{2-}$, with average values of 230 ± 27.0 mg/L and 141 ± 14.5 mg/L, respectively. Among them, $Cl^-$ and $Na^+$ have a larger range of variation, while pH and $K^+$ show less variation. The pH range of the Yellow River tributaries is similar to that of the main stream, between 7.09 and 8.14, with an average of 7.65, also showing weak alkalinity. The main cations and anions are consistent with the main stream, being $Na^+$ and $Ca^{2+}$ for cations and $HCO_3^-$ and $SO_4^{2-}$ for anions, with concentrations similar to those in the main stream ions. However, the coefficient of variation for all hydrochemical components in the tributaries is greater than those in the main stream, indicating that the hydrochemical components of the tributaries are more influenced by human activities and different geological backgrounds. The predominant hydrochemical classifications for both the main stem and the tributaries of the Yellow River are characterized as $HCO_3 \cdot SO_4$—Ca·Na·Mg-type water (Figure 2).

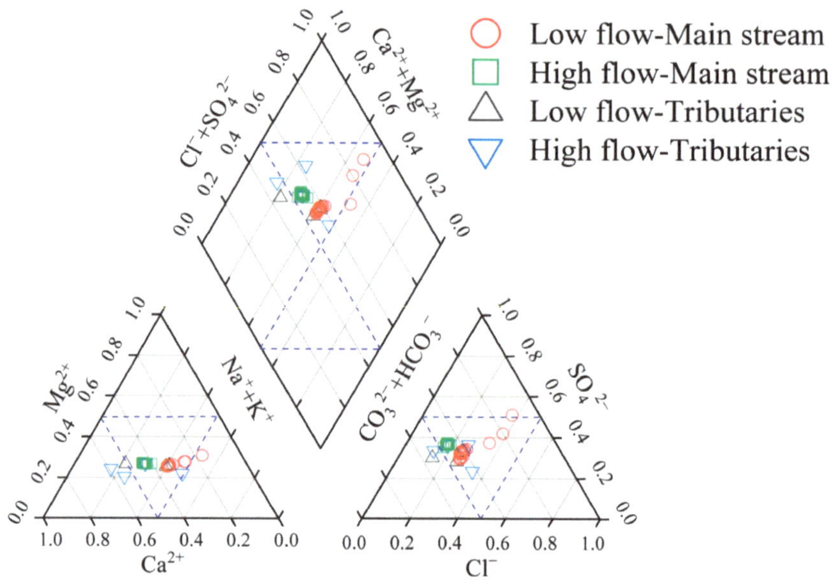


**Figure 2.** Piper map of river water in the Lower Yellow River Water.

### 3.2. *Temporal Variation Characteristics of Hydrochemical Components in the Lower Yellow River*

The temporal variation in hydrochemical components in the lower Yellow River mainly exhibits two patterns (Supplementary Figure S1). The first pattern is that the concentration of hydrochemical components during the low-flow season is greater than that during the high-flow season, with representative indicators being $K^+$, $Na^+$, $Ca^{2+}$, $Mg^{2+}$, $Cl^-$, $SO_4^{2-}$, and $HCO_3^-$. This variation is primarily influenced by the dilution effect of rainfall. The second pattern shows that the concentration during the high-flow season is greater than that during the low-flow season, with the representative indicator being $NO_3^-$. This variation is mainly related to the flushing effect of rainfall, which is primarily attributable to the fact that rainfall runoff washes nitrogen fertilizer from the soil into the rivers, leading to an increase in the concentration of $NO_3^-$ in the river during the high-flow season [1].

### 3.3. *Spatial Variation Characteristics of Hydrochemical Components in the Lower Yellow River*

During the low flow season, the spatial variation in hydrochemical components in the lower Yellow River is primarily influenced by human activities within the basin, showing a significant spatial fluctuation (Supplementary Figure S1). There are two specific patterns observed: The first pattern is that hydrochemical components show a fluctuating upward trend from upstream to downstream, with representative indicators being TDS, $K^+$, $Ca^{2+}$, $HCO_3^-$, $SO_4^{2-}$, and $NO_3^-$. These indicators are significantly impacted by anthropogenic influences along the river's course, such as the discharge of domestic sewage. The second pattern is that the concentration of hydrochemical components shows a slight downward trend from upstream to downstream, with representative indicators being $Na^+$, $Mg^{2+}$, and $Cl^-$.

During the high-flow season, the spatial variation in hydrochemical components in the lower Yellow River is influenced by a combination of human activities within the basin, rainfall-runoff, and the drainage and sediment flushing from the Xiaolangdi Reservoir. Overall, there is less spatial fluctuation, which is related to the large discharge volume of the Xiaolangdi Reservoir (1500 $m^3/s$), which has a significant dilution effect on the spatial variation in hydrochemical components. There are two specific patterns observed: The first pattern is that hydrochemical components show a fluctuating slightly upward trend from upstream to downstream, with representative indicators being $Ca^{2+}$, $HCO_3^-$, and $NO_3^-$. The second pattern is that hydrochemical components show a slight downward trend from upstream to downstream, with representative indicators being TDS, $K^+$, $Na^+$, $SO_4^{2-}$, $Mg^{2+}$, and $Cl^-$.

## 4. Discussion

### *4.1. Comparison of Hydrochemical Components between the Lower Yellow River and Other Major Rivers, Both Domestically and Internationally*

The Yellow River is renowned for its unique high-sediment-laden flow, and its hydrochemical composition significantly differs from other rivers around the world. The water quality characteristics of the Yellow River are mainly influenced by the geological features of the Loess Plateau within its basin and intense human activities, leading to the particularity of its hydrochemical components [19]. Compared to the Yangtze River, Lancang River, and Yarlung Tsangpo River, which all also originate from the Qinghai–Tibet Plateau and share a similar geological background, the hydrochemical components of the Yellow River are generally higher than these three rivers, especially the concentrations of nitrate and sulfate, which are an order of magnitude higher (Table 2). This is mainly related to the intense human activities in the lower reaches of the Yellow River [14]. Chen et al. [20] found that the increase in nitrogen concentration in the Yellow River basin is closely related to regional population density, farmland area, and the degree of industrialization. The Yarlung Tsangpo River mainly flows through the Tibet Autonomous Region, and the Lancang River basin mainly flows through Qinghai Province, Tibet Autonomous Region, and Yunnan provinces in China. The population densities of these three provinces were 8, 3, and 123 people per square kilometer in 2021, respectively. However, the population densities in the downstream provinces of the Yellow River basin, such as Henan and Shandong Province, are respectively as high as 577 and 638 people/km$^2$. Moreover, the development of industry and agriculture in the Lancang River and Yarlung Tsangpo River basins is significantly less than that in the Yellow River basin. In addition, although human activities in the Yangtze River basin are also intense, agricultural activities in the Yellow River basin are more intense than those in the Yangtze River basin. For example, the Yellow River flows through the Hetao Plain—an agricultural irrigation area, and the largest grain storage of China—Henan Province, whose agricultural activities have a stronger impact on the river's hydrochemistry than those of the Yangtze River. Therefore, the concentrations of hydrochemical components in the Yellow River are higher than in these three rivers. Compared to the Nile Delta (20.1 mg/L) and the Thames River (21.9 mg/L), the concentration of nitrate in the Yellow River (2.15 mg/L) is slightly lower, but the concentration of sulfate in the Yellow River is significantly higher. This is mainly related to the high background of sulfate in the Yellow River basin (the strata contain a large amount of sulfide minerals) and the impact of industrial activities [20].

**Table 2.** Comparison of major ion concentrations in the lower Yellow River with other domestic and international Rivers.

| River | $K^+ + Na^+$ | $Ca^{2+}$ | $Mg^{2+}$ | $SO_4^{2-}$ | $Cl^-$ | $NO_3^-$ |
|---|---|---|---|---|---|---|
| This study | 60.1 | 64.1 | 26.2 | 106 | 64.2 | 15.1 |
| Yangtze River [19] | 8.20 | 34.1 | 7.60 | 11.7 | 2.90 | 3.44 |
| Lancang River [1] | 20.6 | 48.17 | 10.72 | 63.01 | 24.8 | 2.52 |
| Yarlung Tsangpo River [21] | — | — | — | 29.6 | 3.29 | 0.54 |
| Nile Delta [22] | 44.8 | 40.0 | 16.6 | 30.1 | 36.4 | 20.1 |
| Thames River [23] | 9.40 | 39.0 | 124 | 5.60 | 73.3 | 21.9 |

### *4.2. Identification of Hydrochemical Component Sources in the Lower Yellow River*

#### 4.2.1. Identification of Hydrochemical Component Sources in the Lower Yellow River Using the Gibbs Diagram

The Gibbs diagram is a useful tool for reflecting the sources of various ions in natural waters (atmospheric precipitation, rock weathering, and evaporation concentration effects) and their evolutionary processes, and it has been widely applied [24]. As shown in Figure 3, in the relationships between TDS and $[Cl^-]/([Cl^-] + [HCO_3^-])$ and TDS with $[Na^+]/([Na^+] + [Ca^{2+}])$, most water samples from both the low- and high-flow seasons fall within the rock weathering and evaporation concentration control zones. During the low-flow season, three water samples are closer to the evaporation concentration control zone, and two points are outside the Gibbs diagram frame; during the high-flow season, one tributary and one main stream water sample are close to the evaporation concentration control zone. This indicates that rock weathering and evaporation concentration are the main controlling factors for the hydrochemical components in the lower Yellow River. Previous research has found that the ion components of surface water are basically located within the Gibbs diagram frame, but when influenced by human activities (such as industrial or mining development), water samples may also fall outside the frame [3], indicating that the variation in hydrochemical components in the lower Yellow River is also affected by human activities.

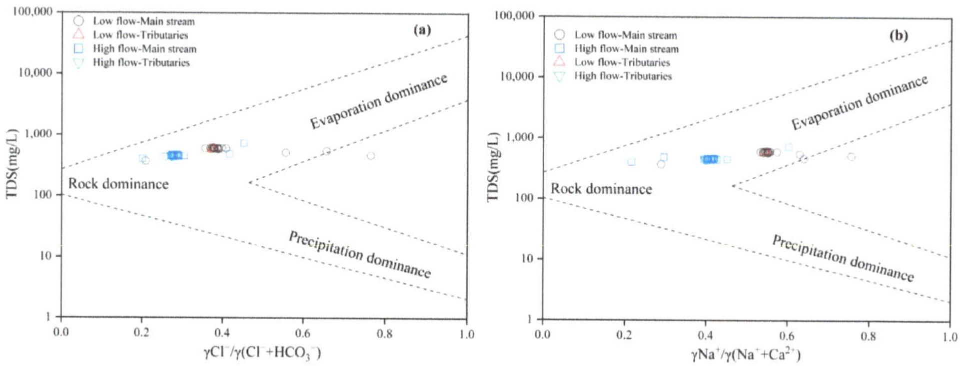


**Figure 3.** Gibbs diagram of water ions from sampling points in the study area: (**a**) $Cl^-/(Cl^- + HCO_3^-)$ vs. TDS; (**b**) $Na^+/(Na^+ + Ca^{2+})$ vs. TDS.

#### 4.2.2. Sources of Hydrochemical Components in the Lower Yellow River—Ion Ratio Method

In natural river water, $Ca^{2+}$, $Mg^{2+}$, and $HCO_3^-$ mainly originate from the dissolution of carbonate rocks, silicate rocks, and evaporite rocks; $SO_4^{2-}$ and $Cl^-$ primarily come from the dissolution of evaporite rocks, and $Na^+$ and $K^+$ are primarily sourced from the dissolution of evaporite and silicate rocks [25]. Therefore, the source of each ion can be inferred based on their ratios.

Typically, the sources of $Ca^{2+}$, $Na^+$, and $Mg^{2+}$ in water bodies are determined by the relationship between $[Mg^{2+}]/[Na^+]$ and $[Ca^{2+}]/[Na^+]$ [26]. As shown in Figure 4a, the main stream of the study area during both the low- and high-flow seasons is primarily located at the carbonate and silicate end-members, while the tributary water samples are close to the carbonate end-member. This is consistent with the geological background of increased carbonate and silicate rock content in the loess of the lower Yellow River region. This suggests that $Ca^{2+}$, $Na^+$, and $Mg^{2+}$ in the river are predominantly a result of the weathering of carbonate and silicate rocks. Meanwhile, the process of evaporite rock dissolution is more pronounced in the regions of the river's tributaries.

Through the ratio of $[K^+ + Na^+]/[Cl^-]$, the sources of $K^+$ and $Na^+$ can be determined. As shown in Figure 4b, the $[K^+ + Na^+]/[Cl^-]$ ratios of the Yellow River water downstream are all greater than one and exhibit a certain correlation, indicating that in addition to being sourced from the dissolution of evaporite rocks, the weathering and dissolution of silicate rocks may also be a source of $K^+$ and $Na^+$. According to the ratio of $[Ca^{2+} + Mg^{2+}]/[HCO_3^-]$, the effects of carbonate and silicate minerals on the formation of hydrochemical constituents can be discerned [27]. Figure 4c demonstrates that the water samples collected from the study area predominantly plot above the 1:1 ratio line and have a good correlation, reflecting the dominant characteristic of carbonate rock weathering [28]. At the same time, $Ca^{2+}$ and $Mg^{2+}$ cannot be completely balanced by $HCO_3^-$, and some $Ca^{2+}$ and $Mg^{2+}$ may come from the dissolution of silicate rocks or evaporite rocks. The ratio of $[Ca^{2+} + Mg^{2+}]/[HCO_3^- + SO_4^{2-}]$ can reflect the source of $SO_4^{2-}$. As shown in Figure 4d, the equivalent concentration of $[HCO_3^- + SO_4^{2-}]$ is higher than that of $[Ca^{2+} + Mg^{2+}]$, and the two are well correlated, indicating that $SO_4^{2-}$ and some $Ca^{2+}$, $Mg^{2+}$ are from the same source. In addition to being sourced from the dissolution of evaporite rocks, they may also be influenced by industrial activities.

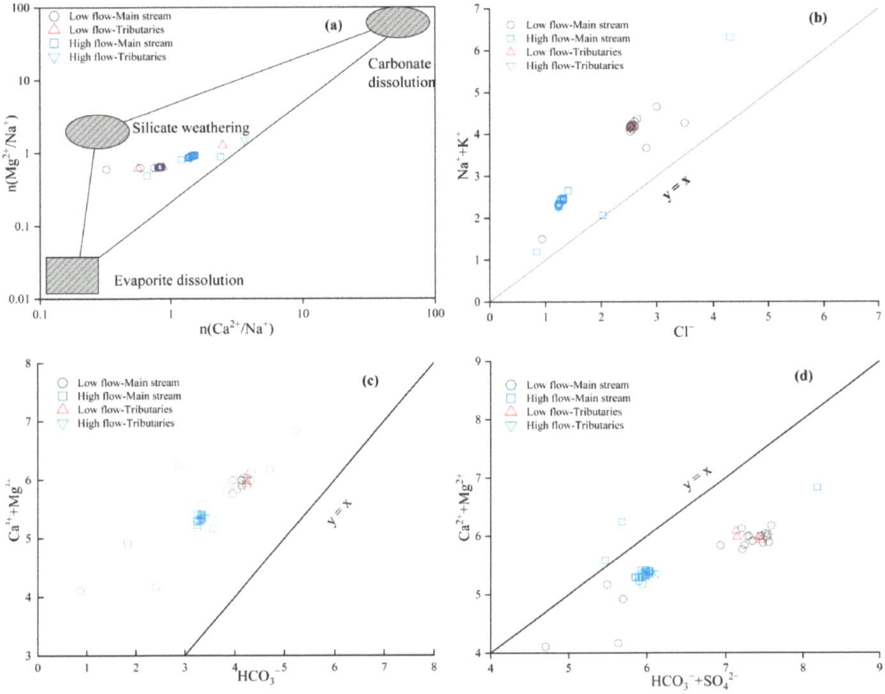


**Figure 4.** Relationship of Major Ion Concentrations in the Lower Yellow River Water: (**a**) $(Ca^{2+}/Na^+)$ vs. $(Mg^{2+}/Na^+)$; (**b**) $Cl^-$ vs. $(Na^+ + K^+)$; (**c**) $[HCO_3^-]$ vs. $[Ca^{2+} + Mg^{2+}]$; (**d**) $[HCO_3^- + SO_4^{2-}]$ vs. $[Ca^{2+} + Mg^{2+}]$.

### 4.3. Impact of Human Activities on Hydrochemical Components in the Lower Yellow River

The $Cl^-$, $SO_4^{2-}$, and $NO_3^-$ in the aquatic environment are often used as characteristic pollutants indicating anthropogenic inputs. $SO_4^{2-}$ primarily originates from industrial activities, while $Cl^-$ and $NO_3^-$ are influenced by agricultural activities and domestic sewage [29]. The source of $NO_3^-$ can be determined based on the ratio of $[NO_3^-]/[Cl^-]$ to $[Cl^-]$. Generally, high $Cl^-$ concentrations and low $[NO_3^-]/[Cl^-]$ indicate fecal and domestic sewage, low $Cl^-$ concentrations and high $[NO_3^-]/[Cl^-]$ indicate agricultural activity, and low $Cl^-$ concentrations with low $[NO_3^-]/[Cl^-]$ indicate the source of soil organic nitrogen [30]. As shown in Figure 5a, the low-flow season has a higher $Cl^-$ concentration and a low $[NO_3^-]/[Cl^-]$, indicating that the river water is affected by domestic sewage during the low-flow season; the high-flow season has a lower concentration of $Cl^-$ and a high $[NO_3^-]/[Cl^-]$, indicating a stronger influence of agricultural activities. In addition, $Cl^-$ and $NO_3^-$ show a clear positive correlation (Figure 5b), suggesting that the sources of the two are both different and the same. The nitrate in the Yellow River comes from domestic sewage and fertilizers, so the $Cl^-$ may mainly come from manure.

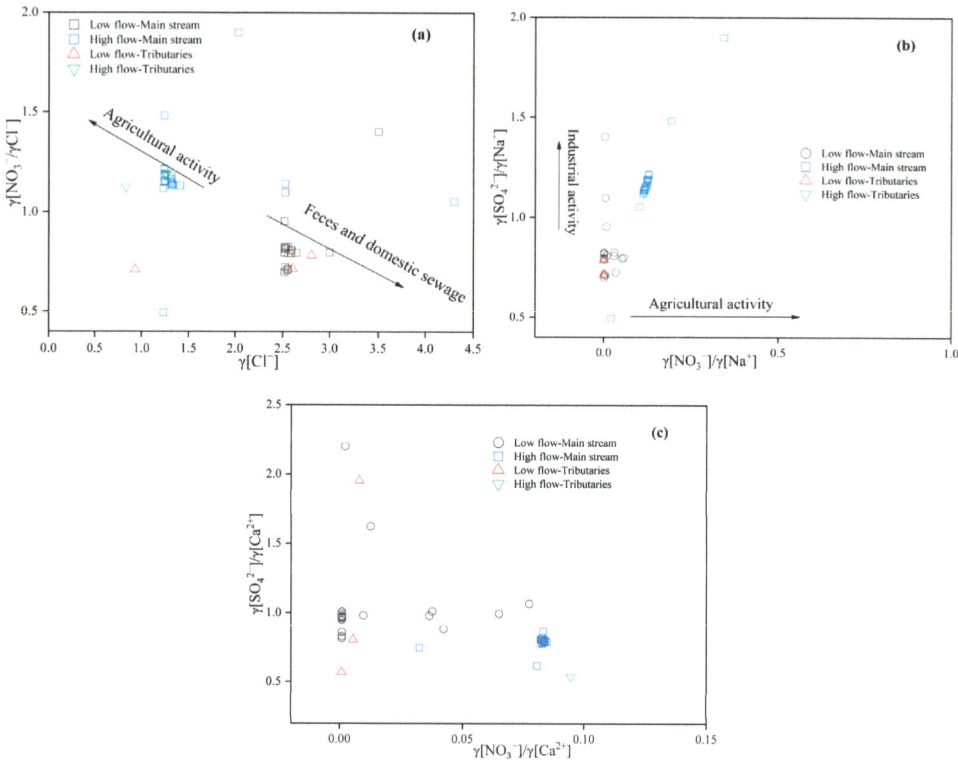


**Figure 5.** Variational Relationships of $[Cl^-]$ and $[NO_3^-]/[Cl^-]$ (**a**), and Variables $[NO_3^-]/[Na^+]$ and $[SO_4^{2-}]/[Na^+]$ (**b**), and $[NO_3^-]/[Ca^{2+}]$ and $[SO_4^{2-}]/[Ca^{2+}]$ (**c**) in the Lower Yellow River Water Body.

Previous analysis has shown that the $SO_4^{2-}$ in the surface water of the study area mainly comes from the dissolution of evaporite rocks and is also affected by the discharge of industrial wastewater. The impact of different human activities on surface water can be further determined through the relationship between $[SO_4^{2-}]/[Ca^{2+}]$ and $[NO_3^-]/[Ca^{2+}]$. Industrial activities have a higher $[SO_4^{2-}]/[Ca^{2+}]$, while agricultural activities have a high $[NO_3^-]/[Ca^{2+}]$ [21]. As shown in Figure 5c, overall, $[SO_4^{2-}]/[Ca^{2+}]$ is significantly higher than $[NO_3^-]/[Ca^{2+}]$, indicating that the $SO_4^{2-}$ in the Yellow River may be affected by industrial activities, and the influence of industrial activities is greater during the high-flow season than during the low-flow season.

### *4.4. Identification of Dominant Factors Controlling Hydrochemistry in the Lower Yellow River*

This study applied principal component analysis (PCA) to further identify the dominant factors controlling the hydrochemistry of the lower Yellow River. Nine water-quality parameters ($K^+$, $Na^+$, $Ca^{2+}$, $Mg^{2+}$, $Cl^-$, $SO_4^{2-}$, $HCO_3^-$, $NO_3^-$, and TDS) were selected for PCA analysis during the low- and high-water periods. Kaiser–Meyer–Olkin (KMO) and Bartlett's test of sphericity were conducted on the study data. The KMO values for the low- and high-water periods were 0.767 and 0.788, respectively, and Bartlett's test values were 228 ($p < 0.001$) and 361 ($p < 0.001$), indicating that the hydrochemical data of the Yellow River met the requirements for principal component analysis for both hydrological periods. Based on the eigenvalue > 1, two principal factors controlling the evolution of hydrochemistry during the low- and high-water periods were identified, which explained

79.5% and 95.1% of all variables, respectively, and could represent all the information of the nine hydrochemical indicators (Tables 3 and 4).

**Table 3.** Principal component analysis of major ions during the low-flow season.

| Parameters | PC1 | PC2 |
|---|---|---|
| $K^+$ | 0.966 | −0.137 |
| $Cl^-$ | 0.962 | 0.099 |
| $HCO_3^-$ | −0.957 | −0.255 |
| $Ca^{2+}$ | −0.759 | −0.607 |
| $SO_4^{2-}$ | 0.707 | −0.209 |
| TDS | −0.674 | −0.516 |
| $Na^+$ | 0.227 | 0.922 |
| $Mg^{2+}$ | −0.155 | 0.875 |
| $NO_3^-$ | 0.001 | 0.621 |
| Eigenvalue | 4.81 | 2.34 |
| Total variance (%) | 53.5 | 26.0 |
| Cumulative (%) | 53.5 | 79.5 |

**Table 4.** Principal component analysis of major ions during the high-flow season.

| Parameters | PC1 | PC2 |
|---|---|---|
| $SO_4^{2-}$ | 0.992 | −0.029 |
| $Mg^{2+}$ | 0.991 | 0.001 |
| TDS | 0.988 | 0.052 |
| $Cl^-$ | 0.987 | −0.116 |
| $Na^+$ | 0.983 | −0.167 |
| $HCO_3^-$ | 0.934 | 0.097 |
| $K^+$ | 0.855 | −0.331 |
| $Ca^{2+}$ | 0.226 | 0.954 |
| $NO_3^-$ | −0.366 | 0.898 |
| Eigenvalue | 6.74 | 1.82 |
| Total variance (%) | 74.9 | 20.2 |
| Cumulative (%) | 74.9 | 95.1 |

During the low-flow season, the first principal component (PC1) explained 37.4% of the total variance (Table 3). The indicators showing a strong positive correlation with PC1 included $K^+$, $Cl^-$, and $SO_4^{2-}$, while those showing a strong negative correlation were $Ca^{2+}$, TDS, and $HCO_3^-$. As mentioned in Section 3, the $Cl^-$ during the low-water period mainly comes from manure, and $HCO_3^-$ mainly comes from the weathering and dissolution of carbonates, while $SO_4^{2-}$ primarily comes from the dissolution of evaporite rocks and industrial activities. Therefore, PC1 represents the control factors of hydrochemical components, which were carbonate weathering, evaporite dissolution, and industrial activities. The indicators showing a strong positive correlation with PC2 were $NO_3^-$, $Mg^{2+}$, and $Na^+$. Generally, $NO_3^-$ in the aquatic environment mainly comes from domestic sewage, fertilizers, soil nitrogen, rainfall, etc. [31]. In this study, it was confirmed that $NO_3^-$ in the Yellow River during the low-flow season mainly comes from domestic sewage, while $Na^+$ and $Mg^{2+}$ mainly come from the weathering and dissolution of silicate rocks. Therefore, PC2 represents the control factors of hydrochemical components, which were silicate weathering and domestic sewage.

During the high-flow season, the PC1 explained 74.9% of the total variance (Table 4). The indicators showing a strong positive correlation with PC1 included $SO_4^{2-}$, $Mg^{2+}$, $HCO_3^-$, TDS, $K^+$, $Cl^-$, and $Na^+$. As mentioned earlier, $K^+$ and $Na^+$ during the high-flow season mainly come from the weathering and dissolution of evaporite rocks, $HCO_3^-$ mainly comes from carbonate dissolution, $Cl^-$ during the high-flow season mainly comes from the discharge of manure, and $SO_4^{2-}$ primarily comes from the dissolution of evaporite rocks and industrial wastewater discharge. Therefore, PC1 represents the control factors of hydrochemical components, which were carbonate weathering, evaporite dissolution, and industrial activities. The indicators showing a strong positive correlation with PC2 were $NO_3^-$ and $Ca^{2+}$. In this study, it was confirmed that $NO_3^-$ in the Yellow River during the high-flow season mainly comes from agricultural activities and domestic sewage. Therefore, PC2 represents the control factors of hydrochemical components, which were agricultural activities and domestic sewage.

## 5. Conclusions

This study, taking the lower reaches of the Yellow River as an example, deeply analyzed the patterns of change in river hydrochemical components and controlling factors under strong human disturbance, revealing the comprehensive impact of natural geological background and human activities on water quality. The study found that the river is influenced by a combination of rainfall and human activities, leading to significant seasonal variations in hydrochemical components between the low- and high-flow seasons. Compared with major rivers domestically and internationally, the concentration of hydrochemical components in the lower Yellow River is relatively high, which may be related to the weathering of evaporite rocks in the upper reaches, agricultural irrigation, and intense human activities in the middle and lower reaches. Through principal component analysis, this study identified the main factors controlling the evolution of hydrochemistry during the low- and high-flow seasons, providing a scientific basis for understanding hydrochemical processes. However, there are some limitations to the study, such as not fully considering the impact of seasonal climate changes on the river's hydrochemical components, not using advanced source tracing methods (such as multiple isotopes combined source tracing technology to quantitatively analyze the sources of hydrochemical components, and not discussing the biogeochemical cycles of the elements). Future research needs to combine advanced isotopic tracing techniques and machine learning methods to conduct high-precision research on the evolution laws, and source identification of hydrochemical substances in the Yellow River Basin. This will provide more comprehensive scientific support for water resource management and ecological and environmental protection in the lower reaches of the Yellow River and the entire Yellow River Basin.

**Supplementary Materials:** The following supporting information can be downloaded at: https://www.mdpi.com/article/10.3390/w16131886/s1, Figure S1: The spatiotemporal variation of the hydrochemical composition of the lower Yellow River.

**Author Contributions:** C.R.: Investigation, Methodology, Software, Data curation, Writing—Original draft preparation. L.L.: Supervision, Methodology, Writing—Reviewing and Editing, Software, Supervision. All authors have read and agreed to the published version of the manuscript.

**Funding:** This work was supported by the Henan Province Science and Technology Development Plan Project (242102321117) and Nanyang Institute of Technology Doctoral Research Startup Fund Project (NGBJ-2023-35) and China Geological Survey, China (Grant No. DD20221773).

**Data Availability Statement:** The data are available from the authors upon reasonable request.

**Acknowledgments:** The authors gratefully acknowledge the editor and anonymous reviewers for their valuable comments on this manuscript. The authors also appreciate the financial support from the different organizations.

**Conflicts of Interest:** The authors declare no conflicts of interest.

## References

1. Zhang, Q.Q.; Zhang, J.; Wang, H.W.; Zhai, T.L.; Liu, L.; Li, G.; Xu, Z. Spatial Patterns in Water Quality and Source Apportionment in a Typical Cascade Development River in Southwestern China Using PMF Modeling and Multivariate Statistical Techniques. *Chemosphere* **2023**, *311*, 137139. [CrossRef] [PubMed]
2. Nguyen, T.A.; Le, D.C.; Nguyen, T.N.; Britta, S.; Tran, L.L. Influences of key factors on river water quality in urban and rural areas: A review. *Case Stud. Chem. Environ. Eng.* **2023**, *8*, 100424.

3. Ren, C.; Zhang, Q.Q. Groundwater Chemical Characteristics and Controlling Factors in a Region of Northern China with Intensive Human Activity. *Int. J. Environ. Res. Public Health* **2020**, *23*, 9126. [CrossRef] [PubMed]
4. Liu, J.; Han, G.; Liu, M.; Zeng, J.; Liang, B.; Qu, R. Distribution, Sources and Water Quality Evaluation of the Riverine Solutes: A Case Study in the Lancangjiang River Basin, Tibetan Plateau. *Int. J. Environ. Res. Public Health* **2019**, *16*, 4670. [CrossRef] [PubMed]
5. Wang, L.Q.; Han, X.X.; Zhang, Y.Y.; Zhang, Q.; Wan, X.M.; Liang, T.; Song, H.; Bolan, N.; Shaheen, S.M.; White, J.R.; et al. Impacts of land uses on spatio-temporal variations of seasonal water quality in a regulated river basin, Huai River, China. *Sci. Total Environ.* **2023**, *857*, 159584. [CrossRef] [PubMed]
6. Ren, X.; Zhang, Z.; Yu, R.; Li, Y.; Li, Y.; Zhao, Y. Hydrochemical Variations and Driving Mechanisms in a Large Linked River-Irrigation-Lake System. *Environ. Res.* **2023**, *225*, 115596. [CrossRef] [PubMed]
7. Zhang, H.Y.; Han, X.; Wang, G.C.; Mao, H.; Chen, X.L.; Zhou, L.; Huang, D.D.; Zhang, F.; Yan, X. Spatial distribution and driving factors of groundwater chemistry and pollution in an oil production region in the Northwest China. *Sci. Total Environ.* **2023**, *875*, 162635. [CrossRef] [PubMed]
8. You, J.; Qi, Y.; Shao, G.; Ma, C. Geochemical Characteristics and Controlling Factors of Groundwater Chemical Composition in the Zihe River Source Area, Shandong, China. *Water* **2024**, *16*, 298. [CrossRef]
9. Xie, C.; Liu, H.; Li, X.R.; Zhao, H.C.; Dong, X.Y.; Ma, K.K.; Wang, N.L.; Zhao, L.J. Spatial characteristics of hydrochemistry and stable isotopes in river and groundwater, and runoff components in the Shule River Basin, Northeastern of Tibet Plateau. *J. Environ. Manag.* **2024**, *349*, 119512. [CrossRef]
10. Zhang, Q.; Zhang, D.; Zhang, F.; Xiao, J. Seasonal Variation in River Water Chemistry of the Middle Reaches of the Yellow River and Its Controlling Factors. *J. Geochem. Explor.* **2015**, *156*, 101–113. [CrossRef]
11. Jiang, H.; Zhang, Q.; Liu, W.; Zhang, J.; Xu, Z. Climatic and Anthropogenic Driving Forces of the Nitrogen Cycling in a Subtropical River Basin. *Environ. Res.* **2021**, *194*, 110721. [CrossRef] [PubMed]
12. Mbaye, M.L.; Gaye, A.T.; Spitzy, A.; Dähnke, K.; Gaye, B. Seasonal and Spatial Variation in Suspended Matter, Organic Carbon, Nitrogen, and Nutrient Concentrations of the Senegal River in West Africa. *Limnologica* **2016**, *57*, 1–13. [CrossRef]
13. Zhang, W.; Jin, X.; Liu, D.; Lang, C.; Shan, B. Temporal and Spatial Variation of Nitrogen and Phosphorus and Eutrophication Assessment for a Typical Arid River—Fuyang River in Northern China. *J. Environ. Sci.* **2017**, *55*, 41–48. [CrossRef] [PubMed]
14. Zhang, Q.Q.; Wang, H.W.; Liu, L.; Zhai, T.L.; Zhang, X.Q. Multiple Isotopes Reveal the Driving Mechanism of High $NO^{3-}$ Level and Key Processes of Nitrogen Cycling in the Lower Reaches of Yellow River. *J. Environ. Sci.* **2024**, *138*, 597–606. [CrossRef] [PubMed]
15. Yue, F.J.; Li, S.L.; Liu, C.Q.; Zhao, Z.Q.; Ding, H. Tracing Nitrate Sources with Dual Isotopes and Long-Term Monitoring of Nitrogen Species in the Yellow River, China. *Sci. Rep.* **2017**, *7*, 506. [CrossRef] [PubMed]
16. Wu, N.; Liu, S.M.; Zhang, G.L.; Zhang, H.M. Anthropogenic Impacts on Nutrient Variability in the Lower Yellow River. *Sci. Total Environ.* **2021**, *755*, 142488. [CrossRef] [PubMed]
17. Wang, H.; Yang, Z.; Saito, Y.; Liu, J.P.; Sun, X.; Wang, Y. Stepwise Decreases of the Huanghe (Yellow River) Sediment Load (1950–2005): Impacts of Climate Change and Human Activities. *Glob. Planet. Change* **2007**, *57*, 331–354. [CrossRef]
18. Zhang, Q.Q.; Wang, X.; Wan, W.; Hou, P.; Li, R.; Ouyang, Z. The Spatial-Temporal Pattern and Source Apportionment of Water Pollution in a Trans-Urban River. *Pol. J. Environ. Stud.* **2015**, *24*, 841–851.
19. Chen, J.; Wang, F.; Xia, X.; Zhang, L. Major Element Chemistry of the Changjiang (Yangtze River). *Chem. Geol.* **2002**, *187*, 231–255. [CrossRef]
20. Wang, H.W.; Zhang, Q.Q. Research Advances in Identifying Sulfate Contamination Sources of Water Environment by Using Stable Isotopes. *Int. J. Environ. Res. Public Health* **2019**, *16*, 1914. [CrossRef]
21. Hu, M.; Wang, Y.; Du, P.; Shui, Y.; Cai, A.; Lv, C.; Bao, Y. Tracing the sources of nitrate in the rivers and lakes of the southern areas of the Tibetan Plateau using dual nitrate isotopes. *Sci. Total Environ.* **2019**, *658*, 132–140. [CrossRef] [PubMed]
22. Taher, M.E.S.; Ghoneium, A.M.; Hopcroft, R.R.; ElTohamy, W.S. Temporal and spatial variations of surface water quality in the Nile River of Damietta Region, Egypt. *Environ. Monit. Assess.* **2021**, *193*, 1–18. [CrossRef] [PubMed]
23. Neal, C.; Neal, M.; Hill, L.; Wickham, H. The water quality of the River Thame in the Thames Basin of south/south-eastern England. *Sci. Total Environ.* **2006**, *360*, 254–271. [CrossRef] [PubMed]
24. Li, P.Y.; Wu, J.H.; Qian, H. Assessment of Groundwater Quality for Irrigation Purposes and Identification of Hydrogeochemical Evolution Mechanisms in Pengyang County, China. *Environ. Earth Sci.* **2013**, *69*, 2211–2225. [CrossRef]
25. Jin, Z.; Qin, X.; Chen, L.; Jin, M.; Li, F. Using Dual Isotopes to Evaluate Sources and Transformations of Nitrate in the West Lake Watershed, Eastern China. *J. Contam. Hydrol.* **2015**, *177–178*, 64–75. [CrossRef] [PubMed]
26. Wang, L.; Zhang, Q.Q.; Wang, H.W. Rapid Urbanization Has Changed the Driving Factors of Groundwater Chemical Evolution in the Large Groundwater Depression Funnel Area of Northern China. *Water* **2023**, *15*, 2917. [CrossRef]
27. Wang, H.W.; Guo, X.J.; Zhang, Q.Q.; Li, B.Y. Evolution of Groundwater Hydrochemical Characteristics and Origin Analysis in Hutuo River Basin. *Environ. Chem.* **2021**, *40*, 3838–3845. (In Chinese)
28. Mayo, A.L.; Loucks, M.D. Solute and Isotopic Geochemistry and Groundwater Flow in the Central Wasatch Range, Utah. *J. Hydrol.* **1995**, *172*, 31–59. [CrossRef]
29. Zhang, Q.Q.; Wang, H.W. Assessment of Sources and Transformation of Nitrate in the Alluvial-Pluvial Fan Region of North China Using a Multi-Isotope Approach. *J. Environ. Sci.* **2020**, *32*, 9–22. [CrossRef]

30. Liu, W.; Jiang, H.; Guo, X.; Li, Y.; Xu, Z. Time-Series Monitoring of River Hydrochemistry and Multiple Isotope Signals in the Yarlung Tsangpo River Reveals a Hydrological Domination of Fluvial Nitrate Fluxes in the Tibetan Plateau. *Water Res.* **2022**, *225*, 119098. [CrossRef]
31. Xue, D.; Botte, J.; Baets, B.D.; Accoe, F.; Nestler, A.; Taylor, P.; Boeckx, P. Present Limitations and Future Prospects of Stable Isotope Methods for Nitrate Source Identification in Surface and Groundwater. *Water Res.* **2009**, *43*, 1159–1170. [CrossRef] [PubMed]

*water*

MDPI

*Article*

# Assessment of Microplastic Pollution in River Ecosystems: Effect of Land Use and Biotic Indices

**David Gutiérrez-Rial [1,*], Iria Villar [1], Romina Álvarez-Troncoso [1], Benedicto Soto [2], Salustiano Mato [1] and Josefina Garrido [1]**

[1] Department of Ecology and Animal Biology, Faculty of Biology, University of Vigo, Campus Lagoas Marcosende s/n, 36310 Vigo, Spain

[2] Department Plant Biology and Soil Science, Faculty of Biology, University of Vigo, Campus Lagoas Marcosende s/n, 36310 Vigo, Spain

* Correspondence: dagutierrez@uvigo.gal

**Abstract:** The proximity of freshwater ecosystems to anthropogenic activities makes them one of the most threatened environments by plastic pollution in the form of microplastics (MPs). Therefore, it is crucial to identify the primary drivers of MP dynamics in rivers to enhance their management. This work analyzed the concentration of MPs in water and sediments and evaluated the influence of land use and its relationship with the main biotic indices employed to assess the water quality of rivers. This research was carried out in four different catchments, with three sampling points established in each river basin. The results revealed that MPs were ubiquitous across all locations, with concentrations ranging from 0.10 to 35.22 items $m^{-3}$ in waters and from 26 to 643 items $Kg^{-1}$ in sediments. The highest concentration of MPs both in water and sediments were found in the Lagares River (35.22 items $m^{-3}$ and 643 items $Kg^{-1}$), while the lowest concentrations were found in the Miñor River for water (0.10 items $m^{-3}$) and Tea River for sediments (138 items $Kg^{-1}$). Urbanization degree was identified as the primary driver of MP pollution in water, whereas population density correlated with sediment pollution levels. These findings explain the elevated MPs abundance in the more urbanized and populated Gafos and Lagares rivers compared to the relatively pristine Miñor and Tea rivers. Furthermore, the presence of MPs in sediments was found to negatively impact the most sensitive benthic macroinvertebrate taxa, as evidenced by lower values of the IASPT and EPT indices at sampling points with higher sediment MPs concentrations (Gafos and Lagares).

**Keywords:** freshwater ecosystems; land use; plastic; EPT; IASPT; IBMWP; biomonitoring

## 1. Introduction

Plastic products have become one of the most demanded materials in modern society; the low manufacturing cost and the versatility of their use led to an exponential increase in their production. Around 400 million tons of plastic are produced at a global scale each year, and it is projected to double by 2050 and more than triple by 2100 [1]. Environmental Protection Agency reported that only 7% of the total produced plastic is recycled annually. Only 8% of the plastic is incinerated, and the remaining is landfilled. However, on many occasions, the high economic and energy costs involved in this process mean that waste ends up accumulating in natural ecosystems [2]. Once in nature, large plastic particles may persist and undergo various degradation processes, such as weathering, photodegradation, or biodegradation, among others, to form microplastics (MPs) [3].

Since the first time that these small particles of plastic were detected, MPs with different shapes, colors, or polymer types were considered by the scientific community as one of the most concerning emerging pollutants for the next decades [4]. The most worrying characteristic of MPs is their ubiquity since these pollutants have been found in the most remote places on earth, such as Lake Hosvgol in Mongolia [5] or in deep-sea sediments from Antarctica [6]. Most of the research efforts were focused on studying the presence and impact of MPs in oceans and marine waters [7,8]. However, terrestrial and freshwater environments may be affected to a higher degree by MPs since all the plastic products are manufactured on land, and most of them are used and disposed of within the continent [9]. Moreover, rivers represent one of the main pathways of efficient transport of MPs from land to the oceans, and the concentration of MPs in freshwater ecosystems is expected to be at least equal to that found in the oceans, with a highly heterogeneous distribution in different areas [10,11].

These pollutants get into rivers through the wind [12], storm sewers, wastewater treatment plants, and as a consequence of direct human activity near riverbanks. The most significant sources of freshwater pollution are wastewater effluents from wastewater treatment plants and runoff from road surfaces caused by the breakdown of road markings and tire debris [13]. Previous studies [14] identified the closeness of urbanization to rivers as sources of MPs from various activities such as effluent discharge, road runoff, littering, and atmospheric deposition to aquatic ecosystems.

Once MPs are present in freshwater ecosystems, the potential threat of contamination may be greater than in the marine environment due to the closer proximity to human activities. The effect of MPs has been studied at various levels (genes, cells, tissues) and both on animals and plants [15]. Benthic macroinvertebrates represent the most abundant group of animals in freshwater ecosystems, and recent studies have evidenced their interaction with these pollutants [16]. As with many other pollutants, the effects depend on many factors, such as the habitat occupied or the MP characteristics. Hence, organisms inhabiting habitats within which MPs are likely to aggregate and be retained will be more susceptible to being affected by them [17]. This would open the possibility of using some taxa as potential indicators of MP pollution in riverine ecosystems, although no research has been conducted on it other than on marine environments where it has already been assessed [18].

Regarding humans, MPs have also entered the food web, thus becoming an emerging problem and a risk to food safety [19]. Recent studies reported the presence of MPs in food and drinking water, indicating that the exposition of MPs is also a fate that threatens human health [20,21]. MPs can cause chemical and physical damage to organisms through oxidative damage and nerve poisoning [22]. In humans, the main effects were mainly manifested in gastrointestinal toxicity and liver toxicity, involving oxidative stress, inflammation, and metabolic disorders [23]. Moreover, the release to water of additives such as bisphenol A (BPA), used as an antioxidant or stabilizing material, may cause endocrine-disrupting effects [24].

Although several studies focused on identifying the major drivers of MP pollution in rivers, and urbanization was reported as one of the most important factors [25], this is not enough to understand the high variability of the data available in the scientific literature. These differences may concern very different magnitude orders even in the same area, indicating that there already exists a problem that could be related to the absence of a unique standardized methodology adopted by the scientific community [26].

It is, therefore, important to obtain data from new regions under different pressures and with different climatic and watershed characteristics. Thus, the main objective of this work was to evaluate the concentration of MPs both in water and in the sedimentary phase in rivers in the NW region of the Iberian Peninsula, an area where the impact of these emerging pollutants has not yet been studied. The specific objectives focused on detecting the main drivers of MP pollution within the study area and evaluating if the current biotic indexes used to assess the water quality could be indicative of MP pollution in rivers.

## 2. Materials and Methods

### 2.1. Study Area

The study area was in the province of Pontevedra (NW Spain), with a total population of 941.772 inhabitants distributed heterogeneously through a surface of 4.459 Km$^2$. The territory has a very particular climatic condition since, in the oriental area, it is considered a transition zone between the Mediterranean and Oceanic climate. According to the Koppen–

Geiger climate classification [27], the study area is characterized by a temperate climate with dry and warm summers and rainy winters (Cfb).

To carry out this work, four rivers were selected, and three sampling points were established, each of them distributed from the headwaters to the mouth in the upper (1), middle (2), and lower (3) sections (Figure 1). Three of the streams, the Gafos River (G), the Lagares River (L), and the Miñor River (M), are managed by the Galicia-Costa basin district. The Tea River is included within the Miño-Sil Hydrographic Demarcation.

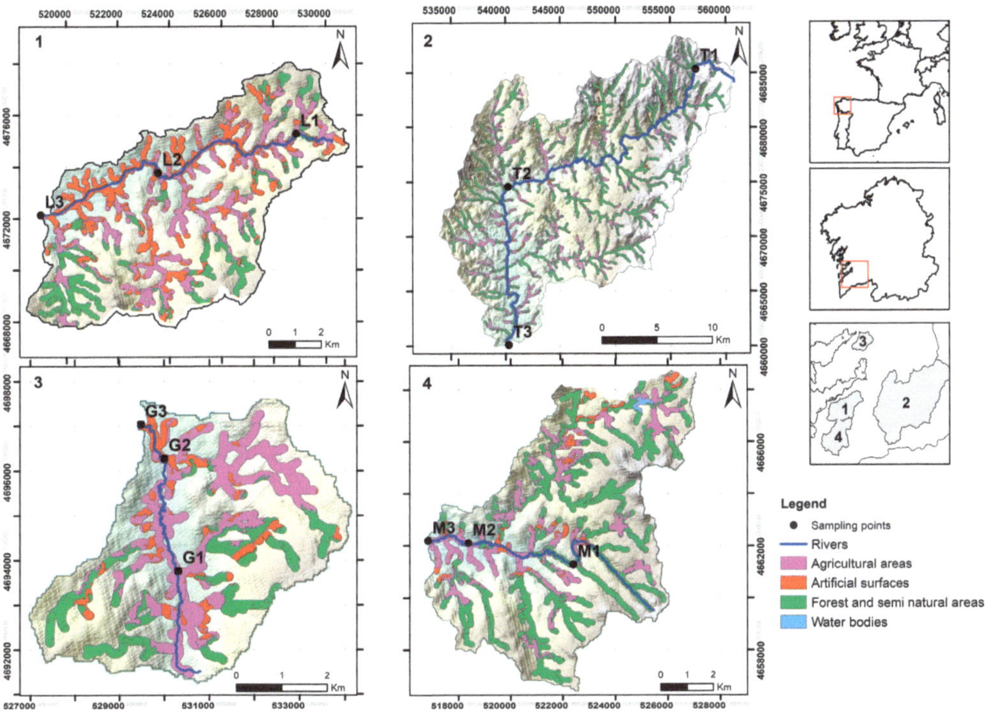


**Figure 1.** Study area and location of the three sampling points established throughout the course of each of the studied rivers: Lagares (1), Tea (2), Gafos (3), and Miñor (4). It represents the catchment and the land use within a buffer zone of 100 extracted from the drainage network.

River Basins Characterization

The hydrological basin of the rivers and each of the sub-basins for the studied points and the drainage network were extracted from the Digital Elevation Models (DEM) with a resolution of 10 m by using the specific hydrology tools of the QGIS Geographic Information System. In each sub-basin was calculated the surface percentage occupied by the land uses of level 1 of the Corine Land Cover classification in a buffer zone of 100 m along the river network up to each sampling point, following the same criteria as Gutiérrez-Rial et al. [28]. Additionally, the population density was calculated based on the population data of the IGE (Instituto Galego de Estatística) and the river basin surface. Characteristics of each of the sub-basins are reflected in Table 1.

**Table 1.** Characteristics of the sub-basin for each of the sampling points in the rivers Tea (T), Miñor (M), Gafos (G), and Lagares (L). CLC parameters represent the percentage of the sub-basin occupied by the Corine Land Cover Classification level 1: **CLC 1:** artificial areas; **CLC 2:** agricultural areas; **CLC 3:** forest and semi-natural areas; **CLC 5:** water bodies.

| | Basin Area (Km$^2$) | Population Density (Hab Km$^{-2}$) | CLC 1 (%) | CLC 2 (%) | CLC 3 (%) | CLC 5 (%) |
|---|---|---|---|---|---|---|
| T1 | 14.03 | 13.79 | 0.00 | 7.80 | 92.20 | 0.00 |
| T2 | 218.79 | 36.03 | 1.14 | 14.01 | 84.85 | 0.00 |
| T3 | 406.62 | 85.02 | 2.01 | 18.95 | 79.03 | 0.00 |
| M1 | 12.66 | 75.43 | 2.46 | 12.51 | 85.03 | 0.00 |
| M2 | 63.74 | 215.38 | 7.70 | 20.28 | 71.62 | 0.41 |
| M3 | 73.92 | 214.67 | 7.83 | 21.05 | 70.76 | 0.35 |
| G1 | 11.56 | 173.46 | 11.26 | 29.16 | 59.58 | 0.00 |
| G2 | 25.82 | 310.94 | 14.55 | 40.20 | 45.25 | 0.00 |
| G3 | 26.94 | 756.02 | 16.94 | 39.68 | 43.38 | 0.00 |
| L1 | 4.27 | 461.93 | 31.74 | 33.58 | 34.68 | 0.00 |
| L2 | 47.33 | 1338.67 | 41.67 | 28.40 | 29.93 | 0.00 |
| L3 | 69.45 | 1958.89 | 42.09 | 25.55 | 32.36 | 0.00 |

### 2.2. MPs Sampling Collection

Two sampling campaigns were carried out in the spring and summer of 2021 in all the mentioned locations. MPs samples from the water column were collected by filtering the water. For this purpose, a specific MP sampling net with a 220 μm mesh size was chosen to avoid the clogging of the net. The sampling net was submerged at a depth of 0–5 cm for five minutes with the open side against the river flow, forcing the MPs into the net. The amount of filtered water was calculated by measuring the flow speed with a flow meter OTT C2 considering the sampling surface of the net ($50 \times 12$ cm/600 cm$^2$). A clean stainless steel was employed to take 1 L of sediment from random locations from the upper 5 cm sediment layer along the river section. Both water and sediment samples were placed in correctly labeled bottles for transport to the laboratory, but first, any significant fragments of organic matter and stones were removed in situ.

### 2.3. MPs Samples Processing

Once in the laboratory, the samples were processed following the NOAA laboratory methods for the analysis of MPs in water and bed samples with some modifications [29]. The water processing consisted of sieving, organic matter elimination, density separation, and finally, vacuum filtration. First, samples were run through a stacked series of metal sieves (5 and 0.55 mm) and cleaned with distilled water. All the organic particles and plastic items >5 mm were discarded. Next, 30 mL of 30% hydrogen peroxide ($H_2O_2$) (Scharlab S.L., Sentmenat, Spain) was added to the fraction smaller than 5 mm to eliminate organic matter. The mixture was heated to 75 °C until it started to boil and left to react for 24 h at room temperature. Then, the samples underwent density separation using NaCl (density 1.2 g cm$^{-3}$) (Labbox Labware S.L., Premià de Dalt, Spain). After 24 h, the supernatant was filtered through 0.45 μm Whatman filters (Cytiva, Global Life Sciences Solutions Operations UK Ltd., Buckinghamshire, UK) under vacuum conditions. Finally, filters were dried at 40 °C before MPs identification under a stereomicroscope.

For the sediments, the procedure was the same but with two previous steps. The wet samples were homogenized by intensive stirring. Then, three sub-samples of 50 g were taken and dried at 90 °C for 2 days. Once dried, the samples underwent the same protocol as for water samples: sieving, organic matter digestion, density separation, and vacuum filtration to retain the MPs in the 0.45 μm Whatman filters. MPs items were identified under stereomicroscope Leica S9D (Leica Microsystems, Wetzlar, Germany) with magnification from 0.6× to 5.5×. The MPs were counted and classified on fragments, spheres, fibers, and foams following the criteria of previous works [30,31].

### 2.4. Quality Assurance and Quality Control

To avoid contamination during sample collection and laboratory processing, all the equipment was carefully washed with distilled water, and all the materials were covered with aluminum foil. During all the steps of the laboratory processing, all the samples were immediately covered, and the exposition time to air was minimized. Moreover, the personnel of the laboratory wore white laboratory coats while the samples were being processed. Whatman filters were placed near the samples exposed to the same contamination as the samples. Filters were explored under a stereomicroscope, and white fibers were found, probably from the laboratory coats, but the amount was negligible compared to the amount found in this study.

### 2.5. Benthic Macroinvertebrate Sampling and Processing

In each of the sampling locations, benthic macroinvertebrates were captured following the methodology established by the Water Framework Directive described in Boonsoong et al. [32]. An entomological hand net with a mesh size of 500 μm was used to collect the benthic macroinvertebrates. The samples were poled and kept in 4% formaldehyde solution to be transported to the laboratory. Once there, all the individuals were identified at the family level using a stereomicroscope and specialized identification keys [33,34], and the abundance (N) of individuals at each site was calculated. After that, several biotic indices were calculated. First, the richness (S) was calculated as the number of different taxa, and then the IBMWP (Iberian Biomonitoring Water Party) and the IASPT (Iberian Average Score Per Taxon) were used to assess the water quality following the protocol of Alba-Tercedor [35]. Additionally, the EPT and PT indexes based on the most sensitive taxa (Ephemeroptera, Plecoptera, and Trichoptera) were calculated considering the number of different taxa of these groups at each sampling station to study the effect of MPs on these individuals.

In addition, at each sampling point, the pH, water temperature, electrical conductivity (EC), redox potential, dissolved oxygen, and the total dissolved solids (TDS) were measured in situ by using the multiparametric sensor Hanna® HI98194 (Hanna Instruments S.L., Eibar, Spain). Moreover, water samples were taken to analyze the Total Organic Carbon (TOC) and the Chemical Oxygen Demand (COD) through the combustion method, and the main inorganic anions ($SO_4^{-2}$, $PO_4^{-3}$, $NO_3^-$ and $Cl^-$) were determined by thermocatalytic decomposition method.

### 2.6. Data Analyses

To analyze the MP concentration in sediments and in the water column, two generalized linear models (GLM) were developed, once for each matrix. Previously, Pearson correlations were calculated to select the variables that influence the concentration of MPs. Initially, the variables considered for each sub-basin were as follows: the basin area (Basin.area), the population density (PopDen), the percentage of agricultural areas (Ag.areas), artificial areas (Art.areas), forest, and semi-natural areas (For.areas) within of the buffers of 100 m; the precipitation accumulated in the previous 15 days (Prec.15) and in the previous 30 days (Prec.30); the total dissolved solids (TDS) and electric conductivity (EC); and two factorial variables, river (Gafos, Lagares, Miñor and Tea) and season (spring or summer). An Exploratory Data Analysis (EDA) was developed in combination with a stepwise process to select the most adequate GLM until no further improvement was possible following the AIC and BIC criteria.

## 3. Results and Discussion

### 3.1. Occurrence and Distribution of Microplastics in Water

The results of this study showed that the presence and concentration of MPs in the surface waters of the studied rivers were highly variable. On the one hand, it varied between rivers and between sampling stations within the same river, and on the other hand, the results showed a marked seasonal pattern, as can be seen in Figure 2. In total, 561

MP particles were identified, of which 81% were fibers, 20% were fragments, and, to a lesser extent, films (1.6%) and spheres (0.17%). In summer, the Lagares River had the highest concentration of MPs in the water (35.22 MPs $m^{-3}$) with a statistically significant difference ($p < 0.05$). The Gafos River had the second-highest concentration (11.36 MPs $m^{-3}$) in summer, while the lowest values were recorded in the Tea River (0.25 MPs $m^{-3}$) in spring and in the Miñor River (0.1 MPs $m^{-3}$) in summer. Despite the differences, the concentrations of MPs found in this study were similar to other European rivers, as can be seen in Table 2, much lower than in rivers from the Asian continent, considered a hot spot for plastic pollution due to rapid economic and demographic growth [36].

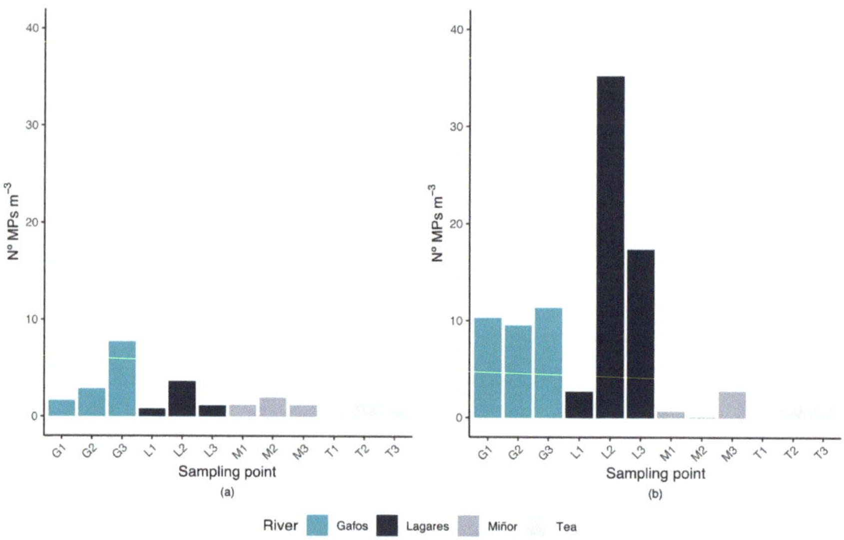


**Figure 2.** Microplastic concentration in water in all the sampling stations in both spring (**a**) and summer (**b**).

Previous works [37] suggested that the difference in MP concentration in rivers depended on different characteristics of the river catchments, such as land use, population density, or the socioeconomic conditions of the area. In this study, the concentration of MPs in rivers varied significantly depending on the artificial surface area in the sub-basin ($p < 0.05$). Previous studies demonstrated that the abundance of MPs increases in urbanized areas with respect to sub-urban [38,39], which would explain the higher concentration of MPs found in the Gafos and Lagares rivers compared to the rest of the rivers. These two rivers, among those studied, have the highest proportion of their catchment area urbanized, as can be seen in Figure 1 and in Table 1. These characteristic conditions were reflected in the higher abundance of MPs found in the most urbanized areas (G2, G3, and L2 and L3). The basin of rivers such as the Miñor or the Tea are characterized by lower population density, and its catchments are much more dominated by forested/natural areas (Table 1). So, anthropogenic activities are less intensive within their basins, and therefore, the potential sources of MPs are less abundant. This is reflected in the lower concentration of MPs found in these rivers.

Another observed trend was the increase in the concentration of MPs from upstream to downstream, something found before by Zhao et al. [40]. Generally, rivers close to their source tend to have steeper gradients and faster flow rates, while downstream sections are flatter. This means that MPs can migrate from upstream to downstream, resulting in higher amounts of MPs as the river flows [41]. However, the dilution capacity of the river may result in a lower concentration of MPs in the lower section of the rivers, where the river flow is higher, so the concentration may be low, as found by Tan et al. [4]. But, in these cases, the total amount of MPs transported by the river continues to be higher than in the upper and middle zones when considering the total volume of water carried by the river. Moreover, human settlements and industrial areas are usually concentrated in downstream sections; in these sites, the high degree of urbanization results in the presence of a high proportion of less permeable soil that facilitates the transport of MPs by runoff to different environmental compartments, including rivers [25].

**Table 2.** Compilation of the abundance of MPs in the water column from different locations available in the recent scientific literature. The list is ordered from the lowest to the highest maximum concentration of MPs.

| Study Area | Continent | MPs Water (MPs $m^{-3}$) | Reference |
|---|---|---|---|
| Paraiba do Sul and Pomba rivers | South America | 0.52 | da Costa et al. [42] |
| Paraiba do Sul and Dois rivers | South America | 0.65 | da Costa et al. [42] |
| Paraiba do Sul and Muriaé rivers | South America | 0.96 | da Costa et al. [42] |
| Tea River. Spain | Europe | 1.54 | This study |
| Miñor River. Spain | Europe | 2.82 | This study |
| Pearl River | Asia | 4.21 | Li et al. [43] |
| Liane River | Europe | 4.52 | Pasquier et al. [44] |
| Ebro River. Spain | Europe | 4.9 | Simón-Sánchez et al. [45] |
| Milwaukee Rivers (Milwaukee) | North America | 5.67 | Lenaker et al. [46] |
| Gafos River. Spain | Europe | 11.36 | This study |
| Elbe River | Europe | 13.24 | Scherer et al. [47] |
| Lagares River. Spain | Europe | 35.22 | This study |
| Ljubljanica | Europe | 45 | Matjasic et al. [48] |
| Kamniška Bistrica Basin | Europe | 75 | Matjasic et al. [48] |
| North Saskatchewan River | America | 88.3 | Ross et al. [49] |
| Neuse River Basin | North America | 131 | Kurki-Fox et al. [50] |
| Neuse River Basin | North America | 221 | Kurki-Fox et al. [50] |
| River Ganga | Asia | 237.9 | Rajan et al. [51] |
| Yangtze River. China | Asia | 258 | Yuan et al. [52] |
| Swat River | Asia | 594 | Bilal et al. [53] |
| Ergene River | Europe | 1206 | Akdogan et al. [54] |
| Yangtze River | Asia | 35,986 | Huang et al. [55] |

The concentration of MPs was also affected by a seasonal pattern; as shown in Figure 2, the number of MPs by cubic meter was higher in all the sampling points in the dry period (summer) rather than in the wet period (spring). During the wet season, with the first rains and floods, surface runoff from the areas surrounding the river increases, resulting in an input of MPs that would increase the abundance and concentration of MPs in the water [25]. However, after a few days of precipitation, the effect would be the opposite; the rainfall would cause a dilution effect of the concentration in the water, and the increased flow would reduce the retention time of the MPs transporting them along the river [56]. Like other types of pollutants, microplastics in dynamic ecosystems such as rivers experience fluctuations over short periods due to the short residence time that water has [47].

### 3.2. Occurrence and Distribution of Microplastic in Sediments

In total, 460 MPs were identified, and, as in the water column, fibers were the most abundant type (70.6%), followed by fragments (26.9%) and, to a lesser extent, spheres (2.4%). The results obtained for the sediments differed markedly from those shown for the water column. Higher concentrations of fragments in sediments rather than in the water column were also found by Lin et al. [57]. Fragments appear more frequently in sediments because of their morphological characteristics; these particles have a higher surface-to-volume ratio and tend to have higher density, which increases the sedimentation of these particles rather than fibers. [58].

Although the concentration of MPs was highly variable between rivers and between sampling points in the same rivers, no clear pattern was observed; nevertheless, the highest concentrations of MPs in the sediments were reported in the same locations as those for the water column as indicated in Figure 3 but differed by a few orders of magnitude between rivers. The highest concentration was found in L3 in summer (643 MPs $Kg^{-1}$) and the lowest in T2 in spring (26 MPs $Kg^{-1}$).

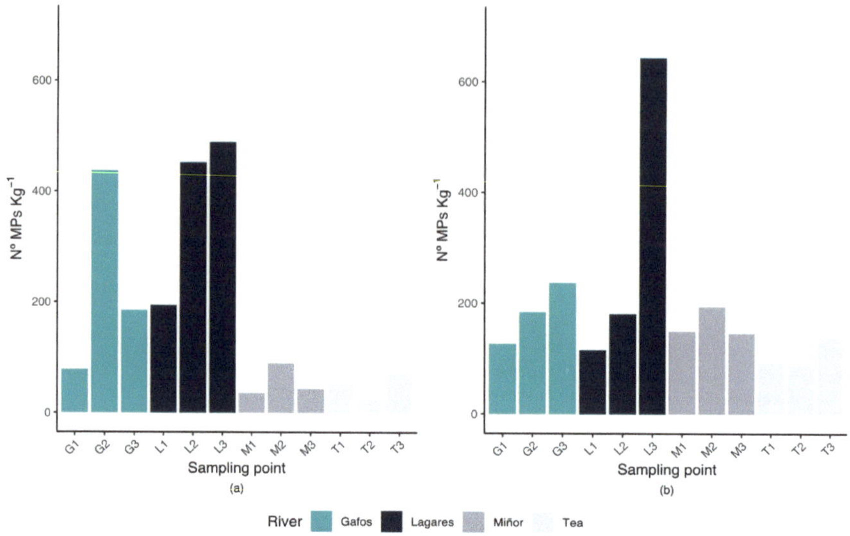


**Figure 3.** Microplastic concentration in sediments in all the sampling stations in both spring (**a**) and summer (**b**).

In contrast to the results obtained for the water column, the concentration of MPs in sediments did not show a spatial or temporal distribution pattern, something common among previous studies [59]. In the case of surface waters, there was a pronounced pattern resulting in Asian rivers being the most affected by MP pollution; the concentration in sediments shows significant spatial variability, as indicated in Table 3. It is expected to be higher in the same areas as for surface waters. The absence of this pattern could be indicative of a lack of consistency in the methodology, previously announced by Lu et al. [56]. Once the differences caused by the methods have been eliminated or at least minimized, the effect of the environmental variables can be quantified with certainty. Logically, river hydrodynamic conditions, as well as river characteristics, should be one of the main drivers of MPs dynamics in rivers.

The results obtained in this study indicated that the concentration of MPs increased from the source of the rivers to the river mouth. This pattern was especially marked in the Gafos and Lagares rivers in summer and in the Lagares River in spring. Other marked pattern was observed in the Miñor River in both seasons and in the Gafos River in spring. In this case, the highest MP concentration in sediments was found in the middle section, similar results to those found by Scherer et al. [47] in the Elbe River. The sampling points in these rivers in the low section are very close to the mouth of the Atlantic Ocean (Figure 1). Thus, these locations are affected by tidal effects, which move marine sediments from the Atlantic Ocean into the rivers, leading to sediment mixture and a further dilution of MP levels [60] compared to that found in the middle section (Figure 3).

**Table 3.** Compilation of the abundance of MPs in river sediments from different locations available in the scientific literature. The list is ordered from the lowest to the highest maximum concentration.

| Study Area | Continent | MPs Concentration (MPs $Kg^{-1}$) | Reference |
|---|---|---|---|
| Nanhuizui tidal flat | Asia | 6 | Peng et al. [57] |
| Ljubljanica | Europe | 44 | Matjasic et al. [48] |
| Kamniška Bistrica Basin | Europe | 48 | Matjasic et al. [48] |
| Yujiabang River | Asia | 53 | Peng et al. [57] |
| Huangpu River branch | Asia | 102 | Peng et al. [57] |
| Shajinggang River | Asia | 104 | Peng et al. [57] |
| Jiangjiagang River | Asia | 117 | Peng et al. [57] |
| Tea River. Spain | Europe | 138 | This study |
| Beishagang River | Asia | 179 | Peng et al. [57] |
| Miñor River. Spain | Europe | 194 | This study |
| Tibet Plateau | Asia | 195 | Jiang et al. [61] |
| Caohejing River | Asia | 230 | Peng et al. [57] |
| Gafos River. Spain | Europe | 437 | This study |
| Lagares River. Spain | Europe | 643 | This study |
| Tisza River | Europe | 5147 | Kiss et al. [62] |
| Milwakee Rivers | America | 6229 | Lenaker et al. [46] |
| Elbe River | Europe | 6750 | Scherer et al. [47] |
| St. Lawrence Rive | America | 7562 | Crew et al. [63] |
| Amazon River | America | 8178 | Gerolin et al. [64] |
| West River | Asia | 10,240 | Huang et al. [65] |

In this study, the concentration of MPs in sediments varied significantly depending on the population density of each sub-basin ($p < 0.05$), something previously found by Matjasic et al. [48], who also reported the size and length of the catchment as one of the major driver of MPs pollution in sediments from freshwater systems. Our results contrast with those found by Klein et al. [66], who did not find any relationship between the MP concentration in the sediments and the population density. But, as for many other previous studies [11,67,68], our model showed that the population density may be a good indicator of MP pollution in the riverbank. This would explain the higher concentration of MPs found in the Lagares and Gafos rivers, whose basins show a higher population density than the rest of the rivers, even in the headwater points (Table 1). Generally, in these highly populated areas, it is expected that high pollution levels may be observed in sediments [69,70]. These rivers, which are much more urbanized, have lower water velocities and less turbulent hydrodynamic conditions. In addition, potential sources of MPs are more abundant, and migration velocity is lower [4]. These conditions facilitate a higher abundance of MPs by increasing sedimentation and retention of MPs in the sediments, compared to much more natural rivers such as the Miñor or Tea, where hydrodynamic conditions increase the transport of MPs along the river course.

Finally, there were no statistically significant differences ($p < 0.05$) between the concentration of MPs found in spring compared to summer. Although some authors consider that seasonal variation is a key factor affecting MP distribution in rivers [71], it should not be applicable in the region of this study, where rainfall is frequent throughout the year. However, in areas where dry and wet periods are much more differentiated, these differences may be more marked. In these cases, rainfall events promote the wet deposition of MPs suspended in the air [72], and the storm events lead to faster river flows that can wash the river sediments and transport them downstream, leading to a lower abundance of MPs in headwater sediments [73]. In contrast, during the dry season, the lower flow velocity may facilitate the accumulation and sedimentation of MPs, which may result in a higher abundance of MPs in the sediments [74].

### 3.3. Biotic Indices and MPs Pollution

The use of benthic invertebrates as indicators to assess different pollution sources has been extended since the last century to guarantee the sustainability of biodiversity and ecosystem services [75]. For this reason, it has been recognized as one of the most useful tools to develop monitoring programs in rivers [76]. Nevertheless, when the biotic indices were developed, threats such as microplastics were not considered, so the potential use to assess the presence of these emerging pollutants remains unknown.

In this study, 21,726 individuals were identified as belonging to 92 different taxa between spring and summer, with insects being the most abundant group (77%). The maximum abundance was found in G2 in spring (N = 3579) and the minimum in L3 in summer (N = 97), while the highest and the lowest values for richness were reported in T1 in spring (S = 40) and M3 (S = 8) and L3 (S = 8) in summer, respectively (Table 4). As shown in Table 5, the correlation test did not show a significant ($p > 0.05$) correlation between the abundance of macroinvertebrates and the concentration of MPs. Although some authors previously found that long-term expositions to these pollutants can result in a decrease in the total abundance of macroinvertebrates [77], this may not happen under natural conditions in rivers. The concentration of MPs in sediments used in experimental conditions usually are much higher than realistic [78], and the presence of MPs in the water column in a dynamic system as a river is not enough to affect these individuals as a consequence of the less prevalence time caused by the water flow. Considering the richness, we found a significant ($p < 0.05$) negative correlation ($\varrho = -0.427$) with the concentration of MPs in sediments (Table 5). So, the presence of high levels of MPs in the sediments could affect the structure of the benthic macroinvertebrate communities by reducing the diversity, as found by Silva et al. [79].

**Table 4.** Diversity and water quality indices calculated based on the benthic macroinvertebrate communities of each point for the rivers Tea (T), Miñor (M), Gafos (G), and Lagares (L) in spring and summer. N: abundance; S: richness; EP: Ephemeroptera and Plecoptera; EPT: Ephemeroptera, Plecoptera, and Trichoptera; IBMWP: Iberian Monitoring Water Party; IASPT: Iberian Average Score Per Taxa.

| | N | | S | | EP | | EPT | | IBMWP | | IASPT | |
|---|---|---|---|---|---|---|---|---|---|---|---|---|
| | Spring | Summer | Spring | Summer | Spring | Summer | Spring | Summer | Spring | Summer | Spring | Summer |
| T1 | 1556 | 269 | 40 | 29 | 8 | 6 | 20 | 14 | 279 | 185 | 6.8 | 6.8 |
| T2 | 417 | 640 | 23 | 32 | 5 | 7 | 13 | 17 | 161 | 210 | 7 | 6.56 |
| T3 | 391 | 429 | 14 | 28 | 5 | 4 | 7 | 13 | 91 | 169 | 6.5 | 6.26 |
| M1 | 1211 | 1660 | 35 | 18 | 11 | 0 | 19 | 7 | 296 | 112 | 7.05 | 5.89 |
| M2 | 2007 | 464 | 32 | 15 | 4 | 2 | 15 | 2 | 203 | 75 | 6.64 | 5 |
| M3 | 792 | 881 | 13 | 8 | 2 | 0 | 5 | 0 | 82 | 32 | 6.31 | 4.57 |
| G1 | 956 | 569 | 31 | 27 | 3 | 4 | 8 | 11 | 151 | 170 | 5.2 | 6.3 |
| G2 | 3579 | 2676 | 25 | 22 | 4 | 3 | 7 | 7 | 141 | 126 | 5.84 | 5.73 |
| G3 | 153 | 792 | 14 | 18 | 1 | 1 | 3 | 3 | 81 | 78 | 4.5 | 4.33 |
| L1 | 482 | 370 | 29 | 30 | 3 | 4 | 9 | 12 | 142 | 182 | 5.68 | 6.07 |
| L2 | 304 | 454 | 25 | 28 | 3 | 4 | 6 | 13 | 128 | 170 | 5.33 | 6.3 |
| L3 | 577 | 97 | 14 | 8 | 2 | 0 | 3 | 1 | 67 | 41 | 4.78 | 5.12 |

The EP index shows the number of different taxa belonging to the orders Ephemeroptera and Plecoptera and, as can be seen in Table 5, also presented a significant ($p < 0.05$) negative correlation ($\varrho = -0.427$) with the concentration of MPs in sediments. When the number of different trichopteran taxa was added to the previous index (EPT index), the correlation was significant ($p < 0.01$) and much higher ($\varrho = -0.427$) (Table 5). These indexes only reflect the most sensitive taxa but are considered good indicators of water quality, especially in altered and urbanized rivers [28]. So, according to the results of this study, the most sensitive taxa may be susceptible to being affected by the presence of MPs in sediments, and as it was shown in this study, the effect can be measured with more severity when considering the trichopterans. Ehlers et al. [80] found that caddisflies can incorporate MPs of different characteristics into the larval cases where they live. This increases the exposition time to

MPs and the risk of being affected by some of the additives, but it also reduces the stability of the cases and the protective function, making the larvae more prone to predation [81,82]. The impact of MPs on some benthic macroinvertebrate groups could affect ecosystem functions by causing changes in community structure, as benthic macroinvertebrates are crucial for ecological parameters such as primary production or leaf decomposition [83,84]. These effects may also affect the energy transfer across trophic levels [85], and the uptake of MPs by different benthic macroinvertebrates represents the pathway for these pollutants to enter food webs [16]. Previous studies evidenced the transfer of MPs from basal to upper levels [86], indicating that further bioaccumulation and biomagnification processes are likely to be produced through food webs [87].

**Table 5.** The Pearson correlation coefficient ($\varrho$) obtained for the concentration of MPs in water and sediments with each of the biotic indices. N: abundance; S: richness; EP: Ephemeroptera and Plecoptera; EPT: Ephemeroptera, Plecoptera, and Trichoptera; IBMWP: Iberian Monitoring Water Party; IASPT: Iberian Average Score Per Taxa.

| | MPs Water | MPs Sediments |
|---|---|---|
| N | −0.125 | 0.024 |
| S | −0.112 | −0.427 * |
| EP | −0.201 | −0.468 * |
| EPT | −0.101 | −0.558 ** |
| IBMWP | −0.130 | −0.500 * |
| IASPT | −0.132 | −0.551 ** |

Note(s): ** 0.01; * 0.05.

The IASPT also showed a significant ($p < 0.01$) negative correlation ($\varrho = -0.551$) with the MP concentration in sediments, supporting the findings mentioned for EP and EPT indices. The IASPT index informs about the mean average degree of tolerance of the taxa found in the benthic macroinvertebrate community, so according to our results, the higher the concentration of MPs in sediments, the lower the IASPT as a consequence of the replacement of the most sensitive by more tolerant taxa [88]. This would explain the lower values for these indexes reported in the rivers Gafos and Lagares with respect to the more pristine rivers (Tea and, to a lesser extent, Miñor rivers), as can be seen in Table 3. Finally, a significant ($p < 0.05$) negative correlation ($\varrho = -0.551$) was found with the IBMWP that indicates that the water quality decreases while the concentration of MPs in sediments increases. It would be expected since the main threats to water quality are caused by anthropogenic activities as equal to the source of MP pollution. However, the use of this index has some disadvantages, as previously shown by several authors [89,90], and as for many other pollutant types, its use is not recommended when considering MPs.

## 4. Conclusions

This study confirmed the ubiquitous presence of MPs in the study area, as these contaminants were detected in all the sampled points, both in water and sediment samples. The study found that urbanization was the primary driver of microplastic pollution in water, while in sediments, the most important factor was the density of the population within the catchment. Finally, the results demonstrated that biotic indices based on the most sensitive taxa, such as EPT, PT, or IASPT, may be suitable for assessing MP pollution in sediments from rivers, although further research is necessary.

**Author Contributions:** D.G.-R.: Conceptualization, Methodology, Investigation, Data curation, Writing—Original Draft, Formal analysis; I.V.: Conceptualization, Methodology, Investigation, Data curation, Formal analysis, Writing—Review and Editing; R.Á.-T.: Conceptualization, Methodology, Investigation, Writing—Review and Editing; B.S.: Conceptualization, Methodology, Investigation, Supervision, Writing—Review and Editing; S.M.: Conceptualization, Methodology, Investigation, Supervision, Validation, Project administration, Writing—Review and Editing; J.G.: Conceptualization, Methodology, Investigation, Supervision, Validation, Writing—Review and Editing, Project administration, Funding acquisition. All authors have read and agreed to the published version of the manuscript.

**Funding:** This work was financially supported by Deputación de Pontevedra. D.G.R was funded by the PhD fellowship of the University of Vigo. I.V.C was supported by the POLENV project (TED2021-131278B-C21) funded by MCIN/AEI/10.13039/501100011033 and by the European Union NextGenerationEU/PRTR.

**Data Availability Statement:** Data will be available upon request to the corresponding author.

**Acknowledgments:** The authors would like to extend their gratitude to Pilar Feijoo Abal for her assistance in processing the samples.

**Conflicts of Interest:** The authors declare no conflicts of interest.

## References

1. Stegmann, P.; Daioglou, V.; Londo, M.; van Vuuren, D.P.; Junginger, M. Plastic Futures and Their $CO_2$ Emissions. *Nature* **2022**, *612*, 272–276. [CrossRef] [PubMed]
2. Kalali, E.N.; Lotfian, S.; Shabestari, M.E.; Khayatzadeh, S.; Zhao, C.; Nezhad, H.Y. A Critical Review of the Current Progress of Plastic Waste Recycling Technology in Structural Materials. *Curr. Opin. Green Sustain. Chem.* **2023**, *40*, 100763. [CrossRef]
3. Mu, J.; Zhang, S.; Qu, L.; Jin, F.; Fang, C.; Ma, X.; Zhang, W.; Wang, J. Microplastics Abundance and Characteristics in Surface Waters from the Northwest Pacific, the Bering Sea, and the Chukchi Sea. *Mar. Pollut. Bull.* **2019**, *143*, 58–65. [CrossRef] [PubMed]
4. Tan, Y.; Dai, J.; Xiao, S.; Tang, Z.; Zhang, J.; Wu, S.; Wu, X.; Deng, Y. Occurrence of Microplastic Pollution in Rivers Globally: Driving Factors of Distribution and Ecological Risk Assessment. *Sci. Total Environ.* **2023**, *904*, 165979. [CrossRef] [PubMed]
5. Free, C.M.; Jensen, O.P.; Mason, S.A.; Eriksen, M.; Williamson, N.J.; Boldgiv, B. High-Levels of Microplastic Pollution in a Large, Remote, Mountain Lake. *Mar. Pollut. Bull.* **2014**, *85*, 156–163. [CrossRef]
6. Cunningham, E.M.; Ehlers, S.M.; Dick, J.T.A.; Sigwart, J.D.; Linse, K.; Dick, J.J.; Kiriakoulakis, K. High Abundances of Microplastic Pollution in Deep-Sea Sediments: Evidence from Antarctica and the Southern Ocean. *Environ. Sci. Technol.* **2020**, *54*, 13661–13671. [CrossRef]
7. Amelia, T.S.M.; Khalik, W.M.A.W.M.; Ong, M.C.; Shao, Y.T.; Pan, H.-J.; Bhubalan, K. Marine Microplastics as Vectors of Major Ocean Pollutants and Its Hazards to the Marine Ecosystem and Humans. *Prog. Earth Planet. Sci.* **2021**, *8*, 12. [CrossRef]
8. Bergmann, M.; Gutow, L.; Klages, M. *Marine Anthropogenic Litter*; Springer Nature: Heidelberger/Berlin, Germany, 2015.
9. Wang, Z.; Zhang, Y.; Kang, S.; Yang, L.; Shi, H.; Tripathee, L.; Gao, T. Research Progresses of Microplastic Pollution in Freshwater Systems. *Sci. Total Environ.* **2021**, *795*, 148888. [CrossRef] [PubMed]
10. Lebreton, L.; Slat, B.; Ferrari, F.; Sainte-Rose, B.; Aitken, J.; Marthouse, R.; Hajbane, S.; Cunsolo, S.; Schwarz, A.; Levivier, A.; et al. Evidence That the Great Pacific Garbage Patch Is Rapidly Accumulating Plastic. *Sci. Rep.* **2018**, *8*, 4666. [CrossRef]
11. Wagner, M.; Lambert, S. Freshwater Microplastics: Emerging Environmental Contaminants? Springer Nature: Heidelberger/Berlin, Germany, 2018.
12. Xiao, S.; Cui, Y.; Brahney, J.; Mahowald, N.M.; Li, Q. Long-Distance Atmospheric Transport of Microplastic Fibres Influenced by Their Shapes. *Nat. Geosci.* **2023**, *16*, 863–870. [CrossRef]
13. Horton, A.A. Plastic Pollution: When Do We Know Enough? *J. Hazard. Mater.* **2022**, *422*, 126885. [CrossRef] [PubMed]
14. Mani, T.; Primpke, S.; Lorenz, C.; Gerdts, G.; Burkhardt-Holm, P. Microplastic Pollution in Benthic Midstream Sediments of the Rhine River. *Environ. Sci. Technol.* **2019**, *53*, 6053–6062. [CrossRef] [PubMed]
15. Zhou, Q.; Zhang, H.B.; Li, Y.; Luo, Y.M. Progress on Microplastics Pollution and Its Ecological Effects in the Coastal Environment. *Chin. Sci. Bullet.* **2015**, *60*, 3210–3220.
16. Windsor, F.M.; Tilley, R.M.; Tyler, C.R.; Ormerod, S.J. Microplastic Ingestion by Riverine Macroinvertebrates. *Sci. Total Environ.* **2019**, *646*, 68–74. [CrossRef] [PubMed]
17. Besseling, E.; Quik, J.T.; Sun, M.; Koelmans, A.A. Fate of Nano-and Microplastic in Freshwater Systems: A Modeling Study. *Environ. Pollut.* **2017**, *220*, 540–548. [CrossRef] [PubMed]
18. Pagter, E.; Nash, R.; Frias, J.; Kavanagh, F. Assessing Microplastic Distribution within Infaunal Benthic Communities in a Coastal Embayment. *Sci. Total Environ.* **2021**, *791*, 148278. [CrossRef] [PubMed]
19. Zhang, Q.; Xu, E.G.; Li, J.; Chen, Q.; Ma, L.; Zeng, E.Y.; Shi, H. A Review of Microplastics in Table Salt, Drinking Water, and Air: Direct Human Exposure. *Environ. Sci. Technol.* **2020**, *54*, 3740–3751. [CrossRef] [PubMed]
20. Cho, Y.; Shim, W.J.; Jang, M.; Han, G.M.; Hong, S.H. Abundance and Characteristics of Microplastics in Market Bivalves from South Korea. *Environ. Pollut.* **2019**, *245*, 1107–1116. [CrossRef]
21. Schymanski, D.; Goldbeck, C.; Humpf, H.-U.; Fürst, P. Analysis of Microplastics in Water by Micro-Raman Spectroscopy: Release of Plastic Particles from Different Packaging into Mineral Water. *Water Res.* **2018**, *129*, 154–162. [CrossRef]
22. Veerasingam, S.; Ranjani, M.; Venkatachalapathy, R.; Bagaev, A.; Mukhanov, V.; Litvinyuk, D.; Mugilarasan, M.; Gurumoorthi, K.; Guganathan, L.; Aboobacker, V.M.; et al. Contributions of Fourier Transform Infrared Spectroscopy in Microplastic Pollution Research: A Review. *Crit. Rev. Environ. Sci. Technol.* **2021**, *51*, 2681–2743. [CrossRef]

23. Chang, X.; Xue, Y.; Li, J.; Zou, L.; Tang, M. Potential Health Impact of Environmental Micro-and Nanoplastics Pollution. *J. Appl. Toxicol.* **2020**, *40*, 4–15. [CrossRef] [PubMed]
24. Halden, R.U. Plastics and Health Risks. *Annu. Rev. Public Health* **2010**, *31*, 179–194. [CrossRef] [PubMed]
25. Cho, Y.; Shim, W.J.; Ha, S.Y.; Han, G.M.; Jang, M.; Hong, S.H. Microplastic Emission Characteristics of Stormwater Runoff in an Urban Area: Intra-Event Variability and Influencing Factors. *Sci. Total Environ.* **2023**, *866*, 161318. [CrossRef] [PubMed]
26. Koelmans, A.A.; Redondo-Hasselerharm, P.E.; Mohamed Nor, N.H.; Kooi, M. Solving the Nonalignment of Methods and Approaches Used in Microplastic Research to Consistently Characterize Risk. *Environ. Sci. Technol.* **2020**, *54*, 12307–12315. [CrossRef]
27. Peel, M.C.; Finlayson, B.L.; McMahon, T.A. Updated World Map of the Köppen-Geiger Climate Classification. *Hydrol. Earth Syst. Sci.* **2007**, *11*, 1633–1644. [CrossRef]
28. Gutiérrez-Rial, D.; González, B.S.; Vázquez, D.G.; Méndez-Martínez, G.; Diego, M.Á.P.; González, J.G. Freshwater Biodiversity Loss in Urbanised Rivers. *Ecol. Indic.* **2023**, *156*, 111150. [CrossRef]
29. Masura, J.; Baker, J.; Foster, G.; Arthur, C. Laboratory Methods for the Analysis of Microplastics in the Marine Environment: Recommendations for Quantifying Synthetic Particles in Waters and Sediments. *Chin. Sci. Bull.* **2015**, *60*, 3210–3220.
30. Zobkov, M.; Esiukova, E. Microplastics in Baltic Bottom Sediments: Quantification Procedures and First Results. *Mar. Pollut. Bull.* **2017**, *114*, 724–732. [CrossRef] [PubMed]
31. Rodrigues, M.O.; Abrantes, N.; Gonçalves, F.J.M.; Nogueira, H.; Marques, J.C.; Gonçalves, A.M.M. Spatial and Temporal Distribution of Microplastics in Water and Sediments of a Freshwater System (Antuã River, Portugal). *Sci. Total Environ.* **2018**, *633*, 1549–1559. [CrossRef]
32. Boonsoong, B.; Sangpradub, N.; Barbour, M.T. Development of Rapid Bioassessment Approaches Using Benthic Macroinvertebrates for Thai Streams. *Env. Monit Assess* **2009**, *155*, 129–147. [CrossRef]
33. Tachet, H.; Richoux, P.; Bournaud, M.; Usseglio-Polatera, P. *Invertébrés d'eau douce: Systématique, biologie, écologie*; CNRS: Paris, France, 2000; ISBN 978-2-271-05745-7.
34. Barrios Barcia, E.; Puig Infante, A.; Fernández Rodríguez, R.; Rodríguez-Carreño Alonso, B.; Corrochano Codorníu, A.; Ruza Rodríguez, J.; Garrido González, J.; Benetti, C.J.; González Martín, J.M.; Pérez Bilbao, A. *Id-Tax: Catálogo y claves de identificación de organismos invertebrados utilizados como elementos de calidad en las redes de control del estado ecológico*; Ministerio de Agricultura, Alimentación y Medio Ambiente: Madrid, Spain, 2012; ISBN 978-84-491-1202-7.
35. Alba-Tercedor, J. Un método rápido y simple para evaluar la calidad biológica de las aguas corrientes basado en el de Hellawell (1978). *Limnetica* **1988**, *4*, 51–66. [CrossRef]
36. Phuong, N.N.; Duong, T.T.; Le, T.P.Q.; Hoang, T.K.; Ngo, H.M.; Phuong, N.A.; Pham, Q.T.; Doan, T.O.; Ho, T.C.; Da Le, N. Microplastics in Asian Freshwater Ecosystems: Current Knowledge and Perspectives. *Sci. Total Environ.* **2022**, *808*, 151989. [CrossRef] [PubMed]
37. Talbot, R.; Chang, H. Microplastics in Freshwater: A Global Review of Factors Affecting Spatial and Temporal Variations. *Environ. Pollut.* **2022**, *292*, 118393. [CrossRef] [PubMed]
38. Nakao, S.; Ozaki, A.; Yamazaki, K.; Masumoto, K.; Nakatani, T.; Sakiyama, T. Microplastics Contamination in Tidelands of the Osaka Bay Area in Western Japan. *Water Environ. J.* **2020**, *34*, 474–488. [CrossRef]
39. Soltani, N.; Keshavarzi, B.; Moore, F.; Busquets, R.; Nematollahi, M.J.; Javid, R.; Gobert, S. Effect of Land Use on Microplastic Pollution in a Major Boundary Waterway: The Arvand River. *Sci. Total Environ.* **2022**, *830*, 154728. [CrossRef] [PubMed]
40. Zhao, X.; Liu, Z.; Cai, L.; Han, J. Occurrence and Distribution of Microplastics in Surface Sediments of a Typical River with a Highly Eroded Catchment, a Case of the Yan River, a Tributary of the Yellow River. *Sci. Total Environ.* **2023**, *863*, 160932. [CrossRef] [PubMed]
41. Tibbetts, J.; Krause, S.; Lynch, I.; Sambrook Smith, G. Abundance, Distribution, and Drivers of Microplastic Contamination in Urban River Environments. *Water* **2018**, *10*, 1597. [CrossRef]
42. da Costa, I.D.; Costa, L.L.; Zalmon, I.R. Microplastics in Water from the Confluence of Tropical Rivers: Overall Review and a Case Study in Paraiba Do Sul River Basin. *Chemosphere* **2023**, *338*, 139493. [CrossRef] [PubMed]
43. Li, T.; Liu, K.; Tang, R.; Liang, J.-R.; Mai, L.; Zeng, E.Y. Environmental Fate of Microplastics in an Urban River: Spatial Distribution and Seasonal Variation. *Environ. Pollut.* **2023**, *322*, 121227. [CrossRef]
44. Pasquier, G.; Doyen, P.; Dehaut, A.; Veillet, G.; Duflos, G.; Amara, R. Vertical Distribution of Microplastics in a River Water Column Using an Innovative Sampling Method. *Environ. Monit Assess* **2023**, *195*, 1302. [CrossRef]
45. Simon-Sánchez, L.; Grelaud, M.; Garcia-Orellana, J.; Ziveri, P. River Deltas as Hotspots of Microplastic Accumulation: The Case Study of the Ebro River (NW Mediterranean). *Sci. Total Environ.* **2019**, *687*, 1186–1196. [CrossRef] [PubMed]
46. Lenaker, P.L.; Baldwin, A.K.; Corsi, S.R.; Mason, S.A.; Reneau, P.C.; Scott, J.W. Vertical Distribution of Microplastics in the Water Column and Surficial Sediment from the Milwaukee River Basin to Lake Michigan. *Environ. Sci. Technol.* **2019**, *53*, 12227–12237. [CrossRef] [PubMed]
47. Scherer, C.; Weber, A.; Stock, F.; Vurusic, S.; Egerci, H.; Kochleus, C.; Arendt, N.; Foeldi, C.; Dierkes, G.; Wagner, M.; et al. Comparative Assessment of Microplastics in Water and Sediment of a Large European River. *Sci. Total Environ.* **2020**, *738*, 139866. [CrossRef] [PubMed]
48. Matjašič, T.; Simčič, T.; Medvešček, N.; Bajt, O.; Dreo, T.; Mori, N. Critical Evaluation of Biodegradation Studies on Synthetic Plastics through a Systematic Literature Review. *Sci. Total Environ.* **2021**, *752*, 141959. [CrossRef] [PubMed]

49. Ross, M.S.; Loutan, A.; Groeneveld, T.; Molenaar, D.; Kroetch, K.; Bujaczek, T.; Kolter, S.; Moon, S.; Huynh, A.; Khayam, R.; et al. Estimated Discharge of Microplastics via Urban Stormwater during Individual Rain Events. *Front. Environ. Sci.* **2023**, *11*, 1090267. [CrossRef]
50. Kurki-Fox, J.J.; Doll, B.A.; Monteleone, B.; West, K.; Putnam, G.; Kelleher, L.; Krause, S.; Schneidewind, U. Microplastic Distribution and Characteristics across a Large River Basin: Insights from the Neuse River in North Carolina, USA. *Sci. Total Environ.* **2023**, *878*, 162940. [CrossRef] [PubMed]
51. Rajan, K.; Khudsar, F.A.; Kumar, R. Urbanization and Population Resources Affect Microplastic Concentration in Surface Water of the River Ganga. *J. Hazard. Mater. Adv.* **2023**, *11*, 100342. [CrossRef]
52. Yuan, W.; Christie-Oleza, J.A.; Xu, E.G.; Li, J.; Zhang, H.; Wang, W.; Lin, L.; Zhang, W.; Yang, Y. Environmental Fate of Microplastics in the World's Third-Largest River: Basin-Wide Investigation and Microplastic Community Analysis. *Water Res.* **2022**, *210*, 118002. [CrossRef] [PubMed]
53. Bilal, M.; Qadir, A.; Yaqub, A.; Hassan, H.U.; Irfan, M.; Aslam, M. Microplastics in Water, Sediments, and Fish at Alpine River, Originating from the Hindu Kush Mountain, Pakistan: Implications for Conservation. *Environ. Sci. Pollut. Res.* **2023**, *30*, 727–738. [CrossRef]
54. Akdogan, Z.; Guven, B.; Kideys, A.E. Microplastic Distribution in the Surface Water and Sediment of the Ergene River. *Environ. Res.* **2023**, *234*, 116500. [CrossRef]
55. Huang, Q.; Liu, M.; Cao, X.; Liu, Z. Occurrence of Microplastics Pollution in the Yangtze River: Distinct Characteristics of Spatial Distribution and Basin-Wide Ecological Risk Assessment. *Water Res.* **2023**, *229*, 119431. [CrossRef] [PubMed]
56. Lu, H.-C.; Ziajahromi, S.; Neale, P.A.; Leusch, F.D.L. A Systematic Review of Freshwater Microplastics in Water and Sediments: Recommendations for Harmonisation to Enhance Future Study Comparisons. *Sci. Total Environ.* **2021**, *781*, 146693. [CrossRef]
57. Lin, L.; Zuo, L.-Z.; Peng, J.-P.; Cai, L.-Q.; Fok, L.; Yan, Y.; Li, H.-X.; Xu, X.-R. Occurrence and Distribution of Microplastics in an Urban River: A Case Study in the Pearl River along Guangzhou City, China. *Sci. Total Environ.* **2018**, *644*, 375–381. [CrossRef] [PubMed]
58. Wang, W.; Ndungu, A.W.; Li, Z.; Wang, J. Microplastics Pollution in Inland Freshwaters of China: A Case Study in Urban Surface Waters of Wuhan, China. *Sci. Total Environ.* **2017**, *575*, 1369–1374. [CrossRef] [PubMed]
59. Gago, J.; Booth, A.M.; Tiller, R.; Maes, T.; Larreta, J. Microplastics Pollution and Regulation. In *Handbook of Microplastics in the Environment*; Springer: Heidelberger/Berlin, Germany, 2022; pp. 1071–1096.
60. Schwartz, R. Aspekte Des Schadstoff-Sedimentmanagements Im Bereich Der Tideelbe. Hydrologie und Wasserbewirtschaftung/BfG–Jahrgang: HW 59. 2015, H.6, ISSN 1439. Available online: https://doi.bafg.de/HyWa/2015/HyWa_2015,6_9.pdf (accessed on 12 January 2024).
61. Jiang, C.; Yin, L.; Li, Z.; Wen, X.; Luo, X.; Hu, S.; Yang, H.; Long, Y.; Deng, B.; Huang, L.; et al. Microplastic Pollution in the Rivers of the Tibet Plateau. *Environ. Pollut.* **2019**, *249*, 91–98. [CrossRef] [PubMed]
62. Kiss, T.; Fórián, S.; Szatmári, G.; Sipos, G. Spatial Distribution of Microplastics in the Fluvial Sediments of a Transboundary River—A Case Study of the Tisza River in Central Europe. *Sci. Total Environ.* **2021**, *785*, 147306. [CrossRef] [PubMed]
63. Crew, A.; Gregory-Eaves, I.; Ricciardi, A. Distribution, Abundance, and Diversity of Microplastics in the Upper St. Lawrence River. *Environ. Pollut.* **2020**, *260*, 113994. [CrossRef] [PubMed]
64. Gerhardt, A.; Janssens de Bisthoven, L.; Soares, A.M.V.M. Macroinvertebrate Response to Acid Mine Drainage: Community Metrics and on-Line Behavioural Toxicity Bioassay. *Environ. Pollut.* **2004**, *130*, 263–274. [CrossRef] [PubMed]
65. Huang, D.; Li, X.; Ouyang, Z.; Zhao, X.; Wu, R.; Zhang, C.; Lin, C.; Li, Y.; Guo, X. The Occurrence and Abundance of Microplastics in Surface Water and Sediment of the West River Downstream, in the South of China. *Sci. Total Environ.* **2021**, *756*, 143857. [CrossRef]
66. Klein, S.; Worch, E.; Knepper, T.P. Occurrence and Spatial Distribution of Microplastics in River Shore Sediments of the Rhine-Main Area in Germany. *Environ. Sci. Technol.* **2015**, *49*, 6070–6076. [CrossRef]
67. Duis, K.; Coors, A. Microplastics in the Aquatic and Terrestrial Environment: Sources (with a Specific Focus on Personal Care Products), Fate and Effects. *Environ. Sci. Eur.* **2016**, *28*, 2. [CrossRef] [PubMed]
68. Shruti, V.C.; Jonathan, M.P.; Rodriguez-Espinosa, P.F.; Rodríguez-González, F. Microplastics in Freshwater Sediments of Atoyac River Basin, Puebla City, Mexico. *Sci. Total Environ.* **2019**, *654*, 154–163. [CrossRef] [PubMed]
69. Eerkes-Medrano, D.; Thompson, R.C.; Aldridge, D.C. Microplastics in Freshwater Systems: A Review of the Emerging Threats, Identification of Knowledge Gaps and Prioritisation of Research Needs. *Water Res.* **2015**, *75*, 63–82. [CrossRef] [PubMed]
70. Rochman, C.M. Microplastics Research—From Sink to Source. *Science* **2018**, *360*, 28–29. [CrossRef] [PubMed]
71. Xia, F.; Yao, Q.; Zhang, J.; Wang, D. Effects of Seasonal Variation and Resuspension on Microplastics in River Sediments. *Environ. Pollut.* **2021**, *286*, 117403. [CrossRef] [PubMed]
72. Brahney, J.; Hallerud, M.; Heim, E.; Hahnenberger, M.; Sukumaran, S. Plastic Rain in Protected Areas of the United States. *Science* **2020**, *368*, 1257–1260. [CrossRef] [PubMed]
73. Hurley, R.; Woodward, J.; Rothwell, J.J. Microplastic Contamination of River Beds Significantly Reduced by Catchment-Wide Flooding. *Nat. Geosci.* **2018**, *11*, 251–257. [CrossRef]
74. Tien, C.-J.; Wang, Z.-X.; Chen, C.S. Microplastics in Water, Sediment and Fish from the Fengshan River System: Relationship to Aquatic Factors and Accumulation of Polycyclic Aromatic Hydrocarbons by Fish. *Environ. Pollut.* **2020**, *265*, 114962. [CrossRef]

75. Feld, C.K. Response of Three Lotic Assemblages to Riparian and Catchment-Scale Land Use: Implications for Designing Catchment Monitoring Programmes: Response of Three Lotic Assemblages to Land Use. *Freshw. Biol.* **2013**, *58*, 715–729. [CrossRef]
76. Armitage, P.; Moss, D.; Wright, J.; Furse, M. The Performance of a New Biological Water Quality Score System Based on Macroinvertebrates over a Wide Range of Unpolluted Running-Water Sites. *Water Res.* **1983**, *17*, 333–347. [CrossRef]
77. Redondo-Hasselerharm, P.E.; Falahudin, D.; Peeters, E.T.; Koelmans, A.A. Microplastic Effect Thresholds for Freshwater Benthic Macroinvertebrates. *Environ. Sci. Technol.* **2018**, *52*, 2278–2286. [CrossRef]
78. Prakash, V.; Dwivedi, S.; Gautam, K.; Seth, M.; Anbumani, S. Occurrence and Ecotoxicological Effects of Microplastics on Aquatic and Terrestrial Ecosystems. In *Microplastics in Terrestrial Environments*; He, D., Luo, Y., Eds.; The Handbook of Environmental Chemistry; Springer International Publishing: Cham, Switzerland, 2020; Volume 95, pp. 223–243. ISBN 978-3-030-56270-0.
79. Silva, C.J.M.; Machado, A.L.; Campos, D.; Rodrigues, A.C.M.; Patrício Silva, A.L.; Soares, A.M.V.M.; Pestana, J.L.T. Microplastics in Freshwater Sediments: Effects on Benthic Invertebrate Communities and Ecosystem Functioning Assessed in Artificial Streams. *Sci. Total Environ.* **2022**, *804*, 150118. [CrossRef] [PubMed]
80. Ehlers, S.; Manz, W.; Koop, J. Microplastics of Different Characteristics Are Incorporated into the Larval Cases of the Freshwater Caddisfly Lepidostoma Basale. *Aquat. Biol.* **2019**, *28*, 67–77. [CrossRef]
81. Ehlers, S.M.; Al Najjar, T.; Taupp, T.; Koop, J.H.E. PVC and PET Microplastics in Caddisfly (Lepidostoma Basale) Cases Reduce Case Stability. *Environ. Sci. Pollut. Res.* **2020**, *27*, 22380–22389. [CrossRef]
82. Nantege, D.; Odong, R.; Auta, H.S.; Keke, U.N.; Ndatimana, G.; Assie, A.F.; Arimoro, F.O. Microplastic Pollution in Riverine Ecosystems: Threats Posed on Macroinvertebrates. *Environ. Sci. Pollut. Res.* **2023**, *30*, 76308–76350. [CrossRef]
83. Campos, D.; Machado, A.L.; Cardoso, D.N.; Silva, A.R.R.; Silva, P.V.; Rodrigues, A.C.M.; Simão, F.C.P.; Loureiro, S.; Grabicová, K.; Nováková, P.; et al. Effects of the Organic UV-Filter, 3-(4-Methylbenzylidene) Camphor, on Benthic Invertebrates and Ecosystem Function in Artificial Streams. *Environ. Pollut.* **2020**, *260*, 113981. [CrossRef] [PubMed]
84. Hasenbein, S.; Lawler, S.P.; Connon, R.E. An Assessment of Direct and Indirect Effects of Two Herbicides on Aquatic Communities. *Environ. Toxicol. Chem.* **2017**, *36*, 2234–2244. [CrossRef] [PubMed]
85. Palmer, M.A.; Covich, A.P.; Lake, S.; Biro, P.; Brooks, J.J.; Cole, J.; Dahm, C.; Gibert, J.; Goedkoop, W.; Martens, K.; et al. Linkages between Aquatic Sediment Biota and Life Above Sediments as Potential Drivers of Biodiversity and Ecological Processes. *BioScience* **2000**, *50*, 1062. [CrossRef]
86. Horton, A.A.; Jürgens, M.D.; Lahive, E.; Van Bodegom, P.M.; Vijver, M.G. The Influence of Exposure and Physiology on Microplastic Ingestion by the Freshwater Fish Rutilus Rutilus (Roach) in the River Thames, UK. *Environ. Pollut.* **2018**, *236*, 188–194. [CrossRef]
87. Krause, S.; Baranov, V.; Nel, H.A.; Drummond, J.D.; Kukkola, A.; Hoellein, T.; Sambrook Smith, G.H.; Lewandowski, J.; Bonet, B.; Packman, A.I.; et al. Gathering at the Top? Environmental Controls of Microplastic Uptake and Biomagnification in Freshwater Food Webs. *Environ. Pollut.* **2021**, *268*, 115750. [CrossRef]
88. Joachim, S.; Roussel, H.; Bonzom, J.; Thybaud, E.; Mebane, C.A.; Van Den Brink, P.; Gauthier, L. A Long-term Copper Exposure in a Freshwater Ecosystem Using Lotic Mesocosms: Invertebrate Community Responses. *Environ. Chem.* **2017**, *36*, 2698–2714. [CrossRef] [PubMed]
89. Morais, M.; Pinto, P.; Guilherme, P.; Rosado, J.; Antunes, I. Assessment of Temporary Streams: The Robustness of Metric and Multimetric Indices under Different Hydrological Conditions. *Hydrobiologia* **2004**, *516*, 229–249. [CrossRef]
90. Yorulmaz, B.; Ertaş, A. Water Quality Assessment of Selendi Stream and Comparative Performance of the Indices Based on Benthic Macroinvertebrates and Physicochemical Parameters. *Biologia* **2021**, *76*, 2599–2607. [CrossRef]

*water*

MDPI

*Article*

# Evaluating the Water Quality of the Keddara Dam (Algeria) Using Water Quality Indices

**Tosin Sarah Fashagba [1], Madani Bessedik [2,3], Nadia Badr ElSayed [4], Chérifa Abdelbaki [1,2,3] and Navneet Kumar [5,6,*]**

1 Pan African University-Institute of Water and Energy Science, Including Climate Change (PAUWES), Pôle Chetouane, P.O. Box 119, Tlemcen 13000, Algeria; fashagbasarah1@gmail.com (T.S.F.); cherifa.abdelbaki@univ-tlemcen.dz (C.A.)

2 Department of Hydraulics, Faculty of Technology, University of Tlemcen, P.B. 230, Tlemcen 13000, Algeria; madani.bessedik@univ-tlemcen.dz

3 Laboratoire EOLE, University of Tlemcen, P.B. 230, Tlemcen 13000, Algeria

4 Environmental Sciences Department, Faculty of Science, Alexandria University, Alexandria 21511, Egypt; nadia.badr@alexu.edu.eg

5 Department of Ecology and Natural Resources Management, Center for Development Research (ZEF), University of Bonn, Genscherallee 3, 53113 Bonn, Germany

6 Global Mountain Safeguard Research (GLOMOS), United Nations University—Institute for Environment and Human Security (UNU-EHS), UN Campus, Platz der Vereinten Nationen 1, 53113 Bonn, Germany

* Correspondence: nkumar@uni-bonn.de

**Abstract:** Dams are regarded as crucial pieces of structure that store water for irrigation and municipal uses. Given their vital role, the dam's water quality assessment is considered to be an important criterion and requires constant monitoring. In this research, we attempted to use two water quality indices (WQIs) methods to assess the water quality of the Keddara Dam, which is located on the Boudouaou River, Algeria, using eleven water quality parameters (temperature, pH, conductivity, turbidity, total suspended solids (TSS), full alkalimetric title (TAC), hydrometric title (TH), nitrite ions ($NO^{2-}$), nitrate ions ($NO^{3-}$), ammonium ions ($NH^{4+}$), and phosphate ions ($PO_4^{3-}$)) for data recorded from 29 December 2018 to 3 June 2021. Application of The Canadian Council of Ministers of the Environment (CCME) WQIs and the Weighted Arithmetic Method (WAM) indicated that the Keddara Dam's water quality parameters were within the WHO's permissible level, except for the conductivity and turbidity values. The results of the CCME WQI ranged from acceptable (81.92) to excellent (95.08) quality, whereas the WAM WQI ranged from 9.52 to 17.77, indicating excellent quality. This demonstrates that the Keddara Dam is appropriate for agriculture and municipal use. The water quality indices (WQIs) methods are recommended as valuable tools that allow both the public and decision-makers to comprehend and manage the water quality of any aquatic environment by providing flexibility in choosing variables.

**Keywords:** water quality index; surface water quality; water quality assessment

## 1. Introduction

Surface water pollution has become a serious environmental issue on a global scale that needs to be continuously assessed and monitored at all scales to ensure the sustainability of ecosystems [1]. Surface water quality monitoring requires a thorough process that involves numerous chemical studies of water [2]. Throughout the world, river systems are significantly impacted by pollution. The primary reason for this is the extremely rapid urbanisation that forces the flow of treated and untreated wastewater into river systems. Harmful chemical pesticides, insecticides, fertilisers, and herbicides are used more frequently as a result of the rapid expansion of industrial facilities along rivers and the increase in agriculture [3]. Water resources, particularly surface water, have gradually declined as a result of this and continuously affect human health, as diseases are largely spread through contaminated water [4]. Every year, water-borne illnesses claim the lives of roughly 1.8 million individuals in developing nations, most of whom are children [4]. Consequently, to protect water resources against harmful contaminants, it is crucial to monitor changes in water quality.

Due to its limited water resources, with annual rainfall of less than 200 mm, Algeria is recognised as an arid and semi-arid region. This results in notable regional and temporal fluctuations in water availability [5]. Approximately 87% of Algeria, covering 2.4 million $km^2$, is composed of desert which sees little to no rainfall, ranking it among one of the driest countries in the world [6]. Over the past two decades, Algeria's water resources have undergone severe stress due to factors such as urbanisation, climate change, and inadequate water usage planning [5] that, in turn, affect the availability and quality of water. Climate change in Algeria has disrupted weather patterns such as precipitation and the water cycle, which has affected groundwater recharge and soil moisture [5]. In addition, the demand for water for various reasons is rising as a result of the expanding population [7] and increased usage of land for agriculture [8]. Furthermore, numerous types of pollution are degrading the condition of rivers in Algeria. The availability of water resources has decreased and they are difficult to use, and are frequently in contact with large volumes of effluent [9]. There have been concerns about diverse pollution sources causing water quality degradation in some dam lakes over the past few decades [10–14].

To assess the water quality and chemistry in aquatic environments, a variety of methods have been used. Water quality indices (WQIs) are the most efficient way to communicate to the public, decision-makers, and user communities to detect threats to many different uses of water, including habitat for aquatic life, irrigation water for agricultural areas, recreation, aesthetics, and drinking water supplies [9,15]. The WQI model, a mathematical method, is frequently used to evaluate aquatic ecosystem quality from various sources using a set of chosen water quality criteria [16]. The model simplifies a large amount of water quality data into a single composite number [16]. The WQI model is an effective tool that makes it easier to comprehend the overall condition of an aquatic ecosystem based on a single index, in contrast to traditional methods of water quality assessment which involve lengthy lists of parametric water quality values. Understanding this index helps to inform the public and the government about the worldwide condition of the water quality in a certain area [9].

Based on ten indicators that were considered critical for the majority of waterbodies, Horton developed the first WQI model in the 1960s [17]. In 1965, Brown, assisted by the National Sanitation Foundation (NSF), developed the NSF-WQI, a more exacting iteration of Horton's WQI model [17]. This model was partially incorporated into the Scottish Research Development Department's method (SRDD). Other specific indices have been developed, such as the SRDD-WQI (1973), Bascaron Index (1979), Dalmatian Index, and House Index (1986) [17]. The British Columbia Ministry for Environment, Lands and Parks created the British Columbia Water Quality Index (BCWQI) in the mid-1990s, which was another significant development. It is used to assess the quality status of numerous waterbodies in the Canadian province of British Columbia [17]. The Canadian Council of Ministers of the Environment Water Quality Index (CCME WQI) was created in 2001 [18]. Recently, many nations and organisations have introduced more than 35 WQI methods to assess surface water quality globally [19]. The lack of a sufficient dataset and the numerous alternatives for choosing different sets of indicators and objective values are the key challenges in assessing the performance of WQIs in surface water bodies around the world [20].

The Keddara Dam is regarded as the region's backbone and provides water to Algiers for municipal and agricultural purposes. Limited research has been carried out on the water quality of the Keddara Dam water using WQIs methods. Thus, this study was conducted using the Weighted Arithmetic Method Water Quality Index (WAM WQI) and the Canadian Council of Ministers of the Environment Water Quality Index (CCME WQI), to assess the Keddara Dam's surface water quality for drinking purposes.

The findings of this research will provide valuable and new information to policymakers and managers of water resources, helping them make decisions about water resource management regarding the area under investigation. This methodology is transferable and applicable to any dry or semi-arid region, making it a practical instrument for environmental management and assessments.

## 2. Materials and Methods

### *2.1. Study Area*

The Keddara Dam is located 6 kilometres (4 miles) northwest of Keddara Town, on the Boudouaou River in Boumerdès Province, Algeria (36°39′03.0″ N 3°24′58.9″ E) (Figure 1). The Keddara Dam is also called the Barrage Keddara. Built between 1982 and 1987, the dam supplies water to Algiers, situated 35 km to the west, for municipal use [21]. With a capacity of 146,500,000 $m^3$ (118,769 acre-ft), the dam's reservoir can store water collected from drainage via a gallery, the Hamiz Dam, 7.6 km to the west, and by pumping water at a rate of 7 $m^3/s$ from the Beni Amrane Dam, 17 km to the east. River Bouira and River Medea are the main sources of the Keddara Dam. The dam is about 100 km east of Algiers. The population density and land use dynamics are rising in the Boumerdès Province. Thus, research in this area will help us to determine if these factors affect water quality.

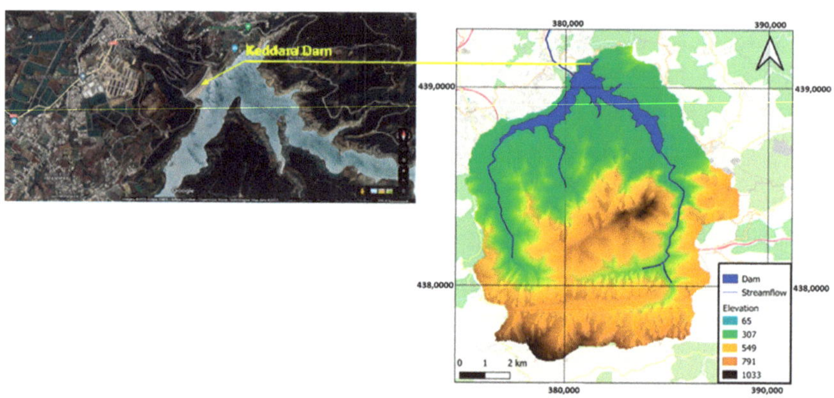


**Figure 1.** Location of the Keddara Dam (36°39′03.0″ N 3°24′58.9″ E).

### *2.2. Data Collection and Analytical Techniques*

This research employed the use of secondary data obtained from the Algiers Water and Sanitation Company (SEAAL). The SEAAL firm adhered to conventional procedures for analysing water and wastewater in all of their analytical methodologies [22]. The sampling period was from 29 December 2018 to 3 June 2021. A total of 338 water samples with 11 water quality parameters for the period of 29 December 2018–3 June 2021 were collected. These were the only data that were accessible. The parameters chosen for this study were based on parameters with sufficient daily measurements. They are Temperature, pH, conductivity, turbidity, total suspended solids (TSS), full alkalimetric title (TAC), hydrometric title (TH), nitrite ions ($NO^{2-}$), nitrate ions ($NO^{3-}$), ammonium ions ($NH^{4+}$) and phosphate ion ($PO_4{}^{3-}$). The data had outliers and missing values, which were considered during the analysis. Days with less than five parameters reported were removed.

### *2.3. Statistical Analysis of the Water Quality Parameters*

Daily time steps are typically used for water quality monitoring in water treatment plants to ensure the system is functioning effectively. This generates time series data for different water quality parameters. Data accuracy is essential for the efficient determination of water quality indices [23,24]. Removing the outliers is crucial to avoid any skewing of the information derived from the raw data. Excel software was used for this analysis.

Detection of Outlier and Normality Testing

According to [24], "An outlying observation (outlier) appears to deviate from other members of the sample in which it occurs and should be identified and removed" [24,25]. To find outliers in this study, the box plot method described by Tukey (1977) was applied [26–28]. The interquartile distance (Q3 − Q1), which is equivalent to 1.5 times the box's height, is typically used to calculate whiskers. In these conditions, a value is considered abnormal if it is more than the interquartile gap either above the third quartile (Q3) or below the first quartile (Q1) [27]. Tukey (1977) [26] asserts that the rule of thumb's pragmatic value of 1.5 is based on a probabilistic basis. The outliers are calculated based on the following equations:

$$\text{Minor outlier (Min. Out.)} = Q1 - 1.5 \times (Q3 - Q1) \tag{1}$$

$$\text{Major outlier (Maj. Out.)} = Q3 + 1.5 \times (Q3 - Q1) \tag{2}$$

Upon detection, outliers were removed to prevent calculation interference. After removal, the raw water samples were reduced to 309 samples. The datasets were then subjected to normality testing. This study used one graphic method (quantile regression) and two numerical methods (Shapiro–Wilk, Kolmogorov–Smirnov) [29]. The different normality tests were implemented to compare different normality methods and account for the small sample size.

a. The Shapiro–Wilk test: If the p-value is greater than the chosen significance threshold (e.g., 0.05), the data are mostly consistent with a normal distribution, and the null hypothesis is still true. If the p-value is less than the significance level, on the other hand, it means that the data deviate significantly from normality, and the null hypothesis is rejected [29].
b. Kolmogorov–Smirnov: By demonstrating that the sample did not originate from the selected distribution, the Kolmogorov–Smirnov test rejects the null hypothesis if the p-value is less than the set significance threshold, which is often 0.05 or 0.01. The null hypothesis cannot be ruled out if the p-value is greater than the significance level, and we may reasonably conclude that the sample may originate from the given distribution [29].
c. Quantile regression approach: Using the quantile regression method, the graph of normal quantiles is produced. A regression model is created for a subset of the response variable's limited distribution. The data have a normal distribution if a linear relation can be inferred from this graph. The closer the $R^2$ is to 1, the better the data distribution [29].

### 2.4. Application of Water Quality Index (WQI)

The WQI is the most effective method for characterising the overall water quality based on several parameters. As previously mentioned, WQI may compress a large amount of data into a single number to communicate the information in a straightforward and accessible manner, making it understandable and accessible to a wide audience. The Keddara Dam's water quality index was assessed using the selected methodologies: CCME WQI and WAM WQIs. The methodology of [30] was followed in this study to employ the WHO standards as suggestions for irrigation and drinking (Table 1).

These two popular techniques were selected for this study, mainly because of their ease of use and ease of calculation, as well as their excellent track record with several experts who have evaluated the quality of water from various sources all over the world (Table 2).

**Table 1.** WHO Standard values used for water quality indices [30–34].

| Parameters | Standard Value |
|---|---|
| Temperature | 26.5 |
| pH | 8.5 |
| Conductivity | 1500 |

**Table 1.** *Cont.*

| Parameters | Standard Value |
|---|---|
| Turbidity | 5 |
| Total soluble solids | 30 |
| Total hardness | 50 |
| Full alkalimetric title | 20 |
| Nitrite ion | 3 |
| Nitrate ion | 50 |
| Ammonia | <0.5 |
| Phosphate ion | 0.5 |

### 2.4.1. Canadian Council of Ministers of the Environment (CCME) WQI

The British Colombia WQI Model (BCWQI) served as the foundation for creating the CCME model in 2001 [35]. Three parameters are included in the CCME-WQI, which offers a measure of the deviation of water quality from water quality guidelines. These parameters are scope, frequency, and amplitude [36]. This model has been used in numerous surface water bodies across the globe.

**Table 2.** The specialised literature used in this study cites case studies using both methods (CCME-WQI and WAM-WQI).

| Study | | Method Used | Reference |
|---|---|---|---|
| Country | Water Resource Type | | |
| Puebla valley, Mexico | Groundwater | CCME-WQI | [15] |
| Reghaia, Algeria | Surface Water | WAM-WQI | [36] |
| Amman-Zaraq, Jordan | Ground Water | CCME-WQI | [37] |
| Ohaozora, Nigeria | Ground Water | WAM-WQI | [38] |
| WWTP of Oran City, Algeria | Wastewater | CCME-WQI<br>WAM-WQI | [39] |

i. The CCME WQI method follows the following steps for parameter selection

The user is free to choose which water quality parameters to utilise. The CCME WQI model only mandates the usage of a minimum of four [35]. The developers advise using the evaluation procedures for expert panels in order to choose model parameters.

ii. Sub-index calculation

There is no sub-index calculation component in the CCME model. This is a significant flaw in this paradigm relative to the others.

iii. Parameter Weightings

The final WQI can be calculated without the use of parameter weight values.

iv. Aggregation

The CCME uses an aggregation function that is different from other models. It is written as follows:

$$WQI = 100 - \left[\frac{\sqrt{F_1^2 + F_2^2 + F_3^2}}{1.732}\right] \quad (3)$$

where

$F_1$ is referred to as the "scope", This is the portion of all parameters that fall short of the desired outcome.

It is expressed as follows:

$$F1 = \left[\frac{\text{Number of failed parameters}}{\text{Total number of parameters}}\right] \times 100 \tag{4}$$

$F_2$ is referred to as the "frequency". This fraction of individual test values that do not match the objective values is known as the "frequency" (failed tests).

It is expressed as follows:

$$F2 = \left[\frac{\text{Number of failed tests}}{\text{Total number of tests}}\right] \times 100 \tag{5}$$

$F_3$: This is calculated in three steps.

1. Calculation of excursion: The number of times an individual concentration exceeds (or falls short of, if the target is at its lowest) the objective is known as the excursion.

   The excursion for a test value that is below the objective value is computed as follows:

$$\text{excursioni} = \left[\frac{\text{Failed test value}_i}{\text{Objective}_j}\right] - 1 \tag{6}$$

On the other hand, the excursion value is determined as follows if the test value is greater than the objective value:

$$\text{excursioni} = \left[\frac{\text{Objective}_j}{\text{Failed test value}_i}\right] - 1 \tag{7}$$

2. Calculation of the Normalised Sum of Excursion (nse).

The total amount that each test deviates from compliance is known as the normalised sum of excursions. The computation involves adding up all of the test deviations from the goals and dividing the result by the total number of tests (including goal- and non-goal-achieving tests).

$$\text{nse} = \left[\frac{\sum_{i=1}^{n} \text{excursion}_i}{\text{Total of number of test}}\right] - 1 \tag{8}$$

3. Calculation of $F_3$

An asymptotic function is used to calculate $F_3$, scaling the normalised total of the deviations from targets to produce a range from 0 to 100.

$$F3 = \left[\frac{\text{nse}}{0.01(\text{nse}) + 0.01}\right] \tag{9}$$

4. WQI evaluation

   The CCME model proposed five water quality classes, as shown in Table 3.

**Table 3.** CCME WQI scale [35].

| CCME WQIs | Quality Range | Water Categories |
|---|---|---|
| 95–100 | Excellent | Natural water quality |
| 80–94 | Good | Water quality is departed from natural or desirable levels |
| 65–79 | Fair | Water quality condition sometimes departs from natural or desirable levels. |
| 45–64 | Marginal | Water quality is frequently threatened or impaired; conditions often depart from natural or desirable level |
| 0–44 | Poor | Water quality is not suitable for use purposes at any level. |

### 2.4.2. Weighted Arithmetic Method (WAM)

This approach, which is popular and straightforward, combines data for many water quality criteria into an equation that evaluates the quality of the water and gives decision-makers and interested individuals a single, easily comprehensible statistic. It achieves this by calculating the unit weight and quality rating scale for each water quality parameter. Furthermore, it requires fewer parameters than all the water quality characteristics for a specific purpose to characterise surface and groundwater sources' quality that is sufficient for human consumption [39]. To categorise water quality according to the level of purity, the most measurable parameters were used [40].

The weighted arithmetic method's computation is as follows:

$$\mathrm{WQI} = \sum \mathrm{QiWi} / \sum \mathrm{Wi} \tag{10}$$

Each water quality parameter's unit weight (Wi) is determined using the following formula:

$$\mathrm{Wi} = \mathrm{K}/\mathrm{Si} \tag{11}$$

where K stands for the proportionality constant and can also be found using the following equation:

$$\mathrm{K} = \frac{1}{\sum(1/S_i)} \tag{12}$$

For every parameter, the quality rating scale (Qi) is computed using the following expression:

$$\mathrm{Qi} = 100 \times [(\mathrm{Vi} - \mathrm{Vo})/\mathrm{Si} - \mathrm{Vo})] \tag{13}$$

where for every parameter, the quality rating scale (Qi) is computed using the following expression:

Vi: the concentration of the ith parameter of the analysed water.

However, the pH = 7.0 and DO = 14.6 mg/L, Vo = 0, which is the ideal value of this parameter in pure water.

Si represents the recommended standard value of the ith parameter.

The weighted arithmetic water quality index method scale [36] is shown in Table 4.

**Table 4.** WAM WQI Scale [36].

| WQI Value | Rating of Water Quality | Grading |
|---|---|---|
| 0–25 | Excellent water quality | A |
| 26–50 | Good water quality | B |
| 51–75 | Poor water quality | C |
| 76–100 | Very poor water quality | D |
| >100 | Unsuitable for drinking | E |

## 3. Results and Discussion

### 3.1. *Descriptive Characteristics of the Keddara Dam*

Table 5 displays the descriptive characteristics of the water quality parameters in the chosen the Keddara Dam over the monitoring period from 29 December 2018 to 3 June 2021 based on daily measurements.

The pH is a measurement of how acidic/basic water is. This is an important water quality parameter that determines the solubility and biological availability of different chemical constituents in water. The pH values of the dam water during the study period ranged from 7.5 to 8.32, indicating that the water was within an alkaline range. The values are within the permissible ranges (6.0 and 9.0) described in the WHO guidelines. According to the research carried out by [41], algae use $CO_2$ in their photosynthetic activity, leading to an increase in pH values due to a shift in the forms of alkalinity present from bicarbonate to carbonate [42].

Table 5. Descriptive characteristics of the Keddara Dam.

| Parameters | Temp. °C | pH | EC µS/cm | Turb. NTU | TSS mg/L | TH $F^0$ | TAC $F^0$ | $NO_2^-$ mg/L | $NO_3^-$ mg/L | $NH_4^+$ mg/L | $PO_4^{3-}$ mg/L |
|---|---|---|---|---|---|---|---|---|---|---|---|
| Minimum | 11.70 | 7.5 | 988 | 1.30 | 1.30 | 29.0 | 15.60 | 0.00 | 2.27 | 0.00 | 0.015 |
| Q1 | 14.50 | 7.82 | 1147 | 3.68 | 3.59 | 39.0 | 17.20 | 0.02 | 3.96 | 0.015 | 0.015 |
| Median | 16.90 | 7.96 | 1231 | 5.90 | 6.40 | 42.8 | 17.80 | 0.02 | 6.31 | 0.015 | 0.015 |
| Mean (Q2) | 17.74 | 7.93 | 1243 | 6.72 | 6.68 | 42.18 | 17.57 | 0.03 | 6.68 | 0.023 | 0.025 |
| Q3 | 20.90 | 8.04 | 1358 | 9.08 | 8.86 | 45.40 | 18.00 | 0.03 | 5.85 | 0.019 | 0.032 |
| Max (Q4) | 25.30 | 8.32 | 1525 | 23.70 | 22.00 | 50.60 | 20.80 | 0.22 | 10.00 | 0.23 | 0.095 |
| Min outlier | 4.9 | 7.49 | 830.5 | −4.42 | −4.32 | 29.4 | 16 | 0.003 | −0.97 | 0.01 | 0.011 |
| Maj outlier | 30.5 | 8.37 | 1674.5 | 17.18 | 16.77 | 55 | 19.2 | 0.049 | 12.18 | 0.025 | 0.058 |
| Permissible limit (WHO) | 26.5 | 8.5 | 1500 | 5 | 30 | 50 | 20 | 3 | 50 | ≤0.5 | 0.5 |

Turbidity is a crucial factor in drinking water, as it can affect consumers' acceptability of water and its utility in some industries. In addition, difficulties in the treatability of water may be encountered in some water treatment plants [43]. The turbidity readings for the Keddara Dam's surface water ranged from 1.3 NTU to 23.70 NTU, exceeding the threshold permitted by the WHO.

The salinity of water can be identified based on electrical conductivity (EC) or the total dissolved solids (TDS). It is an expression of the ability of water to carry an electric current and is directly related to major ions and the amount of TDS. Moreover, it is important as it impacts the taste and user acceptance of the water as potable [44]. High conductivity can originate from human activity, like sewage and industrial effluent, or it can occur naturally as a result of the weathering of some sedimentary rocks [44]. The Keddara Dam's water has conductivity values ranging from 1015 µS/cm to 1523 µS/cm, slightly exceeding the recommended levels set by the WHO (Table 5).

The hardness of water is indicative of a large concentration of dissolved minerals leading to an increase in conductivity. It is a crucial indicator of water quality that determines whether it is suitable for use in agriculture, industry, and residential settings or not. For area under investigation, the WHO criterion of 50 $F^0$ was met by the dam water, as the TH ranged from 29 $F^0$ to 50.60 $F^0$. The water's overall alkalinity is influenced by these ions. In addition to this, the geology of the region the river passes through determines the hard water level. The mean TAC for dam water ranged from 15.60 $F^0$ to 20.80 $F^0$, within the WHO standard limit.

$NH_{4+}$, $NO_3^-$, $NO_2^-$ and $PO_4^{3-}$ totally referred to as essential nutrients for algal growth. They recorded low concentrations in dam water and none of them exceeded the standard limit for WHO standards (Table 5). The low mean values of nutrients in the Keddara Dam's water indicated less contamination from the catchment area.

### 3.2. Outlier Detection Normality Testing

The distribution of the raw data of water from the Keddara Dam (before outliers were removed) for water quality parameters is represented in Figure 2.

After the outliers were removed, normality tests were performed (Table 6). The box plots show that the distribution of the measured values is asymmetric, except for TSS. For temperature, EC, turbidity, $NO_2^-$, $NH_4^+$, and $PO_4^{3-}$, the distribution spreads towards the large values. On the other hand, for pH, TH, TAC and $NO_3^-$ values, the distribution spreads towards small values. This means that for all parameters, the data are not normally distributed, except for TSS. In this context, removing outliers proves to be an effective data processing operation to determine significant WQIs.

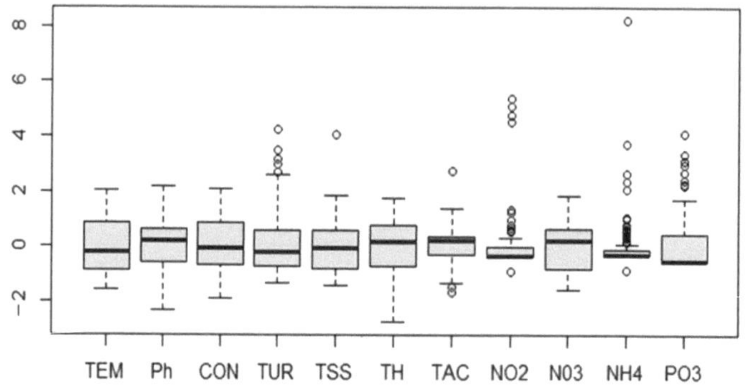


**Figure 2.** Boxplot of the Keddara data before removal of outliers.

**Table 6.** Normality test of the Keddara water quality parameters.

| Water Parameters | Sample Size | Shapiro–Wilk Test | Kolmogorov Test | $R^2$ |
|---|---|---|---|---|
| Temperature | 309 | $1.12 \times 10^{-10}$ | 0.8681 | 0.94 |
| pH | 309 | $4.853 \times 10^{-6}$ | 0.9982 | 0.97 |
| Conductivity | 309 | $6.243 \times 10^{-6}$ | 0.4846 | 0.98 |
| Turbidity | 309 | $2.877 \times 10^{-9}$ | 0.9826 | 0.95 |
| Total dissolved solids (TSS) | 55 | 0.1075 | 0.7854 | 0.98 |
| Hydrometric title (TH) | 24 | 0.1599 | 0.6574 | 0.95 |
| Full alkalimetric title (TAC) | 23 | 0.3749 | 0.83 | 0.96 |
| Nitrogen oxide ($NO_2^-$) | 104 | $<2.2 \times 10^{-16}$ | 0.4472 | 0.29 |
| Nitrate ($NO_3^-$) | 28 | 0.2789 | 0.9841 | 0.97 |
| Ammonia ($NH_4^+$) | 110 | $<2.2 \times 10^{-16}$ | 0.4472 | 0.34 |
| Phosphate ion ($PO_4^{3-}$) | 104 | $1.325 \times 10^{-14}$ | 0.4472 | 0.63 |

Using Shapiro–Wilk test analysis, the *p*-value of all parameters except TSS, TH, TAC, and $NO_3^-$ was found to be statistically significant at level (0.05) (Table 6). This means that data is not normally distributed. This might be because the Shapiro–Wilk test is used to analyse the normality of small data sizes [33]. On the other hand, the Kolmogorov–Smirnov test of the raw water showed that all parameters were normally distributed (>0.05) except $PO_4^{3-}$, $NH_4^+$, and EC. In contrast to the Shapiro–Wilk test, this test is well suited to a long series of data [33]. Aside from $NO_2^-$, $NH_4^+$, and $PO_4^{3-}$ values, the quantile regression approach showed that all other parameters represented a good linear relationship with an $R^2$ that varied from 0.94 to 0.98.

### *3.3. Assessment of the Keddara Dam's Surface Water Using Water Quality Indices*

The outcomes of the monitoring program at the Keddara Dam involve a complex matrix of physical and chemical factors. Each of these factors alone is not sufficient for a dependable assessment of the quality of the water over time. To overcome these challenges, two water quality indices (WQIs) were applied to summarise many monitored parameters into one simple term (Table 7).

According to the computed WQI values obtained using the CCME, the water quality of the Keddara Dam water was good (81.92) during the study period, except during the COVID-19 lockdowns when the dam's water turned out to be of "excellent" quality (95.09) (Table 7). The excellent characteristics during this period can be attributed to low human activities. When the other method (WQI WAM) was used, the results showed that the seasonal surface water quality ranged from approximately 9.52 to 17.77 (excellent condition) and complied with WHO standards (Table 7). The excellent conditions for all seasons compared to the CCME WQI can be attributed to the rating scale used by this method. Consequently, according to both methods, the water quality of the Keddara Dam is suitable for irrigation usage and indicated that the Algerian government have set effective regulations to stop the illegal dumping of waste by industrial and municipal activities since the establishment of Law No. 83-03 on 5 February 1983 relating to environmental protection [45]. This, in turn, helped increase the purity of the Keddara Dam water reaching the water treatment plant.

**Table 7.** Results for CCME WQI and WAM WQI methods.

| | Years | 29 December 2018–29 March 2019 | 3 April 2019–28 September 2019 | 2 October 2019–31 March 2020 | 4 April 2020–29 September 2020 | 2 October 2020–31 March 2021 | 3 April 2021–3 June 2021 |
|---|---|---|---|---|---|---|---|
| | **Season** | **Wet** | **Dry** | **Wet** | **Dry** | **Wet** | **Dry** |
| CCME WQI | Total number of tests | 177 | 383 | 383 | 297 | 348 | 96 |
| | Number of field parameters | 1 | 1 | 1 | 1 | 1 | 2 |
| | Number of failed tests | 31 | 76 | 43 | 21 | 63 | 20 |
| | $F_1$ | 9.09 | 9.09 | 9.09 | 9.09 | 9.09 | 18.18 |
| | $F_2$ | 7.51 | 7.15 | 11.23 | 7.07 | 18.10 | 20.83 |
| | $F_3$ | 12.69 | 2.54 | 3.19 | 2.06 | 16.99 | 14.68 |
| | WQI | 86.45 | 93.16 | 91.46 | 95.09 | 88.34 | 81.92 |
| | Quality range | Good | Good | Good | Excellent | Good | Good |
| WAM WQI | WQI | 14.68 | 9.52 | 11.23 | 12.45 | 17.77 | 15.99 |
| | Quality range | Excellent | Excellent | Excellent | Excellent | Excellent | Excellent |

Currently, there is no research evaluating the water quality of the Keddara Dam using the water quality index methods. However, some researchers have evaluated the dam's water quality using laboratory techniques. The results presented in this paper align with those presented in research carried out by Ghemmit-Doulache and Ouslimani [46], who measured the physicochemical, biological, and heavy metal parameters at three sites at the Keddara Dam using laboratory techniques. They revealed that the sites' physical qualities met WHO standards, while the chemical quality showed increased hardness. Another study was performed by Guettache et al., [47] to determine the water quality of the Keddara Dam before and after treatment using laboratory analysis. Based on their findings, they highlighted that the maximum and minimum temperature, conductivity, pH, and nitrate were within the Algerian standards, while the turbidity and ammonia values were above the norm.

A comparison between the Keddara Dam WQI and other Algerian case studies is presented in Table 8.

As shown in Table 8, the quality of the water in our case study is different compared to that of the other studies. This can be explained by the fact that the water mobilised by the Keddara Dam is not overly affected by human activities, especially as the Boudouaou catchment area has neither large urban areas nor industrial activity zones upstream of the dam.

It is worth noting that every WQI method used in this study could be readily implemented and serve as a guide for similar projects aiming to assess the performance of any dam or treatment plant.

**Table 8.** Comparison between the Keddara Dam WQIs and other Algerian case studies.

| Study | | Method Used | Water Quality | Reference |
|---|---|---|---|---|
| Case Study | Lambert Coordinates | | | |
| Keddara Dam | 36°39′03″ N, 3°24′59″ E | CCME-WQI | Acceptable to excellent | This study |
| | | WAM-WQI | Excellent | |
| Reghaia lake | 36°46′17″ N, 3°20′38″ E | CCME-WQI | Unsuitable | [36] |
| | | WAM-WQI | Poor | |
| Beni Haroun Dam | 36°33′19″ N, 6°16′11″ E | CCME-WQI | Poor | [48] |
| Boukourdane Dam | 36°31′40″ N, 2°18′14″ E | NSF-WQI | impaired | [49] |
| Hammam Boughrara Dam | 34°53′03″ N, 1°38′51″ W | CCME-WQI | Poor | [50] |
| Sikkak Dam | 35°02′42″ N, 1°20′27″ W | CCME-WQI | Marginal | [50] |
| Cheurfa Dam | 35°24′11″ N, 0°15′14″ W | CCME-WQI | Marginal | [50] |

## 4. Conclusions and Recommendations

This study will be the first to apply WQI to the Keddara Dam. It allowed us to evaluate the Keddara Dam's water quality using the CCME WQI and WAM WQI techniques. The dam's water quality was assessed using the concentrations of 11 physicochemical parameters such as temperature, pH, conductivity, turbidity, total suspended solids (TSS), full alkalimetric title (TAC), hydrometric title (TH), nitrite ions ($NO^{2-}$), nitrate ions ($NO^{3-}$), ammonium ions ($NH^{4+}$), and phosphate ions ($PO_4^{3-}$)). The results revealed that all parameters' concentrations were within the permissible ranges stated in the WHO guidelines, except for turbidity and salinity parameters, for which values exceeding these ranges were recorded. Before establishing the WQI, the measured values were cleansed of outliers and their normal distribution was examined. When using the CCME, the water quality of the Keddara Dam ranged from good to excellent (81.92 to 95.09), while the WAM quality ranged from 9.52 to 17.77 (excellent condition). These results indicated that raw water from this dam is of good quality. Nevertheless, before it can be used for municipal purposes, it still needs to undergo more standard treatment (pre-treatment, clarity, and disinfection).

This study's applied methodology may be relevant to areas with similar characteristics in Algeria and elsewhere. This will help decision-makers manage water resources. In prospective, other analyses such as heavy metals and bacteriological analysis are necessary to develop a control approach to limit any negative impacts of the dam's water. Besides, it is advisable to construct an artificial intelligence model for future analysis of temporal and spatial changes, as well as predictions of water quality in the context of sustainable development and preservation of water resources.

**Author Contributions:** Conceptualisation, T.S.F., M.B., N.B.E., C.A. and N.K.; methodology, M.B., N.B.E., C.A. and N.K.; software, T.S.F. and M.B.; validation, M.B., N.B.E., C.A. and N.K.; formal analysis, M.B., N.B.E. and N.K.; investigation, M.B. and N.B.E.; resources, N.B.E., C.A. and N.K.; data curation, T.S.F. and C.A.; writing—original draft preparation, T.S.F., M.B. and N.B.E.; writing—review and editing, M.B., N.B.E., C.A. and N.K.; visualisation, M.B., N.B.E., C.A. and N.K.; supervision, M.B., N.B.E. and C.A.; project administration, M.B., N.B.E., C.A. and N.K. All authors have read and agreed to the published version of the manuscript.

**Funding:** This research received no external funding.

**Data Availability Statement:** Data is contained within the article.

**Acknowledgments:** The authors would like to express their sincere gratitude to the dedicated team at the Water and Sanitation Society (SEAAL Algiers) for generously providing invaluable data for this study.

**Conflicts of Interest:** The authors declare no conflicts of interest.

## References


1. Tay, C.K. Integrating water quality indices and multivariate statistical techniques for water pollution assessment of the Volta Lake, Ghana. *Sustain. Water Resour. Manag.* **2021**, *7*, 71. [CrossRef]
2. Trach, R.; Trach, Y.; Kiersnowska, A.; Markiewicz, A.; Lendo-Siwicka, M.; Rusakov, K. A study of assessment and prediction of water quality index using fuzzy logic and ANN models. *Sustainability* **2022**, *14*, 5656. [CrossRef]
3. Lazizi, A.; Laifa, A. Assessment of the Surface Water Quality: A Case of Wadi El-Kébir West Watershed, Skikda, North-East Algeria. *Nat. Environ. Pollut. Technol.* **2020**, *19*, 1977–1985. [CrossRef]
4. Ahmed, I.; Summers, J.K. (Eds.) *Promising Techniques for Wastewater Treatment and Water Quality Assessment*; BoD–Books on Demand: Schleswig-Holstein, Germany, 2021.
5. Nadir, M.; Boualem, R. Study of Beni Haroun dam pollution (Algeria). *Desalination Water Treat.* **2016**, *57*, 2766–2774. [CrossRef]
6. Hamlat, A.; Guidoum, A.; Koulala, I. Status and trends of water quality in the Tafna catchment: A comparative study using water quality indices. *J. Water Reuse Desalination* **2017**, *7*, 228–245. [CrossRef]
7. Sadigov, R. Rapid growth of the world population and its socioeconomic results. *Sci. World J.* **2022**, *2022*, 8110229. [CrossRef] [PubMed]
8. Djelita, B.; Bouzid-Lagha, S.; Nehar, K.C. Spatial and temporal patterns of the water quality in the Hammam Boughrara Reservoir in Algeria. In *Energy, Transportation and Global Warming*; Grammelis, P., Ed.; Springer International Publishing: Cham, Germany, 2016; pp. 635–653.
9. Guettaf, M.; Maoui, A.; Ihdene, Z. Assessment of water quality: A case study of the Seybouse River (North East of Algeria). *Appl. Water Sci.* **2017**, *7*, 295–307. [CrossRef]
10. Hadji, R.; Limani, Y.; Baghem, M.; Demdoum, A. Geologic, topographic and climatic controls in landslide hazard assessment using GIS modeling: A case study of Souk Ahras region, NE Algeria. *Quat. Int.* **2013**, *302*, 224–237. [CrossRef]
11. Boudoukha, A.; Boulaarak, M. Waters pollution of Hammam Grouz Dam by nutrients (East of Algeria). *Bull Serv. Geol.* **2013**, *24*, 23–34.
12. Demdoum, A.; Hamed, Y.; Feki, M.; Hadji, R.; Djebbar, M. Multi-tracer investigation of groundwater in El Eulma Basin (Northwestern Algeria), North Africa. *Arab. J. Geosci.* **2015**, *8*, 3321–3333. [CrossRef]
13. Hamad, A.; Baali, F.; Hadji, R.; Zerrouki, H.; Besser, H.; Mokadem, N.; Legrioui, R.; Hamed, Y. Hydrogeochemical characterization of water mineralization in Tebessa-Kasserine karst system (Tuniso-Algerian Transboundry basin). *Euro-Mediterr. J. Environ. Integr.* **2018**, *3*, 1–15. [CrossRef]
14. Hamed, Y.; Redhaounia, B.; Sâad, A.; Hadji, R.; Zahri, F.; Zighmi, K. Hydrothermal waters from karst aquifer: Case study of the Trozza basin (Central Tunisia). *J. Tethys* **2017**, *5*, 33–44.
15. Ferahtia, A.; Halilat, M.T.; Mimeche, F.; Bensaci, E. Surface water quality assessment in semi-arid region (El hodna watershed, Algeria) based on water quality index (WQI). *Stud. Univ. Babes-Bolyai Chem.* **2021**, *66*, 127–142. [CrossRef]
16. Iqbal, M.M.; Shoaib, M.; Farid, H.U.; Lee, J.L. Assessment of Water Quality Profle Using Numerical Modeling Approach in Major Climate Classes of Asia. *Int. J. Environ. Res. Public Health* **2018**, *15*, 2258. [CrossRef]
17. Uddin, M.G.; Nash, S.; Olbert, A.I. A review of water quality index models and their use for assessing surface water quality. *Ecol. Indic.* **2021**, *122*, 107218. [CrossRef]
18. Zandagba, J.E.B.; Adandedji, F.M.; Lokonon, B.E.; Chabi, A.; Dan, O.; Mama, D. Application use of Water Quality Index (WQI) and multivariate analysis for Nokoué lake water quality assessment. *Am. J. Environ. Sci. Eng.* **2017**, *1*, 117–127.
19. Egun, N.K.; Oboh, I.P. Surface Water Quality Evaluation of Ikpoba Reservoir, Edo State, Nigeria. *Int. J. Energy Water Resour.* **2022**, *6*, 509–519. [CrossRef]
20. Şener, Ş.; Şener, E.; Davraz, A. Evaluation of water quality using water quality index (WQI) method and GIS in Aksu River (SW-Turkey). *Sci. Total Environ.* **2017**, *584*, 131–144. [CrossRef] [PubMed]
21. Ammari, A. Vulnérabilité à l'Envasement des Barrages (cas du Bassin Hydrographique des Côtiers Algérois). Thèse de Doctorat en Sciences en Hydraulique. Ph.D Thesis, Université Mohamed Khider, Biskra, Algérie, 2012.
22. APHA. *Standard Methods for the Examination of Water and Wastewater*, 22nd ed.; Rice, E.W., Baird, R.B., Eaton, A.D., Clesceri, L.S., Eds.; American Public Health Association (APHA); American Water Works Association (AWWA); Water Environment Federation (WEF): Washington, DC, USA, 2012.
23. Doherty, E.; McNamara, G.; Fitzsimons, L.; Clifford, E. Design and implementation of a performance assessment methodology cognisant of data accuracy for Irish wastewater treatment plants. *J. Clean. Prod.* **2017**, *165*, 1529–1541. [CrossRef]
24. Hoffman, J.I. Chapter 9-Outliers and Extreme Values. *Basic Biostat. Med. Biomed. Pract.* **2019**, *2*, 149–155.
25. Chantarangkul, V.; Peyvandi, F.; Tripodi, A.; Group, I. Effect of different methods for outlier detection and rejection when calculating cut off values for diagnosis of lupus anticoagulants. *Thromb. Res.* **2020**, *190*, 20–25. [CrossRef]
26. Tukey, J.W. *Exploratory Data Analysis*; Springer: Berlin/Heidelberg, Germany, 1977; Volume 2, pp. 131–160.
27. Le Guen, M. TUKEY's boxplot, a tool for learning about statistics. *Stat. Your-SFDS* **2001**, *4*, 1–3.
28. Berenguer-Rico, V.; Nielsen, B. Normality testing after outlier removal. *Econom. Stat.* **2023**. [CrossRef]
29. Razali, N.M.; Wah, Y.B. Power comparisons of shapiro-wilk, kolmogorov-smirnov, lilliefors and anderson-darling tests. *J. Stat. Model. Anal.* **2011**, *2*, 21–33.

30. World Health Organization. Nitrate and Nitrite in Drinking-Water. Background Document for Development of WHO Guidelines for Drinking-Water Quality. Available online: https://www.who.int/docs/default-source/wash-documents/wash-chemicals/nitrate-nitrite-background-document.pdf (accessed on 10 July 2023).
31. Shaibur, M.R.; Hossain, M.S.; Sony, S.J. Drinking water quality of hand tube well water at sub-urban areas of Jashore Municipality, Bangladesh. *J. Jessore Univ. Sci. Technol.* **2019**, *4*, 11–22.
32. Edori, O.S.; Nna, P.J. Determination of physicochemical parameters of effluents at discharge points into the New Calabar River along Rumuolumeni axis, Niger Delta, Nigeria. *J. Environ. Anal. Toxicol.* **2018**, *8*, 2161-0525.
33. Jehan, S.; Ullah, I.; Khan, S.; Muhammad, S.; Khattak, S.A.; Khan, T. Evaluation of the Swat River, Northern Pakistan, water quality using multivariate statistical techniques and water quality index (WQI) model. *Environ. Sci. Pollut. Res.* **2020**, *27*, 38545–38558. [CrossRef]
34. Mohammed, A.Q.; Mouhcine, E.Q.; Nabil Darwesh, M.S.; Hamdaoui, F.; Kherrati, I.; El Kharrim, K.; Belghyti, D. Hydrogeochemical study of groundwater quality in the west of Sidi Allal Tazi, Gharb area, Morocco. *J. Mater. Environ. Sci.* **2017**, *9*, 293–304.
35. CCME (Canadian Council of Ministers of the Environment). *Canadian Environmental Quality Guidelines for the Protection of Aquatic Life, CCME Water Quality Index: Technical Report*, Canadian Council of Ministers of the Environment: Winnipeg, Manitoba, 2001.
36. Bouhezila, F.; Hacene, H.; Aichouni, M. Water quality assessment in Reghaia (North of Algeria) lake basin by using traditional approach and water quality indices. *Kuwait J. Sci.* **2020**, *47*, 57–71.
37. Fathi, P.; Ebrahimi Dorche, E.; Zare Shahraki, M.; Stribling, J.; Beyraghdar Kashkooli, O.; Esmaeili Ofogh, A.; Bruder, A. Revised Iranian Water Quality Index (RIWQI): A tool for the assessment and management of water quality in Iran. *Environ. Monit. Assess.* **2022**, *194*, 504. [CrossRef]
38. Ojukwu, C.K.; Okeah, G.O.C.; Mmom, P.C. A Comparative Analysis of the Weighted Arithmetic and Canadian Council of Ministers of the Environment Water Quality Indices for Water Sources in Ohaozara, Ebonyi State, Nigeria. *Int. J. Eng. Res. Technol. (IJERT)* **2021**, *10*, 498–506.
39. Bessedik, M.; Abdelbaki, C.; Badr, N.; Tiar, S.M.; Megnounif, A. Application of water quality indices for assessment of influent and effluent wastewater from wastewater treatment plant of Oran City, Algeria. *Desalination Water Treat.* **2021**, *236*, 306–317. [CrossRef]
40. Zahedi, S. Modification of expected conflicts between drinking water quality index and irrigation water quality index in water quality ranking of shared extraction wells using multi criteria decision making techniques. *Ecol. Indic.* **2017**, *83*, 368–379. [CrossRef]
41. Varol, M.; Gökot, B.; Bekleyen, A.; Şen, B. Spatial and temporal variations in surface water quality of the dam reservoirs in the Tigris River basin, Turkey. *Catena* **2012**, *92*, 11–21. [CrossRef]
42. Sawyer, C.N.; McCarty, P.L.; Parkin, G.F. *Chemistry for Environmental and Engineering Science*, 5th ed.; McGraw Hill Inc.: New York, NY, USA, 2003; pp. 587–590.
43. Ireland-EPA. *Parameters of Water Quality—Interpretation and Standards*; Environmental Protection Agency: Wexford, Republic of Ireland, 2001.
44. Karakaya, N.; Evrendilek, F. Water quality time series for Big Melen stream (Turkey): Its decomposition analysis and comparison to upstream. *Environ. Monit. Assess.* **2010**, *165*, 125–136. [CrossRef]
45. People's Democratic Republic of Algeria. Law No. 83-03 of February 5, 1983 Relating to Environmental Protection. Official Journal of the Democratic and Popular Algerian Republic No. 6, February 8, 1983, p. 250.(pdf). (This Document Is in French). Available online: https://leap.unep.org/en/countries/dz/national-legislation/loi-n-83-03-relative-la-protection-de-lenvironnement (accessed on 15 March 2024).
46. Ghemmit-Doulache, N.; Ouslimani, N. Water Quality Control of Three Sites At Kedarra Barrage. *J. Fundam. Appl. Sci.* **2021**, *13*, 965–981.
47. Guettache, A.; Bettayeb, S.; Ililes, M. Caractérisation et Étude D'efficacité du Système de Traitement de l'eau de Barrage Keddara. Available online: http://dspace.univ-bouira.dz:8080/jspui/handle/123456789/11443 (accessed on 15 March 2024).
48. Soltani, A.A.; Bermad, A.; Boutaghane, H.; Oukil, A.; Abdalla, O.; Hasbaia, M.; Oulebsir, R.; Zeroual, S.; Lefkir, A. An integrated approach for assessing surface water quality: Case of Beni Haroun dam (Northeast Algeria). *Environ. Monit. Assess.* **2020**, *192*, 1–17. [CrossRef]
49. Arab, S.; Bouchelouche, D.; Hamil, S.; Arab, A. Application of water quality index for surface water quality assessment Boukourdane Dam, Algeria. In *Advances in Sustainable and Environmental Hydrology, Hydrogeology, Hydrochemistry and Water Resources, Proceedings of the 1st Springer Conference of the Arabian Journal of Geosciences (CAJG-1), Tunisia, 2018*; Springer International Publishing: Berlin/Heidelberg, Germany, 2019; pp. 385–387.
50. Hamlat, A.; Tidjani, A.E.B.; Yebdri, D.; Errih, M.; Guidoum, A. Water quality analysis of reservoirs within Western Algeria catchment areas using water quality index CCME WQI. *J. Water Supply Res. Technol.—AQUA* **2014**, *63*, 311–332. [CrossRef]

MDPI AG
Grosspeteranlage 5
4052 Basel
Switzerland
Tel.: +41 61 683 77 34

*Water* Editorial Office
E-mail: water@mdpi.com
www.mdpi.com/journal/water

www.ingramcontent.com/pod-product-compliance
Lightning Source LLC
Chambersburg PA
CBHW041405210926
51850CB00013B/3186

* 9 7 8 3 7 2 5 8 4 0 8 8 5 *